LASER APPLICATIONS IN MEDICINE AND BIOLOGY

Volume 4

LASER APPLICATIONS IN MEDICINE AND BIOLOGY

Volume 4

Edited by
M. L. Wolbarsht
Professor of Ophthalmology and Biomedical Engineering
Duke University
Durham, North Carolina

PLENUM PRESS · NEW YORK AND LONDON

ISBN-13:978-1-4612-8061-3   e-ISBN-13:978-1-4613-0745-7
DOI: 10.1007/978-1-4613-0745-7

Library of Congress Catalog Card Number 77-128514

© 1989 Plenum Press, New York
Softcover reprint of the hardcover 1st edition 1989

A Division of Plenum Publishing Corporation
233 Spring Street, New York, N.Y. 10013

CONTRIBUTORS

Ralph G. Allen, *USAF School of Aerospace Medicine, Brooks Air Force Base, Texas 78235*

William T. Ham, Jr., *Department of Biostatistics, Biomedical Engineering Division, Virginia Commonwealth University, Richmond, Virginia 23298-0001*

Franz Hillenkamp, *Institute for Medical Physics, University of Münster, D-4400 Münster/FRLY, Federal Republic of Germany*

R. Kaufmann, *Department of Clinical Physiology, University of Düsseldorf, D-4000 Düsseldorf, Federal Republic of Germany*

Harold A. Mueller, *Department of Biostatistics, Biomedical Engineering Division, Virginia Commonwealth University, Richmond, Virginia 23298-0001*

Richard M. Osgood, Jr., *Department of Electrical Engineering and Applied Physics, Columbia University, New York, New York 10027*

Garret D. Polhamus, *USAF School of Aerospace Medicine, Brooks Air Force Base, Texas 78235*

Claude Reiss, *Institut Jacques Monod, CNRS and Université Paris VII, 75251 Paris, France*

Leonid B. Rubin, *Department of Quantum Radiophysics, Faculty of Physics, Moscow State University, Moscow, USSR*

PREFACE

The diversity of the chapters presented in this volume illustrates not only the many applications of lasers, but also the fact that, in many cases, these are not new uses of lasers, but rather improvements of laser techniques already widely accepted in both research and clinical situations. Biological reactions to some special aspects of laser exposure continue to show new effects, which have implications for the ever-present topic of laser safety. Such biological reactions are included in fields of research which depend on properties of electromagnetic radiation exposure only possible with lasers, for example, the short pulses necessary for the temperature-jump experiments reviewed by Reiss.

Speciality lasers, such as the transverse excitation atmospheric (TEA) or excimer lasers, add new wavelengths and pulse domains to those already available for biological application. A description of these new types of lasers by Osgood is included to indicate new possibilities for future use and to avoid limiting our coverage to well-developed present-day applications.

Hillenkamp and Kaufmann describe a microprobe mass spectrograph for analysis of the minute amounts of material evaporated by a laser pulse. The analytical possibilities of this instrument are far-reaching, and some of the various results are described to illustrate the power of their method, as well as to show the types of problems that are suitable for it.

The initial steps in photosynthesis have become the subject of intensive investigation. High-speed laser techniques have assisted in unlocking some of the puzzles of the energy transduction involved, and the present review by Rubin shows both the current state of the art and the possibilities for future experiments.

Much of the clinical work using lasers remains in ophthalmology, the first specialty to accept any type of laser exposure as a standard therapy. In addition to its use in the treatment of an ever-increasing number of retinal problems, laser coagulation has become the standard treatment for glaucoma, for the lens capsule cutting necessitated by the complications of lens implant surgery, and bids fair to proceed even further as lasers are applied to vitreous surgery. The analyses of bioeffects of laser exposures in the eye by Ham and Mueller, and by Allen and Pohlhamus, give some of the scientific background for those applications that have already proven successful and indicate the bases for future clinical applications of lasers in ophthalmology.

Each clinical acceptance of a laser instrument broadens the experience of all in the general medical field with regard to laser use. Also, the increasingly larger number of clinicians involved in research on laser surgery adds even more to the diversity of laboratory projects that will in turn become the basis for techniques that in the future will gain clinical acceptance.

As before, suggestions for subjects for critical reviews in future volumes are most welcome.

M. L. Wolbarsht

Durham, North Carolina

CONTENTS

Chapter 1
The Laser as a Tool in the Study of Photosynthesis
Leonid B. Rubin

Chapter 2

**Requirements and Technical Concepts of
Biomedical Microprobe Analysis**

Franz Hillenkamp and R. Kaufmann

Chapter 3

Ultrashort Laser Pulses in Biomedical Research

Claude Reiss

Chapter 4

The Excimer Laser: A New Ultraviolet Source for Medical, Biological, and Chemical Applications

Richard M. Osgood, Jr.

Chapter 5

**The Photopathology and Nature of the Blue Light
and Near-UV Retinal Lesions Produced by
Lasers and Other Optical Sources**

William T. Ham, Jr., and Harold A. Mueller

Chapter 6

Ocular Thermal Injury from Intense Light

Ralph G. Allen and Garret D. Polhamus

CHAPTER 1

The Laser as a Tool in the Study of Photosynthesis

Leonid B. Rubin

Department of Quantum Radiophysics, Faculty of Physics
Moscow State University
Moscow, USSR

1. INTRODUCTION

We can now say with certainty that photosynthesis is the area that has benefited most from laser techniques. In our efforts to comprehend nature and to gain insights into mechanisms of fundamental biological processes, the great advantages offered by laser equipment have been clearly demonstrated by striking progress in photosynthetic research. The laser came into existence at the time when there was the greatest demand for new research methods that would enable us to solve a number of fundamental questions concerning the mechanisms of the ultrafast primary events of photosynthesis.

It is now well established that the prime function of the photosynthetic apparatus is to transform the energy of absorbed light into chemical potential. The process takes place in several stages:

1. Absorption of the incident light photons by the light-harvesting antenna complex (LHA) and the transfer of the excitation energy to a specialized trap or reaction center (RC).

1

   2. Trapping of the electronic excitation energy at the RC.

   3. Translation of the trapped energy into the energy of stabilized separated charges.

   4. Transfer of electrons along the electron transport chain to sites of formation of ATP, a strong reductant and oxidant.

In contrast to other bionergetic processes, the specificity of the operation of the photosynthetic apparatus lies in the most primary events that occur in the picosecond (10^{-10}–10^{-12} s) time domain and lead to charge separation. Studies of such ultrafast photoreactions became a reality only after the development of new optical methods that use all the unique advantages of laser techniques: excitations of ultrashort duration and high intensity, monochromaticity, directionality, and coherence of the light.

The present review focuses on the basic principles of the laser research methods in photosynthesis and on new insights that laser studies have provided in this area of biology.

2. TECHNICAL ASPECTS OF THE LASER RESEARCH METHODS USED IN PHOTOSYNTHETIC RESEARCH

What is surprising about the laser methods used in photosynthetic studies is that their general concepts were worked out long before the development of the laser itself. The use of the laser techniques and nonlinear optics was achieved though fundamental advances that are still only improvements of the well-known optical methods. In the 1930s Gaviola developed the method of pulse fluorometry to study molecular excitation quenching kinetics. The principles of flash photolysis that were developed in the 1940s were successively employed in studies of the mechanisms and kinetics of the primary steps of photochemical reaction. The key to the success of those two methods has been the availability of laser flashes of very short duration as an excitation source, which are several orders of magnitude shorter than the time scale of the processes being measured. Fluorescence decay kinetics and absorption changes are suitable and valuable means to monitor kinetics of the elementary primary steps in a cascade of coupled photochemical reactions. Since experimental time resolution is determined, in large measure, by the duration of an excitation pulse, it was not until the availability of ultrashort laser pulses that time-resolved direct kinetic observation of photoreactions occurring in the 10^{-12} time domain became a reality.

As far back as the 1920s the Raman concept came into existence (Raman, 1928). With the aid of this spectroscopic method, much new valuable information concerning the structure of molecules and their interactions with the surrounding medium has been obtained. The high monochromaticity of the laser emission, along with high intensities, were attractive features that opened the door for the rapid development of this method and its successful application in different areas of biology and notably in photosynthesis. Meanwhile, new research methods have emerged in recent years, based on a specific interaction of laser irradiation with biological systems. In the following sections, an outline of general principles involved in the laser technique and nonlinear optics is given and details of their application in photosynthetic research are presented.

2.1. The Sources of Short and Ultrashort Pulses

2.1.1. Q-Switched Lasers

Soon after the development of the first laser, new techniques were designed leading to the availability of pulse widths on the order of 10 ns (Hellwarth, 1961). The method, termed Q-switching, consists of placing a shutter inside the laser cavity. The function of the shutter is to build up a much higher level of population inversion when there is no optical feedback. A sudden opening of the shutter improves the cavity Q to a significant extent and the energy stored in the laser is rapidly emitted in a single short, strong pulse. The shutter function can be accomplished mechanically by rapid rotation of the back mirror, electro-optically by placing a Kerr cell in the cavity, or optically by placing a dye cell containing a photobleaching dye in the cavity. Typical pulse widths thus obtained were on the order of from 10^{-8} to 3×10^{-8} s, and as a consequence, Q-switched lasers were used to study photosynthetic processes soon after their development (DeVault, 1964).

2.1.2. Mode-Locked Lasers

In the late 1960s, with the advent of mode-locked lasers (De Maria et al., 1966) the duration of laser pulses was decreased from 10^{-8} to 10^{-12} s. Without going into details on mode locking, which has been treated at length in the literature (De Maria, 1971; Eckardt et al., 1974), note that the output of a mode-locked laser is a train of 50–200 periodic pulses occurring every 5–10 ns (T), the pulse spacing being dependent on the pathlength of the cavity (L): $T = 2L/c$, where c is the speed of light. The duration of each pulse (Δt) is restricted by the fluorescence emission bandwidth (Δv) where $\Delta v \cdot \Delta t$ is 1. Thus, in a Nd:glass laser (the luminescence bandwidth Δv is 100 cm^{-1}), pulses of 0.3 ps in duration can, in principle, as predicted by theory, be generated. However, in practice the pulses are significantly broader (10 ps) owing to the long relaxation time of the bleaching dyes that are used. The synthesis of special dyes Nos. 3282U and 3323U, with a relaxation time of 1.5 ± 0.3 ps, opened the way for laser pulses as short as 2 ps (Babenko *et al.*, 1977). The durations of pulses that are now available are 5–10 ps from Nd:glass lasers, 15–25 ps from ruby lasers, and 30–50 ps from YAG lasers.

Since the fluorescence emission bandwidths of a dye are broader than solid state lasers, one may expect shorter pulses from mode-locked dye lasers. Efforts in this area led to construction of lasers operating in the very high frequency domain with an output of a series of subpicosecond pulses with a duration up to 0.3 ps separated by 10 ns (Ippen *et al.*, 1972; Shank and Ippen, 1974; Ippen and Shank, 1975; Ruddock and Bradley, 1976).

Picosecond laser trains can be used as an excitation source in picosecond spectroscopy if the relaxation time of the process under study is short compared with the time between pulses. Moreover, because pulses from different positions in the pulse train are sometimes different in intensity, width, and energy, most modern lasers are provided with a Pockels cell as a means to select a single pulse out of the train.

2.1.3. Frequency Tuning

Optical methods techniques, including picosecond spectroscopy, need not only ultrashort pulses but also a tunable wavelength capability within a broad spectral region. A search for more convenient equipment led to several methods. Harmonic generation is now most frequently used. By using various nonlinear crystals such as potassium dihydrogen phosphate (KDP), the fundamental wavelength of the Nd:glass laser can be doubled, tripled, or quadrupled to 530, 353, and 265 nm, and that of the ruby doubled to 347 nm.

On passing a strong laser pulse through some liquids, such as benzene, ethanol, and gasoline, stimulated Raman emission occurs as a strongly directional monochromatic radiation. Both methods have high efficiency of energy conversion but provide only discrete, separate irradiation lines.

Dye lasers provide continuous frequency tuning over the band of fluorescence. The use of different dyes makes it possible, in principle, to have laser radiation varied over the entire spectral region.

Among recent new laser technologies of great interest are parametric light generators, which offer the advantages of high stability, broad spectral tuning, directionality, and monochromaticity (Tanaka *et al.*, 1978; Akhmanov and Khokhlov, 1966). New technologies on the basis of YAG–Nd^{3+} laser and nonlinear LiNbO$_3$ crystals opened the door to parametric oscillators, which provide continuous tuning of 30–40-ps pulses over a region from 240 to 3600 nm.

The passage of a high-power picosecond pulse through some compounds, such as water, polyphosphate acid, tetrachloromethane, and various glasses, is attended by directional radiation of white light "continuum" at wavelengths from 300 to 1200 nm, lasting picoseconds (Alfano and Shapiro, 1970a,b; Smith *et al.*, 1977). The development of such intense white light picosecond sources has made picosecond absorption studies possible.

2.2. Picosecond Spectrometers for the Absorbance Change Measurements

There are only two concepts, echelon and optical delay, being used nowadays in any instruments of this sort. The main difference between them lies in the principles that provide measurement of the absorption spectrum of an intermediate product at a certain time interval after a photoexcitation.

2.2.1. Echelon Method

The key to the echelon method that has been developed at the laboratory of Professor Rentzepis is the use of the simplest echelon optical element, a stack of microscopic slides of different lengths (Fig. 1). A single 1060-nm pulse is selected (by means of a Pockels cell) from the output train of a mode-locked Nd:glass laser, then amplified and frequency doubled. The second harmonic 530-nm pulse is used as the activating light at the sample cuvette. The residual 1060-nm pulse is focused onto a cell containing *n*-octanol and produces a broadband continuum of light. The continuum (still a single pulse at this point) is then passed through the echelon. The resulting pulse train consists of picosecond pulses separated by preselected discrete time intervals, determined by the slide thickness (d), refraction (n), and angle of an incident light θ:

$$\Delta t = d/c [(n^2 - \sin^2 \theta)^{1/2} - \cos \theta] \tag{1}$$

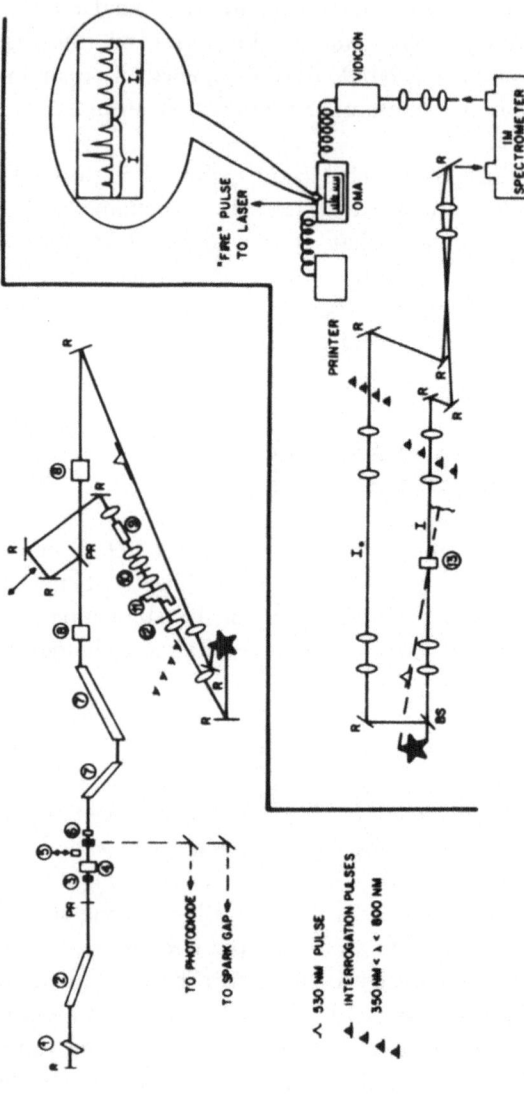

FIGURE 1. Double-beam picosecond spectrometer utilizing a silicon vidicon detector. Components: 1, mode-locking dye cell; 2, laser oscillator rod; 3, calcite polarizer; 4, Pockels cell; 5, translatable 90° polarization for 1060 nm radiation; 6, fixed position 90° polarization rotator; 7, laser amplifier rod; 8, second harmonic (530-nm) generating crystal (KDP); 9, 20-cm octanol cell for generating the interrogation wavelengths; 10, ground glass diffuser; 11, index matched glass echelon for producing picosecond optical delay between the stacked interrogation pulses; 12, vertical polarizer; 13, sample cell; R, reflector; PR, partial reflector; BS, beam splitter; OMA, optical multichannel analyzer. (After Leigh *et al.,* 1974.)

This pulse train is split into a measuring beam, I, and a reference beam, I_0. The measuring beam I passes through the sample cell simultaneously with the single 530-nm activating pulse. Since the measuring beam I is a series of successive pulses spaced at different time intervals, the I pulses monitor the absorbance of the sample at different times before and after the activating pulse. I_0 pulses are directed around the sample. Both I and I_0 are recombined behind the sample cuvette and imaged beside each other on the entrance slit of the spectrometer; I and I_0 are then detected on a silicon vidicon screen and the image stored in an optical multichannel analyzer (OMA). The image consists of two trains, I and I_0, following one after the other at certain times. Figure 1 gives a schematic representation of the data at the OMA output. The activating pulse in this case hits the cell at the beginning of the second or third time segment in the I train. Changes in absorbance for each time segment can be calculated from the intensities of the measuring and reference beams with (I^e, I_0^e) and without (I^u, I_0^u) the activating pulse, using the following equation (Leigh $et\ al.$, 1974):

$$A = -\log(I^e \cdot I_0^u / I_0^e \cdot I^u) \qquad (2)$$

The advantage of this method is that absorbance changes can be monitored in time from a single excitation flash. At the same time, sampling times depend entirely on the geometry of the echelon. To measure slow processes lasting more than 1 ns, the stepped delay of the echelon must be as long as several tens of centimeters in the loading edge of the pulse.

2.2.2. Optical Delay Line Method

The disadvantage of a long stepped delay in the echelon can be avoided by employing a variable optical delay line (Magde and Windsor, 1974; Rockley $et\ al.$, 1975; Moskowitz and Malley, 1978). One of the versions of this equipment is described in Fig. 2. A single 1060-nm pulse from a mode-locked Nd:glass laser is frequency doubled and split into its 530-nm and 1060-nm components. The 530-nm pulse which serves as an activating source is passed through a variable length optical line into the sample cuvette. The residual 1060-nm pulse is focused on a CCl_4 continuum generating cell. This pulse provides the measuring light. The resultant white continuum measuring pulse passes along another optical delay line, and is collimated and focused onto the sample cuvette. Both the activating and measuring pulses overlap physically. The measuring beam from the sample cuvette is delivered to the entrance slit of a spectrograph and then to a silicon vidicon detector and an OMA. Images of a sample both above and

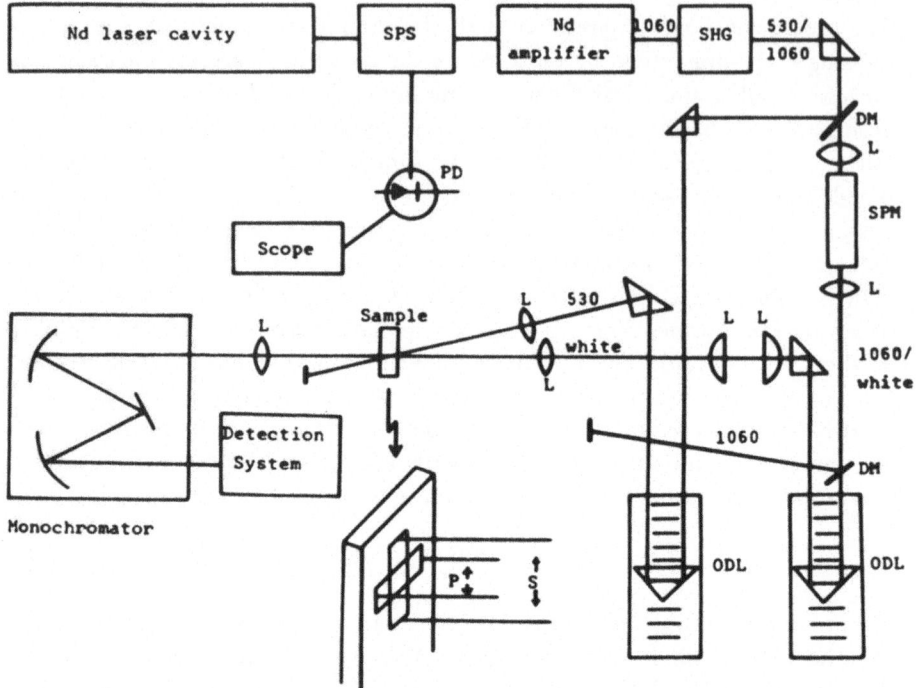

FIGURE 2. Picosecond absorption and kinetic spectroscopy apparatus. SPS, single pulse selector; SHG, potassium dihydrogen phosphate crystal for second 3 harmonic generation; DM, dichroic mirror; L, lens; SPM, carbon tetrachloride cell for continuum production by self-phase modulation; ODL, optical delay line; PD, photodiode; Scope, oscilloscope and camera. Insert shows expanded view of the beam geometry at the sample; P, 530-nm excitation or pump pulse; S, white continuum or spectroscopic pulse. (After Magde and Windsor, 1974.)

below the excited region are used to measure the initial absorbance of the sample. By varying the time between the measuring pulse and an exciting one, the absorbance of the sample can be monitored on a point-by-point basis. This method has the advantage that absorbance changes can be monitored over a broad time range, from a subpicosecond to nanoseconds, with the same equipment. This method, however, cannot provide comprehensive information about kinetics of reactions after a single pulse excitation. However, for studying some irreversible photochemical and photobiological reactions, such information is urgently needed.

Both the ethelon and optical delay methods are widely used, and have been significantly modified in many laboratories. The modifications involved, first of all, changes in the sources of measuring and exciting light, that is dye lasers, tunable parametric oscillators, stimulated Raman scattering frequencies, as well as subpicosecond excitation sources. Advances in these techniques have made possible selective excitation of pigments or biochemical components in complex biological systems. The rapid progress in this direction promises further improvement of the equipment time resolution, which is already now as small as 0.5 ps (Martin *et al.*, 1979).

2.3. Picosecond Fluorometry

Subsequent to the advent of picosecond lasers, great progress was made in picosecond fluorometry, even with simple conventional recording methods such as a photodiode-oscilloscope system (Arsenjev *et al.*, 1973; Paschenko *et al.*, 1975). Improvements in the sensitivity and time resolution of such equipment enabled investigators to observe fluorescent emission in the 100-ps time scale. To investigate more rapid fluorescent phenomena in various photosynthetic organisms, the biological community was faced with the need for specialized instrumentation. Two kinds of equipment are now available for this purpose: an optical gate and a streak camera.

2.3.1. Optical Kerr Gate Method

In 1969, Duguay and Hansen introduced an ultrafast shutter, based on an optical Kerr gate (Duguay and Hansen, 1969). The key to this shutter is the placement of a 1-cm path length CS_2 cell to induce birefringence (similar to that encountered in a Kerr cell exposed to high voltage) when a high-power pulse hits it, thus allowing the fluorescent light from the sample cuvette to pass through the cross Polaroid configuration only during the induced birefringence. A 1060-nm laser pulse may, for instance, be used to open the light gate. The duration of this pulse and the CS_2 relaxation (1.8 ps) determine the response time of the shutter. It is the order of 2–10 ps. The development of a new fast shutter gave an impetus to the development of a new, delicate method of ultrafast spectrophotometry as shown in Fig. 3 (Rentzepis *et al.*, 1970). The 1060-nm pulse from a Nd:glass laser is split into its component 530-nm and 1060-nm wavelengths. The 530-nm pulse serves as an excitation source and the 1060-nm pulse is used to control the opening of the CS_2 light gate. The gate is a 0.1 × 20-mm cuvette, which allows the fluorescent emission from the sample to pass at different times. At first the fluorescent light is allowed to pass from the upper and then from the low portion of the cuvette. The key to the setup is that the gate allows the fluorescence to pass only at a time when a control pulse arrives. When the 530-nm and 1060-nm pulses coincide in time and space, a two-dimen-

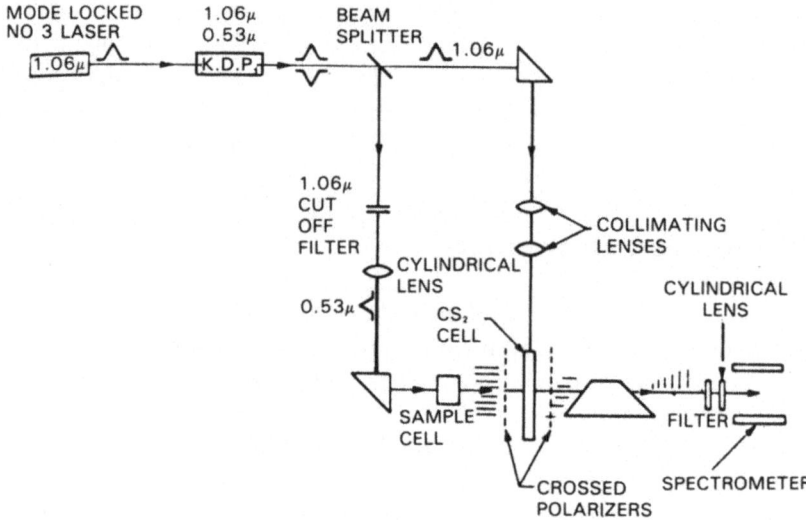

FIGURE 3.   Method for time–frequency resolving picosecond spectra. (After Rentzepis, 1970.)

sional output of the fluorescence is produced, with the length of the cuvette serving as a time coordinate. Improvements of this technique led to wide use of it in many laboratories.

Seibert and his co-workers were the first to apply this method to photosynthetic research (Fig. 4) (Seibert *et al.*, 1973; Yu *et al.*, 1975; Pellegrino *et al.*, 1978). The fluorescent output can be measured at various times with respect to the activating pulse by adjusting the optical delay.

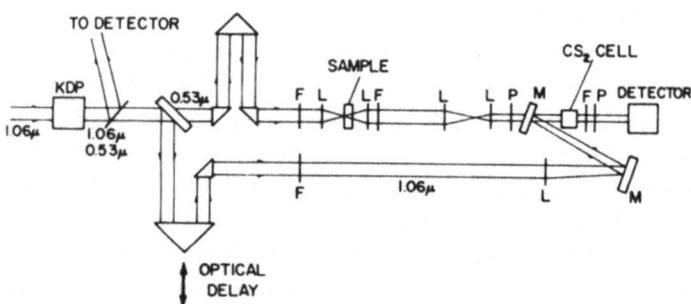

FIGURE 4.   Schematic view of the experimental setup for fluorescence lifetime measurements. At the Kerr cell (CS_2 cell), the activating 1.06-nm beam is aligned colinearly with the collected fluorescence light. L, lens; F, filter; M, mirror; P, Polaroid; KDP, potassium dihydrogen phosphate crystal. (After Pellegrino *et al.*, 1978.)

Point-by-point measurements of fluorescent emission kinetics are thereby possible. To correct the data for differences in laser pulses the fluorescence output from the monochromator is coupled with the 530-nm signal on a double-beam oscilloscope. A major advantage of this setup is a high time resolution—up to 2 ps. The method, however, suffers from two serious disadvantages. Because of the great optical signal losses at the CS_2 gate, the sensitivity is low. Besides, a complete fluorescence decay kinetics scheme can be obtained only on a point-by-point basis by multiple laser flash excitations.

2.3.2. Streak Camera Techniques

A streak camera has the advantage over the Kerr cell gate technique of allowing continuous recording of fluorescence excited by a single laser pulse (Paschenko *et al.*, 1975a,b; Shapiro *et al.*, 1975). An instrumental setup employing a streak camera detecting system is shown schematically in Fig. 5. A single harmonic 530-nm pulse from a Nd:glass laser is split into a reference and excitation beam. The reference pulse goes directly to the streak camera to monitor the shape and energy of the laser excitation pulse. Other Nd^{3+}-laser harmonics as well as pulses from different lasers (dye, ruby, etc.), or parametric light oscillators, may be used to excite fluorescence. Fluorescent light, after going through the spectrograph, is focused onto a photocathode in the camera. Photoelectrons, produced in proportion to the light intensity, are accelerated in an electric field and then deflected by a voltage ramp onto a phosphorescent screen such that electrons released from the cathode at different times hit the screen at different

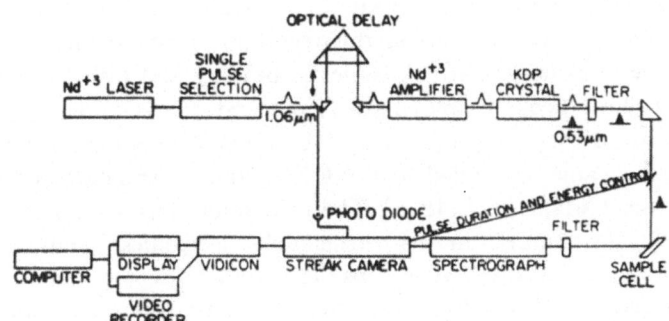

FIGURE 5. Picosecond fluorometry apparatus. (After Rubin and Rubin, 1978.)

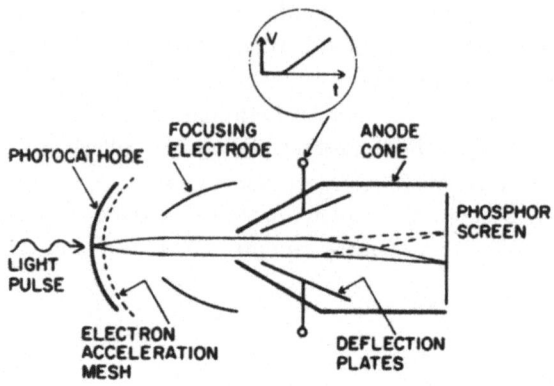

FIGURE 6. Schematic diagram of streak tube operation. (After Campillo and Shapiro, 1978.)

locations (Fig. 6). This provides a means to calibrate the screen linearly in time. With a very fast sweep, a streak camera can have a time resolution of the order of 10^{-12} s, while its sensitivity is not worse than that of photomultipliers (Bradley and Sibbet, 1975; Babenko *et al.*, 1977). The phosphorescence can be photographed directly on film or detected on a silicon vidicon screen and the image can be stored in an OMA. After data processing, the output represents the fluorescence decay and laser excitation pulse as a function of time. Relative quantum yields of fluorescence can be obtained as a ratio of an area under the fluorescent decay curve and a laser pulse which in turn can be calibrated in absolute units of energy.

Chlorophyll fluorescence *in vivo*, because of its low yield, is difficult to measure from low-energy laser pulses (5×10^{12} photons cm^{-2}). In recent times these experiments have become the subject of special interest in connection with studies of nonlinear fluorescence quenching processes (see Section). Recent modifications in this technique promise further improvements in the sensitivity of streak cameras. Bradley and his co-workers have designed a chronoscope employing a picosecond dye laser and providing a pulse repetition frequency (PRF) of 140 MHz. It operates in conjunction with a streak camera coupled to the OMA, the latter being synchronized with the laser (Adams *et al.*, 1978). The time resolution of 10 ps is restricted largely by instabilities in its associated electronics, and with new modifications can be brought to 1 ps. The high pulse repetition frequency and good reproducibility from shot to shot make recording, storage, and averaging of low-level fluorescent emissions possible. An important point to note is that high excitation frequencies can be used in investigating systems lacking long-lived photoproducts, such as triplet states. The pulse repetition frequency can, in principle, be notably reduced by using a Pockels cell (Beddard *et al.*, 1979).

2.3.3. Photon Counting Fluorometry

The latest advances in fast photodiodes and time-to-amplitude converters have led to rapid progress in a new pulse fluorometric technique, based on the photon counting concept (Sauer, 1979). The method employs 10-kHz flashes from a flash lamp as an excitation source. Because of the high sensitivity of the equipment, the energy of an excitation pulse may be as low as 10^5 photons. Further, by substituting the flash lamp for a dye laser with a high PRF, the time resolution was improved to 10 ps (Beddard *et al.*, 1979). These authors used a quasicontinuous rhodamine 6G laser (10-ps pulses) excited by a picosecond Ar-laser with a PRF of 75 MHz. By including a Pockels cell, the PRF was reduced to 33 kHz. The instrument was equipped with a photomultiplier tube coupled to a time-to-amplitude converter, which was frequency-locked to the excitation laser. The equipment is a suitable tool to use for observation of *in vivo* chlorophyll fluorescence decay kinetics with excitation pulses whose energy is as low as 10^9–10^{11} photons cm^{-2}.

2.3.4. Frequency Mixing in Nonlinear Crystals

Before ending the instrumentation part of this review, it seems appropriate to briefly describe here another method that has recently become available. Thus far it has been used only to observe the fluorescence kinetics of bacteriorhodopsin; photosynthetic studies by this technique have not been undertaken as yet. The method is based on frequency mixing in a nonlinear $LiNbO_2$ crystal. The frequencies of a fluorescence emission, W_f, and of a picosecond excitation pulse, W_L (from a Rhodamine 6G laser), are mixed in the nonlinear crystal to give a resulting frequency W, composed of $W = W_f + W_L$ which mixed frequency is recorded with a photomultiplier tube (Hirsch *et al.*, 1976). Since a signal with frequency W is produced only when both beams arrive at the crystal simultaneously, the fluorescence output can be sampled at various times by adjusting a delay line. The time resolution of 4 ps that is now available is far from being the least possible value. It can, in principle, be reduced to 0.5 ps. An important feature that must be stressed is high sensitivity of the equipment. Since signal W occurs within the UV spectral region, i.e., has a shorter wavelength than either W_L

or W_f, the signal-to-noise ratio is very great. Devoid of disadvantages of sophisticated electronic devices, the equipment offers high time resolution and sensitivity and is extremely attractive to investigators concerned with photosynthetic research.

2.4. Raman Spectroscopy

Raman (1928) was the first to demonstrate that inelastic light scattering is accompanied by new frequencies originating from the natural vibrational frequency of molecules. In contrast to IR absorbance, which is associated with changes in dipole moments due to molecular vibrations, Raman scattering (RS) is associated only with such vibrations as lead to changes in polarization. For this reason only symmetrical vibrations are involved in this phenomenon. Raman spectroscopy has the advantage of allowing investigations on small amounts of compounds in any phase, gaseous, liquid, or solid, including aqueous solutions. Even in a system with a great number of vibrations in the range from 1 to 4000 cm^{-1} the Raman spectrum falls within the visible region and can be recorded on a single spectrophotometer.

If the laser excitation frequency falls within the absorption band of the molecule being investigated, there occurs a sharp increase (by a factor of up to 10^7) in vibrational amplitudes of atoms and groups of molecules involved in an excited electron transfer. The phenomenon, known as resonance Raman scattering (RRS), has the advantage of greater sensitivity and selectivity. Varying the laser frequency makes it possible to selectively excite both molecules in complicated, multicomponent systems and electronic transitions within a single molecule.

The widespread application of Raman spectroscopy has become possible after the development of powerful monochromatic lasers. Now available are industrial Raman spectrometers with high resolution and sensitivity which allow excitation by different laser wavelengths. The difficulty with the Raman technique is the handling of weak scattering signals, which are often masked by intense fluorescent emissions. Since the fluorescence occurs mainly in the Stokes spectral region, the methods that provide the possibility of observing the Raman signal in the anti-Stokes region are most intriguing. Recently a new method for such a purpose was developed: the coherent anti-Stokes Raman scattering (CARS) based on four-photon parametrical interaction carried out in the cubic nonlinear sensitivity of a material (Akhmanov and Koroteev, 1977).

Excitation of a sample by two laser beams with a certain angle between them and a difference in their frequencies that corresponds to a frequency of a certain vibration of a molecule produces a strong signal in the anti-Stokes region. The frequencies of the optical and wave vectors of the interacting waves must satisfy the following conditions:

$$\omega_3 = 2\omega_1 - \omega_2 \tag{3}$$

$$\mathbf{K}_3 = 2\mathbf{K}_1 - \mathbf{K}_2 \tag{4}$$

When one of the two laser frequencies falls within the electron transition band, resonance occurs that leads to a dramatic increase in the anti-Stokes scattering.

The CARS technique has found a wide application in different fields of physics; attempts to use it in biological research have also been made. An observation of anti-Stokes coherent scattering from some biological molecules has been reported (Nitsh and Kifer, 1977; Carriera *et al.*, 1977, 1978). The extremely high sensitivity of the method was seen in studies of carotene, which allowed an observation of carotene concentrations as low as $10^{-8}\,M$.

As is known, all chemical reactions involve structural changes in the reagents. Biological processes, including photosynthesis, are attended by conformational changes both in individual molecules and macromolecular complexes, as well as in parts of biomembranes. Clearly, a knowledge of such reactions and processes is impossible without understanding the mechanisms and kinetics of the conformational alterations. In solving those problems, dynamic Raman spectroscopy seems most promising. The pulsed Raman spectrometers that are now available made it possible to obtain the RR spectra of intermediates of rhodopsin and CO–hemoglobin photoreactions with a time resolution as high as 3 ns. The key to the method is the use of two successive pulses, the first as a source of excitation that initiates a photochemical reaction and formation of photoproducts, and the second, shifted in frequency, as a measuring beam. By varying the time interval between the two pulses, the dynamics of the conformational changes of the photoproducts can be followed.

General considerations suggest that, for photosynthetic research, RR technique is a very suitable method: Antenna pigments display intense absorbance in the spectral region, with the absorbance band coinciding with the major lines of Ar and He–Cd lasers; meanwhile, the fluorescence from Chl molecules does not mask the true picture, because it is characterized by a sizable Stokes shift. So far, little work has been done in photosynthesis using this technique, although there is no reason to have doubts about its benefits. Perhaps, the bar to work is the complication of interpreting RR spectra. Moreover, from among hundreds of Chl molecules that contribute to an RR spectrum, only one or two are from the reaction center complex and participate in the charge separation process (see below). Obviously, what is needed now is a better scheme for interpretating the complicated, multicomponent RR spectra of the photosynthetic apparatus.

2.5. Laser Monitoring Techniques

The laser spectroscopic methods that have been discussed above were devised for scientific research under laboratory conditions. Meanwhile, photosynthetizing organisms are very widespread in nature. For surveillance of ecosystems as to their productivity, or in pollution control, the functional activities of these key components of our environment are most important. A suitable means for this is a new method of laser probing (Fadeev, 1978) that employs a lidar concept. The lidar uses a continuously tunable laser and a receiver. The latter consists of a telescope, spectrum analyzer, photo-electronic receiver, and data processing equipment.

It has been found in work done at the University of Moscow that laser diagnostic techniques can provide, at distances as long as 100 m, infor-mation about the contents of photosynthetizing organisms and their chlorophyll concentration. The qualitative analysis involves monitoring both the spectra of excitation and of fluorescence, as well as RRS. In determining the chlorophyll content of aqueous media, for instance that of phyto-plankton in oceans, it is suitable to compare the fluorescence intensity with that of the Stokes component of the RS signal from a sample of water at a frequency of 3.440 cm^{-1} (Fig. 7). By this approach, there is no need to adjust and keep under control many of the parameters (such as a general geometry, initial hydraulic and optical characteristics, light path length in a water layer, etc.), because they have equal effects on both signals (Klyshko and Fadeev, 1978). In this method, the concentration of fluorescing particles can be adequately expressed as

$$n = F_0 \frac{X_{RS}}{\sigma_{fl}} \tag{5}$$

where n is the concentration of particles, in cm^{-3}; X_{RS} is $4\sigma_{RS}n_{H_2O}$, with σ_{RS} a cross section of Raman scattering of H_2O in cm^2/cr; n_{H_2O} is $3.3 \times 10^{22} \text{ cm}^{-3}$, the concentration of H_2O molecules; σ_{fl} is a cross section of fluorescence in cm^{-2}; it is a parameter that is a function of hydrooptical properties of water at a wavelength of fluorescence and Raman scattering (by suitably choosing laser excitation wavelength, its value can be obtained

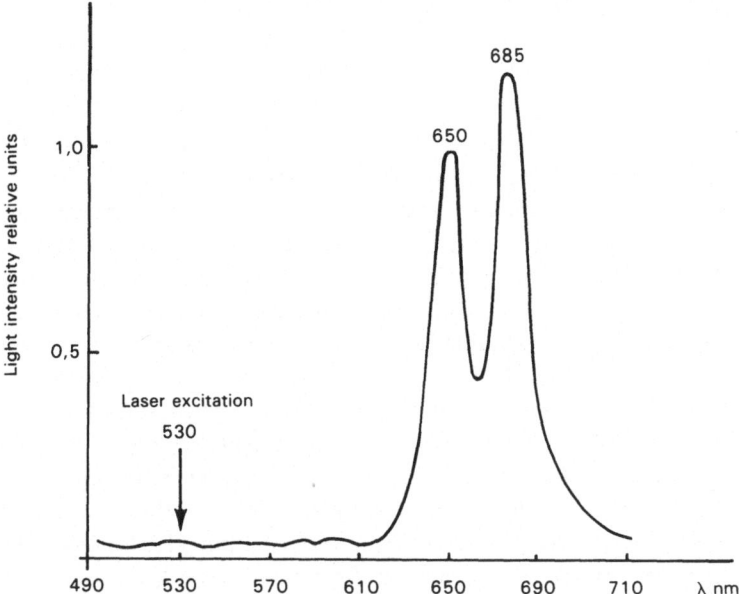

FIGURE 7. Spectra of the Raman scattering of water and the phytoplankton fluorescence obtained by laser monitoring of sea water. Excitation wavelength 530 nm. The maximum at 650 nm corresponds to the Raman scattering band of water, and the 685-nm maximum is the chlorophyll fluorescence. (After Fadeev, 1978.)

to be nearly 1); F_0 is $(\mathcal{N}_{fl}/\mathcal{N}_{RS})\delta$, where δ is a factor of fluorescence saturation, and can be written $I + \sigma_{abs}\tau \cdot F$, and where $\mathcal{N}_{fl}$ and $\mathcal{N}_{RS}$ are the integrals of a scattering and fluorescence (see Fig. 7), σ_{abs} is the cross section of the absorbance at the excitation wavelength, τ is the lifetime of the excitation, F is the photon density in photons $cm^{-2} s^{-1}$.

Experimental verification of Eq. (5), using a 532-nm laser excitation and a large variety of samples, and correlative analysis yielded the following empirical relation (Fadeev, 1978):

$$\beta_a! \, \mu g \, l^{-1}1 = (2.6 \pm 0.3) F_0$$

$$\beta_{ac}! \, \mu g \cdot l^{-1}1 = (1.2 \pm 0.3) F_0$$

$$(6)$$

Here β_a and β_{ac} are concentrations of Chl, and accessory pigments which absorb at 532 nm and deliver energy to Chl molecules, respectively.

3. LASER STUDIES OF PHOTOSYNTHESIS

In the past few years, as a consequence of the aggravation of the energy crisis, much scientific research has been directed toward searches for new alternative energy sources. In this regard, there is heightened interest in the use of sunlight as a source of energy. The most valuable prospects for practical solar energy conversion might be based on the photosynthetic model, since photosynthesis is an extremely effective means by which nature captures solar energy, converts it into separated charges, and stores it as useful chemical energy. New valuable insights have been gained recently into the mechanisms of photosynthesis by the application of new physical methods such as laser spectroscopy and nonlinear optics. The literature covering the primary events of photosynthesis is now extensive—more than several hundreds of publications. For this reason, this chapter is not meant to be a comprehensive review on the mechanism of photosynthesis; a few of the latest reviews on the subject are Holton and Windsor (1978), Sauer (1979), and Witt (1979). Rather, this chapter focuses on some leading questions in order to show the important role of laser techniques in this area of biology.

3.1. The Primary Events

In the last 10–20 years, the conception of the primary photosynthetic reactions has been considerably revised. In the 1950s and 1960s, electron transport (ET) reactions lasting 10^{-6} s were considered primary, while experimental research to date is concerned with reactions occurring in the 10^{-10}- to 10^{-12}-s time domain. Naturally enough, these steps are now thought to be the primary ones.

It is now well established that the primary process of excitation energy conversion takes place in the following several stages. The excitation quanta of light are absorbed by aggregates of pigments, called photosynthetic units (PSU) (Emerson and Arnold, 1932). The PSU consists of several hundred pigment molecules, mainly chlorophyll (Chl), or bacterichlorophyll (BChl), carotenoids, and phycobilins, which together act as a light-harvesting antenna (LHA) system. The role of the antenna is to transfer energy to a specialized trap or reaction center (RC). This process takes less than 1 ns. Charge separation and stabilization of the separated charges then occurs within the RC in about 10^{-10} s. It is now universally recognized that the PSUs form close associations among themselves, and are capable of exchanging energy with one another (the multitrap, or "lake" model).

In green plants, photosynthesis is driven by two different light reactions, sensitized by photosystems I and II (Emerson, 1958). PS II is responsible for oxygen evolution from water and for the reduction of plastoquinone (PQ); PS I sensitizes the reduction of nicotineamidadenindinucleotide phosphate (NADP), and the oxidation of PQ. As a consequence of two photosystems, the pigment complex in green plants has a complicated organization (Fig. 8) and its LHA channels the energy either to the RC of PS II (RC II) or to a focusing antenna of PS I to be then trapped by its RS (RC I) (Butler and Kitajima, 1975; Butler, 1978). The two photosystems have different spectral properties. PS I fluoresces at 735 nm; the fluorescence maximum of PS II is at 685 nm ($77\frac{1}{4}$ K). The fluorescence of PS II is believed to be due to the LHA Chls; the PS I fluorescence is usually ascribed to the long-wavelength Chls in its own antenna system (Satoh and Butler, 1978). Photosynthetic bacteria have only a single photosystem and therefore only one type of RC.

The overall efficiency of the primary photosynthetic events is very high —about 0.8–0.9 for energy migration from the LHA to the RC (Clayton

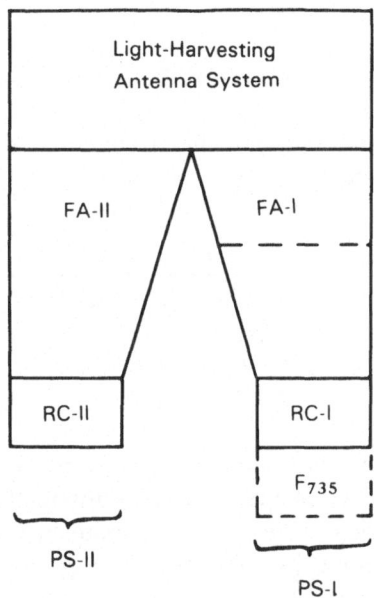

FIGURE 8. Schematic diagram of the pigment apparatus organization in green plants. FA-II, FA-I, focusing antenna of PSII and PSI; F-735, fluorescence of far-red form of Chl of PSI.

et al., 1972), and 0.97–0.99 for charge separation (Wraight and Clayton, 1974). Obviously, knowledge of the physical mechanisms that provide such unique characteristics might be very helpful in attempts to create better artificial electric converters for solar energy.

The section that follows gives a closer look at different stages of photosynthesis, placing emphasis on the progress that has been made in the understanding of them with the aid of laser techniques.

3.1.1. Energy Migration

The structure of the photosynthetic apparatus is arranged so as to provide a surprisingly high efficiency of energy absorption and energy migration from the PSUs to the RCs. Since electronic excitations of pigments are the direct result of light absorption by the light-harvesting pigments (Chl, BChl, carotenoids, phycobilins), fluorescence measurements have, thus far, played an important role. As is known, the rate equation describing the decay of excited singlets within a PSU is expressed as

$$- \frac{dn}{dt} = (K_f + K_h + K_{st})n \tag{7}$$

where K_f, K_h, K_{st} are the constants of fluorescence and internal and inter-combination conversion to a triplet state, respectively. The solution of Eq. (7) gives the relationship for the exponential decay time ($\tau_{1/e}$) and quantum yield of fluorescence (ϕ):

$$\tau = \frac{I}{K_f + K_h + K_{st}} = \frac{I}{\sum_i Ki} \tag{8}$$

$$\phi = \frac{K_f}{\sum_i Ki} \tag{9}$$

where $\sum_i Ki$ is the sum of intermolecular constants for the excitation decay processes. In the presence of intermolecular interactions, when an excited molecule, D, can either enter into a photochemical reaction with a quencher, Q, or transmit its excitation energy to an acceptor, A, the relations for τ and ϕ [Eqs. (8) and (9)] take the form

$$\tau_{DQ} = \frac{I}{\sum_i Ki + K_{DQ}(Q)} \tag{10}$$

$$\tau_{DA} = \frac{I}{\sum_i Ki + K_{DA}} \tag{11}$$

$$\phi_{DQ} = \frac{K'_D}{\sum_i Ki + K_{DQ}(Q)} \tag{12}$$

$$\phi_{DA} = \frac{K'_D}{\sum_i Ki + K_{DA}} \tag{13}$$

where τ_{DQ}, ϕ_{DQ}, and τ_{DA}, ϕ_{DA} are the lifetimes and quantum yield of the fluorescence emitted by D for cases when it interacts with Q and when the excitation migrates to A, respectively; K_{DQ} and K_{DA} are constants of photochemical interaction and energy migration, while K'_D is the fluorescence constant in the absence of an intermolecular interaction.

There are three possible resonance mechanisms by which the energy can migrate from D to A, depending on the energy of their interactions, diffusion of coherent or noncoherent excitons, and induction resonance (Agranovitch and Galanin, 1978). The molecular organization of the photosynthetic apparatus suggests that diffusion of noncoherent excitons and resonance induction are the most probable mechanisms for energy migration within the LHA. The probability for energy migration by resonance induction can be expressed as

$$W_M = \frac{K}{\tau} \frac{X}{R^6} \int \frac{D(\omega) A(\omega)}{\omega^4} d\omega \tag{14}$$

Here $D(\omega)$ and $A(\omega)$ are the normalized fluorescence spectrum of D and absorption spectrum of A; X is a factor for mutual orientation of the D and A dipoles; R is the distance between D and A; K is a dielectric constant of the medium; and τ is the lifetime of the singlet excitation in the absence of energy migration. As is seen, W_M depends not only on the overlap integral between the fluorescence spectrum of D and the absorption spectrum of A, but also on the sixth power of the distance between them, R^6, as well as on the mutual orientation of their dipole moments. Since biological systems are capable of active conformational rearrangements, one may expect that mutual orientations and positions of components of the photosynthetic apparatus play an important role in regulation of the excitation energy distribution.

In accordance with Eq. (7), the fluorescence decay of excited molecules D must be exponential, even when photochemical quenching and energy migration take place. The time course of the fluorescence from A molecules can be written as

$$\mathcal{N}^A(t) = \frac{K_{DA}\mathcal{N}_0^D(e^{-K_A t} - e^{-K_D t})}{K_D K_A} \tag{15}$$

where $\mathcal{N}^A(t)$ represents in time the population of A in the singlet excited state; $\mathcal{N}_0^D$ is the initial number of excited D molecules, t_0; and K_A and K_D are the sums of the deexcitation constants for A and D, respectively (Paschenko *et al.*, 1977). As is seen from Fig. 9, deduced on the basis of Eq. (15), $\mathcal{N}^A(t)$, and therefore the time course of the A fluorescence, is non-exponential. The two are identical under two boundary conditions, namely, when $K_D \gg K_A$ and $K_D \ll K_A$. As a rather good approximation, one can get K_A from the decay segment of the curve $\mathcal{N}^A(t)$. For the case where $K_D \ll K_A$ ($\tau_d \gg \tau_A$), the value of K_A refers to the rise, and the value of K_D to the decay, portion of the curve $\mathcal{N}^A(t)$. When $K_D \sim K_A$, the rise and decay of $\mathcal{N}^A(t)$ are nonexponential. Analysis makes it clear that determinations of K_A and K_D in binary systems require a knowledge of at least one constant, K_A or K_D. This must be taken into consideration in any investigations of the primary events of photosynthesis in which donor–acceptor interactions are dominant.

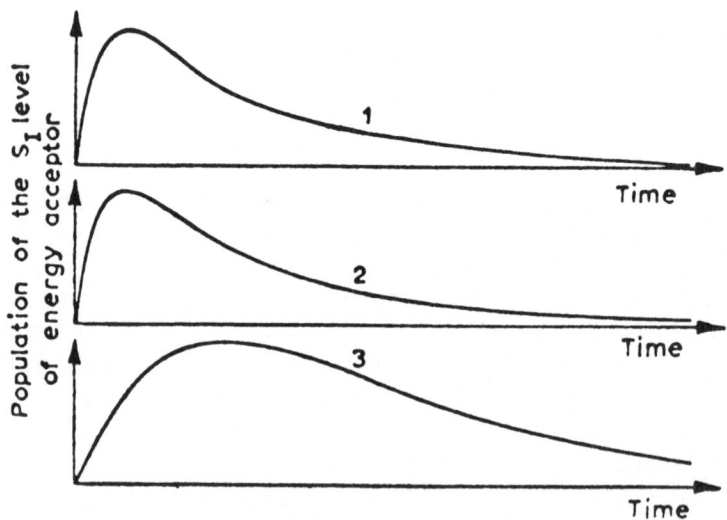

FIGURE 9. Dependence of the population of the lowest singlet excited state of an energy acceptor, $\mathcal{N}^A(t)$, on the relationship of K_D and K_A, the constants for deactivation of the energy donor and acceptor, respectively. 1, $K_A \gg K_D$; 2, $K_A \ll K_D$; 3, $K_A \sim K_D$. For details see the text. (After Paschenko *et al.*, 1977.)

From Eqs. (10)–(13) it is clear that photochemical interactions and energy transfer processes cause shorter excited state lifetimes, and lead to a lowering in fluorescence quantum efficiency. On the other hand, the decrease in τ and ϕ may also be an indication of the involvement of fluorescing molecules in photochemical reactions or in excitation transfer processes. In systems containing only identical molecules, there should be direct proportionality between the duration and quantum efficiency of fluorescence:

$$\phi = \tau/\tau_0 \tag{16}$$

$$\tau = \phi/\phi_0 \tag{17}$$

where τ_0 and ϕ_0 are the intrinsic lifetime and quantum efficiency, respectively, of the fluorescence in the absence of intermolecular interactions. It would appear from Eqs. (16) and (17) that for simple homogeneous systems τ can be deduced (at given values of τ_0 and ϕ_0) from ϕ and vice versa. Nevertheless, such an approach is far from being the best way to use in an effort to gain insights into ϕ and τ phenomena, since the photosynthetic apparatus consists of hundreds of Chl molecules having different spectral properties and different functional assignments in the energy transfer processes. The decrease in quantum efficiency may not necessarily be correlated with the shortening of excited state liketimes of photoactive antenna Chls.

Brody and Rabinowitch (1957), in an attempt to elucidate the role of excited Chls in the primary light reactions, measured the duration of Chl fluorescence in solution and *in vivo*. It appeared to be much shorter in green algae *Chlorella* as compared with that in ethyl ether (1.6 ns and 5 ns, respectively). The findings of those investigators were the first direct physical demonstration of the involvement of an excited pigment in the primary act of photosynthesis. Later, similar observations were made in the USSR (Rubin and Osniskaya, 1963).

The advent of high-resolution picosecond fluorometers made possible measurements with a resolution of 10^{-11} s of fluorescence kinetics in higher plant chloroplasts and subchloroplast fragments (Table I). The first picosecond fluorometric observations were done with high-intensity picosecond pulse train excitation, with pulse energies of 5×10^{14} photons cm^{-2} or even more. Subsequent studies have demonstrated that, under high-intensity excitation conditions, nonlinear decays of fluorescence take place with a dramatic drop in quantum efficiency and lifetime. Mauzerall (1976) showed that the quantum efficiency of fluorescence from *Chlorella* decreased markedly at high intensities for 7-ns excitation pulses. Campillo *et al.* (1976a, b), in similar picosecond measurements, found that the decrease in the quantum efficiency of chlorella fluorescence occurred simultaneously

TABLE I. Room Temperature *in Vivo* Chl a Lifetime Measurements in Dark-Adapted Samples: Early Results with Pulse Trains of High Intensity

Sample	Lifetime (ps)	Reference
Spinach chloroplast	10 and 320 for PS I and PS II	Siebert *et al.* (1973)
Pea chloroplast	80 and 200 for PS I and PS II	Paschenko *et al.* (1975)
PS I- and PS II-enriched preparations of spinach chloroplasts	60 and 200 for PS I and PS II	Yu *et al.* (1975)

with a dramatic shortening of its lifetime as the pump intensity was increased. These observations have been explained in terms of bimolecular singlet–singlet and singlet–triplet annihilation processes (Swenberg *et al.*, 1976, 1978; Geacintov *et al.*, 1977; Campillo *et al.*, 1977; Campillo and Shapiro, 1978; Tredwell *et al.*, 1978). It is thought that at high-intensity excitation several singlet excitons (probably triplets too) are produced within the PSU. As a consequence, alongside normal monomolecular excited state deactivation, bimolecular annihilation processes begin to take place:

$$S_1 + S_1 \xrightarrow{\sigma_{SS}} S_n + S_0 \to S_1 + S_0 + \text{heat} \qquad (18)$$

$$S_1 + T_1 \xrightarrow{\sigma_{ST}} S_0 + T_n \to S_0 + T_1 + \text{heat} \qquad (19)$$

S_0, S_1, S_n represent ground, first excited, and higher excited singlet states, respectively; T_1 and T_n represent lowest and higher excited triplet states; σ_{SS} and σ_{ST} are the rate constants for the $S_1 + S_1$ and $S_1 + T_1$ annihilations, respectively. It is probable that because of such interactions the excess energy is released as heat, and there is a decrease in τ and ϕ. The significance of the works listed above is not so much in providing true estimates for τ_{fl} (Table II), as in the impetus they gave to singlet diffusion studies of the photosynthetic pigment apparatus. In fact, σ_{SS} and σ_{ST} [Eqs. (18) and (19)] reflect not only the activity of the annihilation process in the system, but also the exciton diffusion. If so, studies of the effects of excitation intensities on fluorescence lifetime and quantum efficiency may provide information about coefficient (D) and diffusion distance (L) for excitons formed within the antenna complexes of PSUs (Campillo *et al.*, 1976a). The creation and destruction of excitons, $n(t)$, for the case of $S_1 + S_1$ annihilation, when the singlets come within their interaction radius, can be adequately described by a standard kinetic equation:

$$\frac{dn(t)}{dt} = C(t) - \beta n(t) - \tfrac{1}{2}\sigma_{SS} n^2(t) \qquad (20)$$

TABLE II. Room Temperature *in vivo* Chl a Lifetime Measurements in Dark-Adapted Samples, Obtained after Excitation by Single Laser Pulse of Low Intensity (10^{13} photon cm^{-2})

Sample	Lifetime (ps)	Reference
1. Chlorella	650 (PS II?)	Campillo *et al.* (1976b)
2. Chlorella	450 (PS II?)	Porter *et al.* (1978)
3. Spinach chloroplasts		Searle *et al.* (1977)
PS II	500	
PS I[a]	100	
4. Pea chloroplast fragments		Korvatovsky *et al.* (1979)
LHA	3000	
PS II (RC II-"open")	450	
PS II (RC II-"closed")	1200	
5. Entire pea chloroplasts		Korvatovsky *et al.* (1979)
RC II-"open"	340	
RC II-"closed"	900	

[a] Pulse energy was 5×10^{14} photon cm^{-2}.

where $C(t)$ is a function of the singlet excitation source; $\beta = 1/\tau_0$ is a deactivation rate constant for low intensity excitation; σ_{SS} is a constant for an $S_1 + S_1$ annihilation process. Equation (20) will be applicable only on some assumptions (Geacintov *et al.*, 1977). First, one must assume that excitons can move free by within the photosynthetic membrane (diffusion in a two- or three-dimensional multitrap PSU). Secondly, the exciton density is uniformly distributed within the PSU, because of its large size, and is uniformly reduced as an $S + S_1$ annihilation process takes place in it. It should also be assumed that these annihilation processes normally commence only after a uniform exciton distribution has been reached, i.e. the condition $\sigma_{SS} \cdot n \ll I/\tau_j$ must hold (τ_j is a jump time; n is an exciton density). At large excitation energies, n increases to a point that the mean separation between excitons becomes comparable with the distance between neighboring molecules. In this case the exciton diffusion mechanism does not work and Eq. (20) is inadequate. It can be shown that for the LHA, which normally consists of about 300 Chl molecules, 10^{16} photons cm^{-2} are a limiting intensity of a 530-nm pulse for which exciton diffusion is still at work.

Assuming that the excitation pulse is a δ function ($\tau \cdot \text{pulse} \ll I/\beta$), Eq. (20) can be rewritten as

$$n(t) = \left\{ \left[-\frac{\sigma_{SS}}{2} + \left(\frac{\zeta I}{n(0)} + \frac{\sigma_{SS}}{2} \right) e^{\beta t} \right] \right\}^{-1} \tag{21}$$

Here $n(0)$ is the amount of absorbed photons. $n(0) = \sigma \mathcal{N} I$; σ is a cross section of an absorbing molecule; $\mathcal{N}$ is the number of pigment molecules involved in the process; and I is an excitation pulse intensity. Clearly, when $\sigma_{SS} = 0$, the decay of $n(t)$ has a monomolecular character. At $\sigma_{SS} \neq 0$ the decay kinetics are nonexponential. From Eq. (21) the excitation intensity dependencies of fluorescence lifetime, $\tau_{I/e}$, and its quantum efficiency, ϕ/ϕ_0, can be found:

$$\tau_{1/e} = \frac{I}{\beta} \ln \left[\frac{e + \sigma_{SS} n(0)/2\beta}{I + \sigma_{SS} n(0)/2\beta} \right] \tag{22}$$

$$\frac{\phi}{\phi_0} = \frac{2\beta}{\sigma_{SS} n(0)} \ln \left[1 + \frac{\sigma_{SS} n(0)}{2\beta} \right] \tag{23}$$

Using Eqs. (22) and (23), one can obtain a family of theoretical curves—$\tau_{I/e}$ and ϕ/ϕ_0 versus excitation intensity—for different values of σ_{SS}. Comparisons of such a curve with experiments will make it possible to estimate the value of σ_{SS} in a certain system of investigation (Geacintov $et\ al.$, 1977; Campillo $et\ al.$, 1976a).

In terms of a structural model for the LHA complex in green plants (Fig. 8), it is now generally accepted that excitation energy absorbed in the antenna migrates to the RCs of PS II and PS I (Butler, 1978). It is believed that excitation trapping at the RCs is a dominant process that accounts for

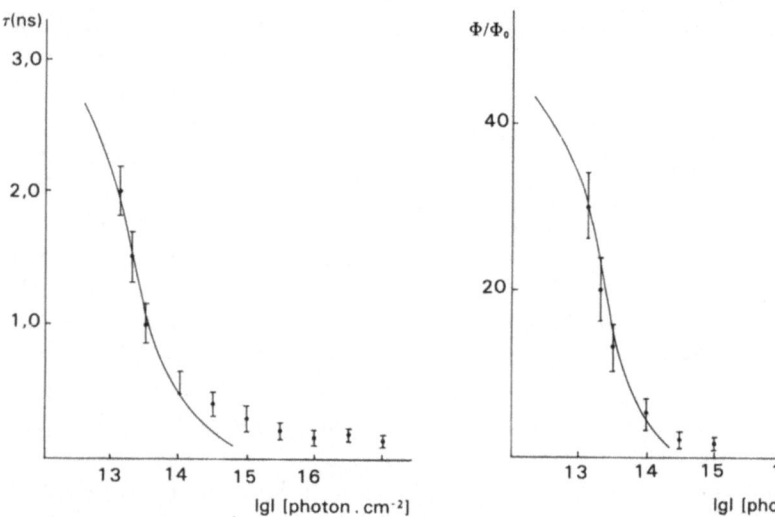

FIGURE 10. Dependence of the lifetime (τ) and the relative quantum yield ϕ/ϕ of the fluorescence of the LHA preparation from pea chloroplast on the energy of the 530-nm picosecond excitation pulse. Theoretical curves, for $\gamma_{SS} = 7 \times 10^{-8}\ cm^3\ s^{-1}$.

short lifetimes of electronic excitation in the LHA (Tables I and II). The possible role of other factors will be discussed later. As the theory predicts, RCs can be regarded as quenchers. Obviously, for the exciton diffusion mechanisms to be understood we need experiments on material free of those quenchers. Our fluorescence investigations have been designed to this end (Rubin *et al.*, 1979; Korvatovsly *et al.*, 1979), and the effects of different excitation intensities starting at 3×10^{12} photons cm^{-2}, at 530 nm on τ and ϕ/ϕ_0 in pea chloroplasts and their fragments were sought. In the first series of experiments, we used RC-free LHA preparations (Guljaev and Tetenkin, 1979). The lifetime of fluorescence at 685 nm was found to be reduced from 3000 to 150 ps, as the excitation intensity was varied from 3×10^{12} to 10^{17} photons cm^{-2} (Fig. 10). Nonexponential decays were observed at high intensities and monoexponential ones at low intensities of excitation. Experiments with entire chloroplasts and PS II-enriched chloroplast particles also showed dramatic changes in τ and ϕ/ϕ_0 at 685 nm with increased intensity (Fig. 11).

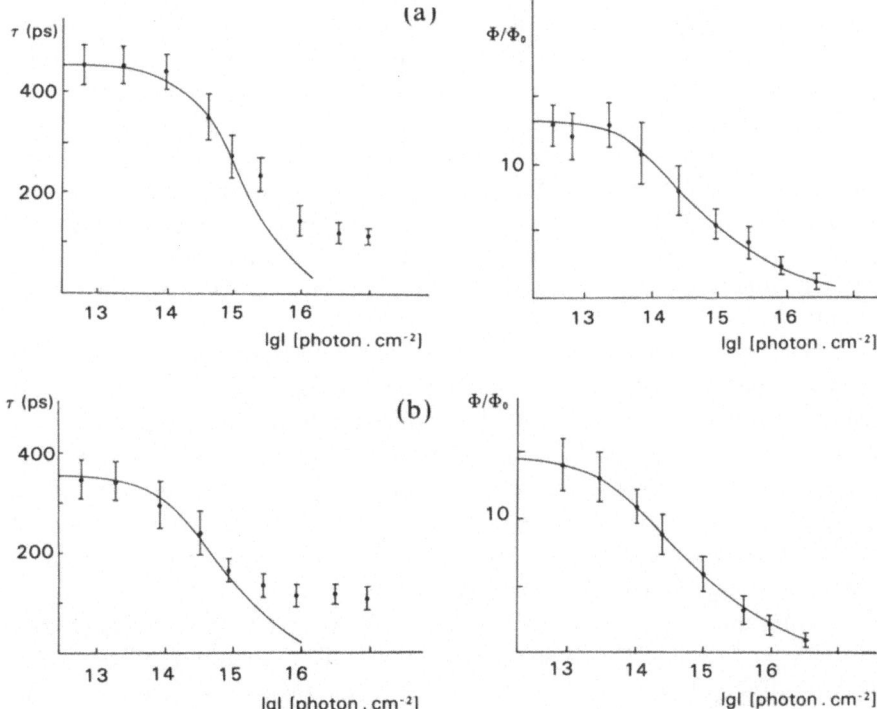

FIGURE 11. Dependence of the lifetime τ and the relative quantum yield ϕ/ϕ of the fluorescence: (a) Entire pea chloroplast; ———, theoretical curves for γ_{SS} of 1.5×10^{-8} cm^3 s^{-1}. (b) PSII particles from pea chloroplasts; ———, theoretical curves for γ_{SS} of 0.7×10^{-8} cm^3 s^{-1}.

According to the well-known Duesence hypothesis, RCs act as quenchers for antenna excitation only when they are in the reduced ("opened") state. One would expect, keeping in mind the high quantum efficiency of photosynthesis, that the oxidation of the RC and its transition into the "closed" state should be attended by a rise in the lifetime and quantum efficiency of LHA fluorescence to values that are normally encountered in Chl solutions. As our experiments showed only a two- to threefold rise in τ_{fl} for PS I and PS II, with the RC driven in the "closed" state (Paschenko et al., 1975, 1977), this negligible rise suggests that even in the "closed" state RCs remain effective quenchers of antenna excitation. As is known, the oxidation of an RC occurs concomitant with a bleaching of its main absorption maximum and the appearance of a new maximum in the far red spectral region. Thus, the 680-nm absorption band of RC II's was found under such conditions to be shifted toward 830 nm (Mathis and Van Best, 1978), and the 890-nm band of bacterial RCs toward 1250 nm (Reed and Clayton, 1968). When such absorbance changes take place (with other conditions unchanged), the overlap integral between the LHA and RC should decrease by a factor of 20–30 and as a consequence τ_{fl} for the LHA should increase by the same factor. The experimental results suggest that the oxidation of the RC is attended by conformational changes in the photosynthetic apparatus, which reduce the separation between the RCs and the LHA by 1.3–1.4 times. Since the energy migration rate constant is the sixth power of the distance over which the excitation migrates, it is clear that even with a negligible decrease in distance between the antenna and trap a 20- to 30-fold decrease in their overlap integral can readily be expected. It is also possible that owing to the conformational changes the mutual orientation of the dipole moments of LHA and RC molecules may change too.

With high-intensity laser excitations in a multitrap domain PSU there occurs a partial oxidation, and therefore "closing," of some of the RCs, leading to changes in τ and ϕ of fluorescence from them. Since σ_{SS} is a function of τ_0 [(22), (23)], failure to control the trap state ("open" or "closed") may introduce confusion in results, leading to misinterpretation of the excitation dynamics. To avoid this uncertainty, we carried out a series of experiments with entire chloroplasts and PS II subchloroplast particles containing only closed RCs (Fig. 12).

Examining the data of Figs. 10 and 11 shows first of all that a τ_{fl} (LHA) of 3000 ps is close to those reported for Chl solutions (Brody and Rabinowitch, 1957; Paschenko et al., 1975), thus suggesting that there is no concentration effect on Chl in vivo fluorescence although high concentration of Chl in solution have strong quenching of fluorescence. In fact, in a Chl b chlorophorm solution the fluorescence lifetime of 572 ps was reduced to

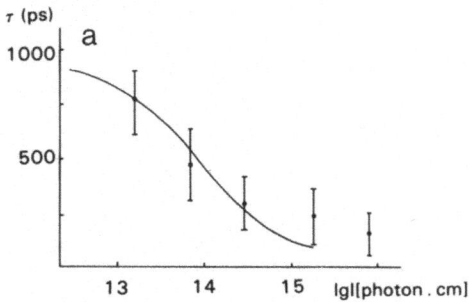

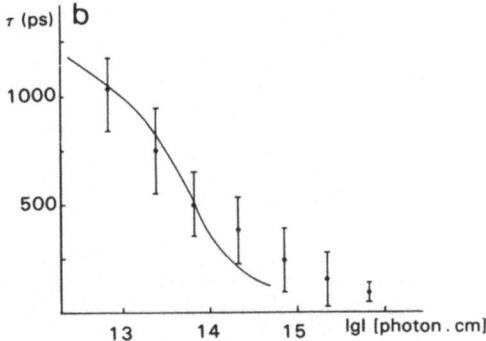

FIGURE 12. Dependence of the lifetime τ of the fluorescence: (a) Entire pea chloroplasts with RCI in "closed" state (RC^+II); ——, theoretical curve for γ_{SS} of 4×10^{-8} cm^3 s^{-1}. (b) PSII particles from pea chloroplasts with RCII in "closed" state (RC^+II); ——, theoretical curve for γ_{SS} of 7×10^{-8} cm^3 s^{-1}.

15 ps when the concentration was increased from 3.8 mM to 0.1 M (Shapiro *et al.*, 1975).

It is probable that the lack of concentration quenching of Chl *in vivo* is the consequence of its specific imbedding and the way it interacts with proteins of the LHA. If this conclusion is true, it is evident that Chl–protein complexes *in vivo* play an extremely important role, since they are the very structures that provide the unique property of the LHA—the efficient light harvesting and transfer of the energy to the RC with no losses associated with concentration quenching (Lutz, 1977; Paillotin, 1977; Korvatovshk *et al.*, 1979).

The variations in lifetimes observed in our preparations (Fig. 13) are obviously due to different quenching activities of RC II's and RC I's. In fact, in samples containing only RC II's (PS II-enriched chloroplast fragments) the lifetime of antenna pigment fluorescence (normally 3000 ps) reduces to 1200 ps when the RC II's were driven to the "closed" state, and

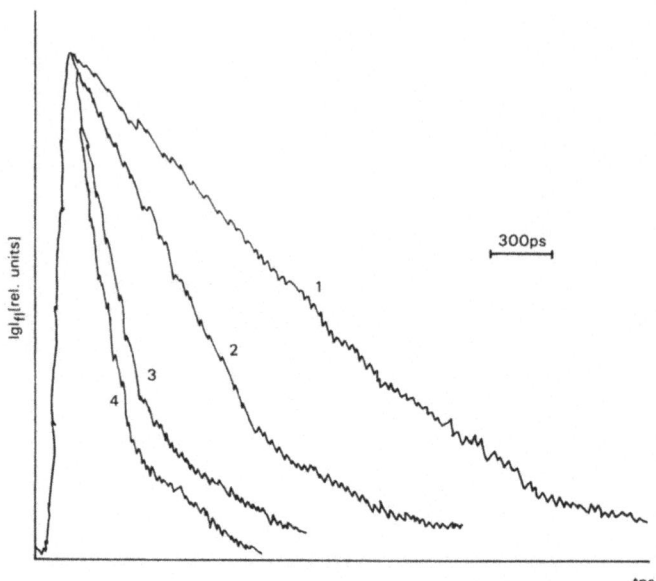

FIGURE 13. Fluorescence decay curves of entire pea chloroplasts and their fragments: 1, LHA τ_0 is 3.000 ps; 2, PSII particles with RCII in "closed" state (RC^+II), τ_0 of 1.200 ps; 3, PSII particles with RCII in "open" state τ_0 at 450 ps; 4, Entire chloroplasts $(RCII + RCI)$; τ_0 at 340 ps.

to 450 ps when they were "opened." When both RC I and RC II quenchers were present (entire chloroplasts) the reduction in τ of the LHA fluorescence was even more pronounced: to 900 ps with "closed" RC II's, and to 340 ps with "open" RCs. From these data, the rate constants for energy migration from the LHA to the RCs with their nearest pigment surroundings K_M^{11} and K_M^1, can readily be calculated:

$$K_M^{11} = \frac{I}{\tau(\text{PS II})} - \frac{I}{\tau(\text{LHA})} \tag{24}$$

$$K_M^1 = \frac{I}{\tau(\text{chloroplasts})} - \frac{I}{\tau(\text{LHA})} - K_M^{11} \tag{25}$$

where $\tau_{\text{PS II}}$, τ_{LHA}, and $\tau(\text{chloroplasts})$ are fluorescence lifetimes for PS II, LHA, and entire chloroplast preparation. Calculations give K_M^{11} as $1.82 \times 10^9 \text{ s}^{-1}$ ("open" RCs), and K_M^1 as $0.64 \times 10^9 \text{ s}^{-1}$. Hence, for normal conditions with "open" RC II's K_M^{11}/K_M^1 is 3, i.e., RC II's trap 3 times as much energy as RC I's. The data of Table II show only mean excitation times for the LHA. Further information on exciton dynamics occurring within the LHA can be obtained by mathematical analysis. By comparing theoretical and experimental curves (Figs. 10, 11, and 12), we found the values of σ_{SS} for singlet–singlet annihilation in the presence of RC II and RC I quenching centers (Table III). In the case of three-dimensional isotropic diffusions, there is the following relation between σ_{SS} and diffusion coefficient D (Suna, 1970):

$$\sigma_{SS} = 8\pi DR \qquad (26)$$

R here is the radius of the interaction between the two excitons with efficiency of $P \leqslant I$. The validity condition for Eq. (26) is

$$L = (D\tau)^{1/2} \gg R \qquad (27)$$

where L is the diffusion distance and R the interaction radius. These conditions can be satisfied only at moderate excitation levels, when exciton diffusion can take place. Assuming that the LHA is a rigid three-dimensional crystal configuration with Chl molecules in the cross links, the crystal constant for a concentration of 0.1 M will be equal to 20 Å. In all likelihood, there is strong exciton–photon coupling in the LHA complex, and the excitation, once localized on a single Chl molecule, moves further by jumps (Paillotin, 1977). The probability for excitation transfer declines as the sixth power of the transfer distance, r^6. Therefore, to a first approximation, exciton jumps over distances greater than the crystal constant may be neglected. In fact this means that the effective radius for exciton–exciton interactions can be taken as a crystal constant (Agranovitch, 1968;

TABLE III. Parameters of the Exciton Diffusion in Green Plants Photosynthetic Apparatus

Sample	γ_{ss} (cm^3 s^{-1} × 10^8)	D (cm^2 s^{-1})	L (Å)	W (s^{-1})	Reference
LHA from pea chloroplasts[a]	7 ± 3	2×10^{-2}	900	10^{12}	Rubin et al. (1979)
PS II particles from pea chloroplasts[a]					Rubin et al. (1980)
RC II–closed	7 ± 3	2×10^{-2}	490	10^{12}	Rubin et al. (1980)
RC II–open	1 ± 5	2.5×10^{-3}	110	10^{11}	Rubin et al. (1980)
Spinach chloroplasts[b]	0.5	10^{-3}	200	3×10^{-10}	Geacintov et al. (1977)

[a] Three-dimensional model.
[b] Two-dimensional model.

Agranovitch and Galanin, 1978), i.e., ~ 20 Å. As theory predicts, the probability, W, of a noncoherent exciton jump is a function of a diffusion coefficient, D (Treifai, 1956):

$$W = DR^2 \qquad (28)$$

Using the values of σ_{SS} given in Table III and taking R as 20 Å, Eqs. (26), (27), and (28) can be solved to obtain D, L, and W (Table III). Note that calculations were made on the assumption that the annihilation probability, P, is unity, while in reality P is 1. Therefore, the estimates of Table III correspond to their lowest probable limits.

Our value for the antenna diffusion constant, D, of 2×10^{-2} cm^2 s^{-1} is somewhat greater than those reported in the literature (Geacintov *et al.*, 1977a, b; Swenberg *et al.*, 1976). We believe this to be due to the material used. No exciton diffusion experiments with RC-free LHA complex preparations have been performed as yet. Obviously, in the presence of RC quenchers exciton diffusion is not homogeneous. Besides, after excitation some of the RC II quenchers become "closed," a fact that has been disregarded thus far. The essential feature is a rather large part, $L \geqslant 900$ Å, for excitation diffusion in the LHA complex preparations, which is taken as evidence for a multitrap domain model for PSU organization.

It is known that lowering the temperature leads to an increased fluorescence lifetime. The fluorescence activation energy, ΔE, was found by us to be 0.02 eV (Guljaev *et al.*, 1979). This energy is in fact a measure of the energetic depth of a RC II quencher with respect to the energy level of the localized excitation. Taking the probability, W, for an excitation jump from an antenna Chl molecule to the CR II as being equal to 10^{12} s^{-1} (Table III), and using the ΔE value of 0.02 eV, the probability of a reverse excitation jump can be estimated as

$$\vec{W} = \vec{W} e^{-\Delta E/kT} \simeq 0.45\vec{W} \qquad (29)$$

Thus, the probability for an exciton to leave the RC II is as large as $0.45W$. The long excitation lifetime (3000 ps) and the great diffusion path ($L = 900$ Å) are probably factors that together provide optimal conditions for multitrap quenching at the RC, and the resulting quantum yield of photosynthesis.

According to the model of Fig. 8, the LHA complex acts as an energy donor for both RC IIs and PS I. In support of this scheme is an observation of nonlinear quenching of the 735-nm fluorescence from the pigment complex of PS I. In a recent fluorescence study (Searle *et al.*, 1977), the lifetimes of fluorescence in PS I and PS II subchloroplasts were investigated at room temperature. The lifetime of fluorescence from PS II-enriched particles was reported as 500 ps for an excitation photon density of 5×10^{13} photons ch^{-2}, and was found to decrease to 150 ps as the photon density was increased to 5×10^{16} photons ch^{-2}. In the PS I, the fluorescent emission was found to last 100 ps without any variation with photon density. Similar observations, which were accompanied with a simultaneous and large drop in the PS I fluorescence quantum efficiency, were reported by several research groups (Rubin *et al.*, 1979; Campillo and Shapiro, 1978). Of principal importance is the fact that the dependency of τ_{fl} and ϕ/ϕ_0 as a function of excitation energy for PS I and PS II appeared to be identical. The results were interpreted (Campillo and Shapiro, 1978) in terms of annihilation processes in the LHA complex (the energy supplier for the two photosystems) that caused the observed short lifetimes and drop in the 685-nm fluorescence quantum efficiency (PS II). The annihilation processes cause the energy flow to PS I to be reduced, a phenomenon seen as a drop in 735-nm fluorescence. Within the antenna system of PS I annihilation is presumably not taking place, and for this reason no changes in its fluorescence lifetime could be observed. If this logic is true, and the LHA complex acts as an energy donor and PS I as energy accepter, one would expect a rise time in level of PS I fluorescence at 735 nm. Meanwhile, there are only a few works (Paschenko *et al.*, 1975; Searle *et al.*, 1978; Porter *et al.*, 1978) in which this phenomenon was observed, with the rise time much shorter than that of 685-nm fluorescence decay. Our recent experiments, using an extremely sensitive fluorometer (Rubin and Rubin, 1978) demonstrated a distinct correlation between the dependences of a τ_{685} and τ_{735} upon the energy of the excitation at room temperature. This led us to conclude that at room temperature the fluorescence at 735 nm is mainly the standard slope of the LHA fluorescence band with a maximum at 685 nm (Tusov *et al.*, 1980). The quantum yield of a fluorescence 735 nm emitted by a far-red form of Chl-705 (705 nm—its maximum of an absorption) is estimated to be 0.002 with a τ of 30 ps at room temperature. That means that at room temperature it is practically impossible to measure the fluorescence of PS I at 735 nm, and all results obtained before must correspond to the emission of the LHA. The origin of the differences in the values of τ_{735} at room temperature (Table I) may be the result of a dependence of the lifetime of the LHA fluorescence upon the energy of the excitation. It seems that the Chl-705 acts as a source of energy for the RC I by the mechanism of an anti-Stokes energy transfer. The value of the activational energy was found to be 0.08 eV and a constant of energy migration $K = 3.3 \times 10^{10}\,\mathrm{s}^{-1}$. The dependence of the τ_{735} upon the temperature can be described by an analytical formula in good agreement with an experimental result

$$\tau_{735} = (A + Be^{-\Delta E/kT})^{-1} = 10^{-9}(0.45 + 830e^{-0.08/kT})^{-1}\,\mathrm{s} \qquad (30)$$

Our recent experiment with Chl-protein complex containing only RC I and focusing antenna (FA) of Chl with the ratio 1:40 gave the lifetime of the Chl in FA of 30 ps, which is independent of temperature. A simple calculation made it possible to estimate the value of the RC I lifetime, $\tau_{RC\,I}$, of 5 ps, and a constant of a charge separation K_e^I of 1.5×10^{11} s^{-1}. As a summary of our new results it is possible to present the following flow scheme for the distribution of an energy of the excitation within the pigment apparatus of green plants (Korvatovsky *et al.*, 1980):

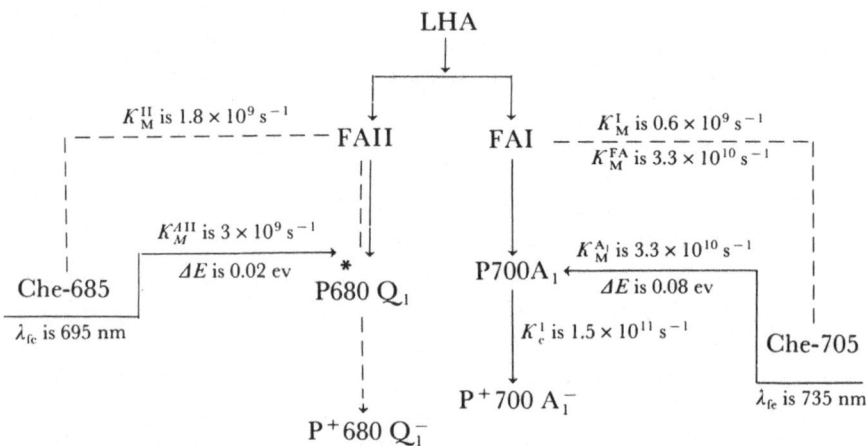

Recently, a new method, based on observations of radiation transitions from higher excited states of Chl molecules, has been modified to study energy distribution in the photosynthetic pigment apparatus (De Vault and Kung, 1978).

Electronic excitation dynamics of bacterial pigment complexes were also investigated in a number of works. Fluorescence lifetimes were measured for different bacterial species (Table IV) and estimates were made of BChl cross section of absorbance *in vivo* and of σ_{SS} for four species of purple bacteria (Campillo *et al.*, 1977; Paschenko *et al.*, 1977, 1978).

In bacterial chromatophores, the fluorescence from antennal BChl appeared to be temperature independent (with the accuracy of the experimental error) in the temperature region from 295 to 80 K. Presumably, the energy transfer and trapping mechanism in bacteria differs somehow from that in higher plants (Rubin *et al.*, 1979a). In bacteria, the decay time of the 200-ps fluorescence rose to 500 ps, when the RCs were driven to the "closed" state (Paschenko *et al.*, 1977, 1978). In low-temperature experiments, fluorescence lifetimes of chromatophores frozen in

TABLE IV. Fluorescence Lifetime of Photosynthetic Bacteria at Room Temperature

Sample	Lifetime (ps)	Investigators
R. sphaeroides PM-8dpl	1110 ± 100	Campillo *et al.* (1977)
R. sphaeroides 2.4.1.	100 ± 50	Campillo *et al.* (1977)
R. sphaeroides Ja	100 ± 25	Campillo *et al.* (1977)
R. sphaeroides R-26	300 ± 50	Campillo *et al.* (1977)
R. sphaeroides 1760-1	200	Paschenko *et al.* (1977)
Chromatophores reaction	$15 \pm 8 (Bchl)_2$	Paschenko *et al.* (1977)
Centers	250 (BPh)	Paschenko *et al.* (1977)

the dark ($80\frac{1}{4}$ K), and under continuous illumination, were measured as 200 and 330 ps, respectively. In dark-adapted chromatophores, after cooling, all the RCs are fixed in the "open" state, whereas cooling of light-adapted samples made it possible to keep most of the RCs in the "closed" form owing to the stabilization of $(BChl)_2^+$ in the oxidized form (Lukashev *et al.*, 1976). This leads to less efficient excitation quenching of the LHA and to a longer lifetime of its fluorescence.

Although the electronic excitation dynamics of the LHA complex in purple bacteria have been studied only slightly, there seems to be a close resemblance between the primary steps of photosynthesis in bacteria and higher plants: the absorption of excitation energy is followed by efficient exciton diffusion within the multitrap domain LHA complex which supplies the excitons further to reaction centers where separation and stabilization of charges occurs.

3.1.2. Charge Separation and Stabilization

The reaction center of the photosynthetic apparatus is a special complex of a primary donor, BChl/Chl dimer and protein, with other intermediate primary and secondary electron donors and acceptors. Normally, the electron level of the BChl/Chl dimer excited state is lower than that of the corresponding light harvesting chlorophylls, which provides an efficient electron trap for it. Excitation of the primary donor in the RC initiates a series of coupled electron transfers via an intermediate electron carrier, I, to a primary acceptor, A_1, and subsequently to a secondary acceptor A_2:

$$PIA_1A_2 \xrightarrow{h\nu} P^*IA_1A_2 \longrightarrow P^+I^-A_1A_2 \longrightarrow P^+IA_1^-A_2 \longrightarrow P^+IA_1A_2^-$$

This scheme represents the main general principle for RC function. However, there are certain specific organizations peculiar to the RCs of individual photosynthetic bacteria and green plants. Since 1973 much scientific research has been directed toward gaining insights into these specific primary events with the aid of picosecond spectroscopy.

3.2. Photosynthesizing Bacteria

The light reactions of the photosynthesizing bacteria have been the most extensively studied and the best known at the present. This is the happy result of the progress in preparative biochemistry that has provided means for extracting active RCs. The first work (Reed and Clayton, 1968) was done with a particular carotenoidless mutant, R-26, of *Rhodopseudomonas sphaeroides*. It was found that the RC consists of a set of four molecules of BChl, two of BPh (BChl without the Mg atom), one non-heme iron atom, two quinones (usually ubiquinones), and three different component proteins of 21, 24, and 28 kdaltons. Two of the BChl molecules have an absorption maximum at 865–885 nm. They form a closely coupled dimer, or "special pair," identified as $(BChl)_2$ or P870 (Katz *et al.*, 1977). Light absorption results in bleaching of the 870-nm band and in an appearance of a weak maximum at 1250 nm, as a consequence of $(BChl)_2$ photooxidation (Reed and Clayton, 1968). The other two BChl molecules absorb at 800 nm. No clear role has, thus far, been assigned to them in the electron transfer process. The two BPhs have absorption maxima in the red region at 760 nm, but differ in absorbance in the visible region, exhibiting, at low temperatures, maxima at 532 and 542 nm, respectively. The molecule of BPh that absorbs at 542 nm was found to be involved as an electron carrier from $(BChl)_2$ to quinone (Parson and Cogdell, 1975; Parson *et al.*, 1975; Fajer *et al.*, 1975; Clayton and Yamamoto, 1976). The reduction of $(BChl)_2^+$ occurs by electron donation from the associated cytochrome. Once $(BChl)_2^+$ is reduced, the reaction center is returned to its "neutral" form and is ready to react again.

The kinetics of bleaching of the 870-nm band with a rise time of 7 ± 2 ps were first observed in 1973 (Netzel *et al.*, 1973). An observation of $BChl_2^+$ absorption at 1250 nm with a similar kinetic pattern was a strong argument for the occurrence of $(BChl)_2$ oxidation in an electron transfer reaction (Dutton *et al.*, 1975):

$$(BChl)_2 \xrightarrow{\ h\nu\ } (BChl)_2^* \xrightarrow[\ 7\,ps\ ]{} (BChl)_2^+ + \bar{e}$$

Once oxidized, the $(BChl)_2^+$ lasts unless it undergoes reduction by electron donation from cytochrome or recombination in a back reaction. The oxidation of $(BChl)_2$ appears to be concomitant with the reduction of BPh in times of $\lesssim 10$ ps (Kaufman *et al.*, 1975; Rockley *et al.*, 1975), suggesting a close coupling between the two processes without involvement of intermediate electron carriers between $(BChl)_2$ and BPh. However, recent work (Shuvalov *et al.*, 1978) has reported data for the involvement of BChl 800 in electron transfer between $(BChl)_2$ and BPh in *Rhodospirillum rubrum*. The charge separated state $(BChl)_2^+ PBh^-$, commonly designated P^F (Parson *et al.*, 1975; Parson and Cogdell, 1975), has a lifetime of 100–120 ps (Kaufmann *et al.*, 1975) or 180 ps (Rockley *et al.*, 1975). The differences between the lifetimes probably lies in the different methods of preparing the material. This is the time in which the electron is transfered to quinone, a secondary acceptor:

$$(BChl)_2^+ \, BPh^- Q_1 Fe \xrightarrow{\;150\ ps\;} (BChl)_2^+ \, PBhQ_1^- Fe$$

If $Q_1^- Fe$ was preliminarily reduced, the lifetime of P^F is found to increase up to 10 ns (Parson *et al.*, 1975). The relaxation of the P^F state under such conditions involves back reactions and the formation of the RC's $(BChl)_2$ triplet state. The transfer of the electron from the primary to secondary acceptor occurs on a slower time scale up to 10^{-4} s (Halsey and Parson, 1974; Chamorovsky *et al.*, 1976). The early electron transfer events in the bacterial RC can be presented schematically as follows:

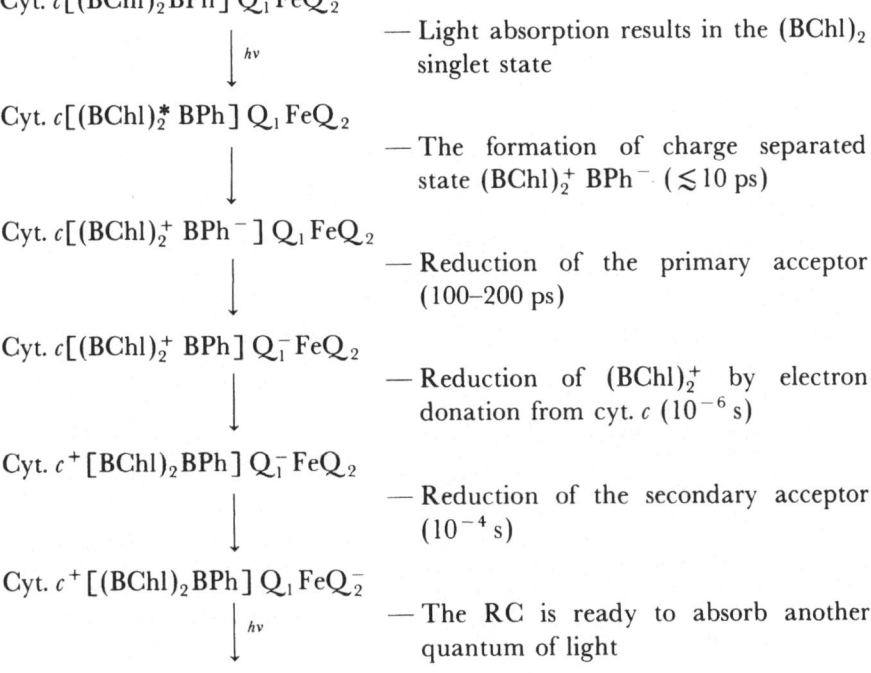

Cyt. $c[(BChl)_2 BPh] Q_1 FeQ_2$

↓ hv — Light absorption results in the $(BChl)_2$ singlet state

Cyt. $c[(BChl)_2^* BPh] Q_1 FeQ_2$

↓ — The formation of charge separated state $(BChl)_2^+ BPh^-$ ($\lesssim 10$ ps)

Cyt. $c[(BChl)_2^+ BPh^-] Q_1 FeQ_2$

↓ — Reduction of the primary acceptor (100–200 ps)

Cyt. $c[(BChl)_2^+ BPh] Q_1^- FeQ_2$

↓ — Reduction of $(BChl)_2^+$ by electron donation from cyt. c (10^{-6} s)

Cyt. $c^+ [(BChl)_2 BPh] Q_1^- FeQ_2$

↓ — Reduction of the secondary acceptor (10^{-4} s)

Cyt. $c^+ [(BChl)_2 BPh] Q_1 FeQ_2^-$

↓ hv — The RC is ready to absorb another quantum of light

As pointed out above, a distinctive feature of RC function is the extremely high quantum efficiency of charge separation—about 1.02 ± 0.04 (Wraight and Clayton, 1974). This means that only one of 100 absorbed quanta is dissipated. Indeed, the quantum yield of fluorescence from the RCs is extremely low, which makes it difficult to use the picosecond fluorometric techniques to study electron transfer reactions in the bacterial RC. We were the first to have obtained insights into these processes (Kononenko *et al.*, 1976; Noks *et al.*, 1976; Paschenko *et al.*, 1977). In *Rps. sphaeroides* R-26 the fluorescence from the RCs exhibited a two-component nature with decay times of 15 ± 8 and 250 ps. On the basis of spectral and kinetic analysis the τ_1 of 15 ± 8 ps was ascribed to the deactivation of $(BChl)_2^*$ associated with the transfer of an electron to BPh. The τ_2 of 250 ps was ascribed to the deactivation of BPh^* (following its direct excitation with a ruby laser flash) in the $BPh^* \xrightarrow{\bar{e}} Q_1^- Fe$ reaction. Note that our observations are in good agreement with those seen with the aid of absorption spectroscopy. Our results suggest that BPh^* can, in principle, act as a primary electron donor for $Q_1 Fe$. The apparent primary donor function of $(BChl)_2$ probably results from the much more efficient energy migration from the antenna to $(BChl)_2$, than to BPh which absorbs in a shorter wavelength region.

Along with rapidity and high quantum efficiency of charge separation

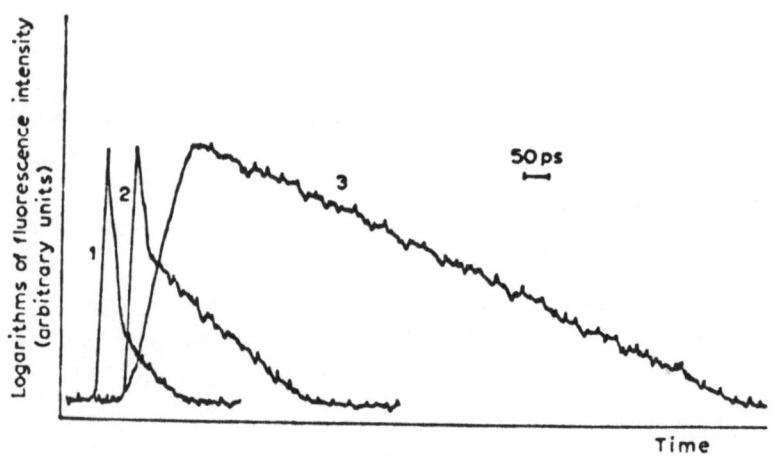

FIGURE 14. Decay kinetics of the fluorescence from photosynthetic reaction center preparations made from *Rps. sphaeroides*, strain 1760-1. λ_{exc} is 694.3 nm; λ_{meas} is 850–1000 nm; 1, 100% reduction of P870; 2, 40%–50% oxidation of P870; 100% oxidation of P870. The redox state of P870 was buffered by adding sodium ascorbate and potassium ferricyanide, and monitored by ESR measurements. (After Paschenko *et al.*, 1977.)

at the RC, the distinguishing feature of electron transfers from $(BChl)_2$ to BPh, and further to Q_1Fe, is the temperature dependence (Peters *et al.*, 1978). As reported by Clayton (Clayton, 1977), the quantum efficiency of the long-wave (900-nm) RC's fluorescent emission (when the Q_1 Fe complex is chemically reduced) increases 2–3 times as the temperature is lowered from 220 to 180, but remains invariant with its further lowering to 40 K. Our low-temperature fluorescence observations (Fig. 14) (Rubin and Rubin, 1978; Kononenko *et al.*, 1978) were similar in this regard. At temperatures

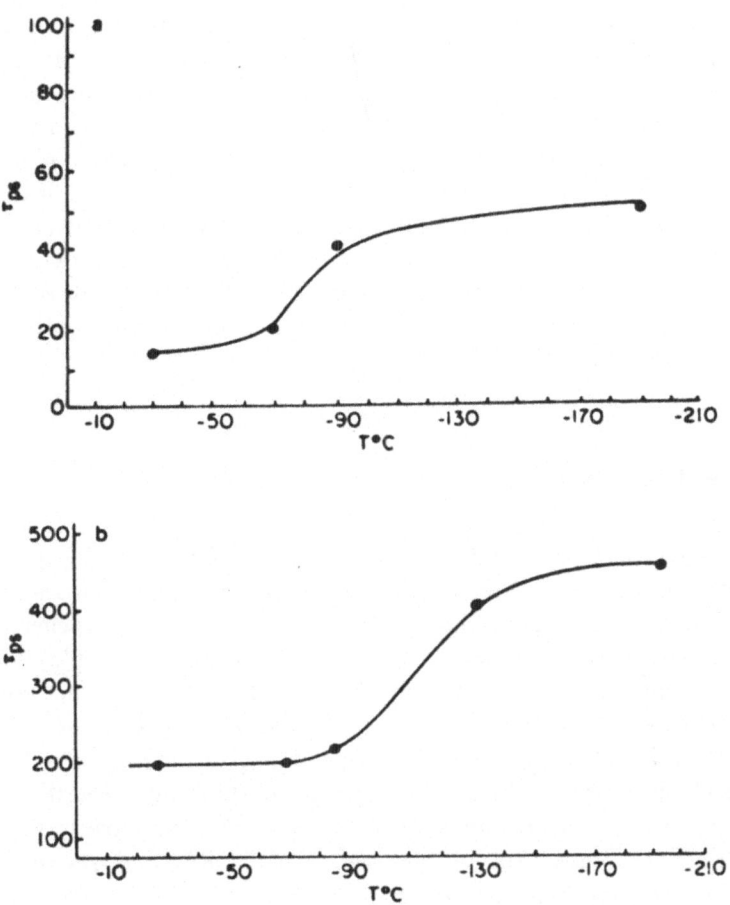

FIGURE 15. Dependence of the lifetime (τ) of the fluorescence of *Rps. sphaeroides* PC_s on the temperature. (a) Fast component; (b) slow component. (After Rubin and Rubin, 1978.)

from 230 to 170 K, we observed a twofold slowing of both fast and slow fluorescent components (Fig. 15). These data give a reasonably good fit with a tunneling model for electron transfer.

A possible mechanism need not be examined in detail, but it should be noted that prevention of backward electron tunneling and charge recombination is a necessary condition for the provision of efficient charge separation and stabilization. We believe that conformational changes are dominant in this mechanism. The radical-ion pair, $(BChl)_2^+ BPh^-$, that results from charge separation is a strong dipole and the field arising from such a dipole at a distance of 20 Å may be as strong as 3×10^{-6} V cm^{-1}, based on calculations from kinetic data of the carotenoid bandshift in chromatophores of *Rps. sphaeroides* (Leigh *et al.*, 1974). Obviously, such strong electric fields may cause electric and nuclear polarization of proteins, the basic components of the RC structural matrix. Photoinduced polarization of this kind may not only induce conformational alterations within carriers by which the stabilization of separated charges is maintained, but may also be implicated in the storage of energy during the primary reaction of photosynthesis (Rubin and Rubin, 1978; Rubin *et al.*, 1979a). Local electric fields arising from electron transfers within the RC may control, in some sense, the feedback in the regulation of charge separation. Support for this idea is provided by experiments (Paschenko *et al.*, 1977; Pellin *et al.*, 1978) that show that variations in redox states of the components within the RC strongly affect ET rates.

3.3. Higher Plants

Progress in understanding the mechanism of the primary photoreactions in the reaction centers of PS I and PS II in higher plants has lagged behind that in the photosynthetic bacteria. In part this deficiency arises from the difficulty in isolating simple reaction center complexes free of associated antenna Chl. Commonly there are 40 Chl *a* molecules associated with each PS I RC (Thornber *et al.*, 1978) and over 100 Chl associated with each PS II RC. Indeed, there is strong masking of the photoconversions of "special" Chl molecules in the RC structure of this kind. Nevertheless, laser pulse spectroscopy, ESR, and some other optical studies have provided new insights in the operation of the higher plant RC and revealed many analogies with the bacterial ones.

The primary light reaction in the RC I of higher plants is thought to proceed via the following successive steps (Sauer *et al.*, 1978; Shuvalov *et al.*, 1979b):

$$P700\ A_1 A_2\ P430 \xrightarrow{\quad hv \quad} P^*700\ A_1 A_2\ P430$$

$$\xrightarrow{\quad <60\ ps \quad} P700^+\ A_1^-\ A_2\ P430$$

$$\xrightarrow{\quad 200\ ps \quad} P700^+\ A_1\ A_2^-\ \overline{P}430$$

$$\xrightarrow{\quad 300\ ps \quad} P700^+\ A_1\ A_2\ \overline{P}430$$

$$\xrightarrow[pc\ \downarrow pc+]{\quad 20\ ms \quad} P700\ A_1 A_2\ \overline{P}430$$

$$\xrightarrow[Fd\ \downarrow Fd]{\quad 5\ ms \quad} P700\ A_1 A_2\ P430 \xrightarrow{\quad hv \quad} \cdots$$

Here P700 is the primary electron donor of PS I; A_1, A_2, and P430 are electron acceptors; PC, plastocyonine, the electron donor for P700; Fd, ferredoxin, the electron acceptor for P430. The molecular species involved in P700 appears to be a dimer of the Chl a molecules with many properties similar to those seen in the bacterial systems (Katz $et\ al.$, 1977). The time required for P700 to undergo oxidation has only recently been measured to be around 10 ps (Fenton $et\ al.$, 1979; Shuvalov $et\ al.$, 1979a). The intermediate acceptor A_1 is believed to be a form of Chl a monomer (Fujita $et\ al.$, 1978) or a dimer (Shuvalov $et\ al.$, 1979b). BPh has this function in the bacterial RCs. It is probable that A_1 is a complex of Chl a and Fe–S protein. Normally, the electron stays on A_1 about 200 ps before transferring to subsequent acceptors (Shuvalov $et\ al.$, 1979a; Friesner $et\ al.$, 1979).

The second electron acceptor A_2 is made up of Fe–S protein of ferredoxin. The rate of electron transfer from A_1 to P430, the final acceptor (bound ferredoxin center), has not been measured as yet, but since the $P700^+\ A_1^-$ back reaction (recombination) occurs for about 700 ps (Sauer $et\ al.$, 1978), one would expect it to be of the same order of magnitude. The normal lifetime of reduced $P430^-$ is about a few milliseconds before the electron is moved to the bound ferredoxin center where the final stabilization of separated charges occurs. The reduction of P700 occurs by electron transfer for a copper-containing protein, PC, and takes about 20 ms. Like the bacterial light reaction, that of PS I has an extremely high quantum efficiency. Recent low-temperature observations using nano- and picosecond absorption spectroscopy have demonstrated that tunneling is the mechanism by which electron transfers occur within the PS I RC (Shuvalov $et\ al.$, 1979b; Ke $et\ al.$, 1979). The time of the electron transfer from P700 to A_1 appears temperature independent over 5–70 K but drastically decreases with further increases in temperature, and gives an activation energy of 1

(Ke *et al.*, 1979). The statistically exponential of pairs ($P700^+$ A_1^-, A_2) that has been found in PS I is characterized by different mutual orientations and distances between the components in each pair. The mean distance between $P700^+$ and A_1^- is believed to be ~ 13.2 Å (Ke *et al.*, 1979).

The sequence of steps in the light reaction sensitized by PS II (Mathis and Van Best, 1978) can be presented as follows:

$$D \xrightarrow{10^{-6}} P680 \xrightarrow[1,5'10\,s^{-4}]{hv < 1\,ns} A_1 \xrightarrow{4,2'10\,s^{-3}} A_2 \longrightarrow PQ$$

The primary donor P680, a $(Chl)_2$ dimer (Van Gorkom *et al.*, 1974), undergoes oxidation within less than 10^{-9} s, donating an electron to an intermediate carrier A_1. Pheophitin seems a very probable candidate for this role (Klimov *et al.*, 1978). The oxidation of A_1 goes concurrently with the reduction of a secondary acceptor A_2, quinone or its derivative, designated $\times 320$ for its optical properties. The acceptor A_2 forms a close association with the pool of endogenous plastoquinones.

It is clear from what has been said so far that there are numerous common features among the molecular structures and operations of reaction centers in evolutionary quite different photosynthetizing organisms such as bacteria and the higher plants. Following the absorption of a photon by antenna Chl the excitation migrates to an adjacent RC where it is trapped by a "special" pair, an (B-) Chl dimer, which is converted after that into an excited singlet state. Within a few picoseconds there occurs charge separation leading to radical pair formation: $(Chl/BChl)_2^+$ and $(BPh/Chl)^-$. The stabilization of separated charges occurs in stages that follow: electron transfers from $(BPh/Chl)^-$ to the primary and then secondary acceptors. The molecules involved in those stages of electron transport show some differences between bacteria (quinone), PS II (plastoquinone), and PS I (Fe–S proteins and ferredoxin), but they are all integral components imbedded into the protein matrix of their RC. In some sense, the reaction center may be regarded as a photoenzyme endowed with a specific activity in regard to charge separation and stabilization. The duties of this enzyme include the provision of a strict coordination between electron transfer processes within its $[(BChl)_2 BPh\, Q_1 FeQ_2]$ center, and conformational changes in its protein matrix which prevent charge recombination.

Indeed, during the charge separation processes some of the energy is lost. As much as 25 % of the energy trapped is lost during the first electron transport (ET) step: $(BChl)_2^+ BPh$–radical pair formation. After the separated charges have been stabilized on the secondary donors and acceptors, the fraction of useful photon energy that is stored is about 30 % for the purple bacteria and 50 % for the higher plant light reactions. In other words, although the bacterial RC traps 1.3 eV, only 0.4 eV is stored in the

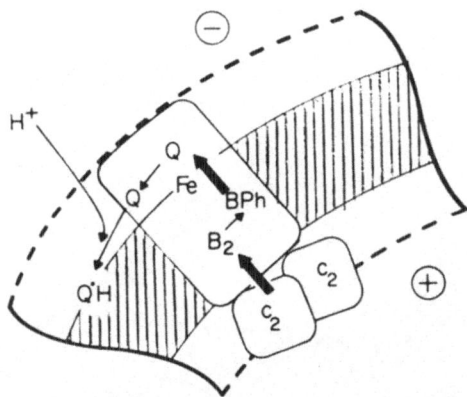

FIGURE 16. Schematic representation of the reaction center of *Rps. sphaeroides* with its associated cytochromes, in the chromatophore membrane. The thick arrows represent transmembrane electrogenic events. B_2, $(BChl)_2$; C_2, cytochromes; BPH, bacteriopheophytin; Q, quinone. (After Dutton *et al.*, 1978.)

form of separated charges, and in higher plants the relation is 0.95 eV out of 1.9 eV. According to Boltzmann's formulation, it is due to such great losses that the separated charges are stable without much probability of recombining in a backward reaction. Clearly, the storage of energy at the RC must invoke a mechanism whereby the rapid dissipation of the excess quanta can be provided. An important energy sink is heat dissipation via molecular vibrations. Meanwhile, the role of protein conformational changes as an energy sink, at least for some of it, cannot be entirely ruled out. This conformational energy may further be utilized for the synthesis of high-energy compounds. In this connection, it is interesting to raise the question of why the primary photochemistry leading to charge separation operates on a step-by-step principle. Is this not an indication that there is possible storage of a small portion 0.2–0.3 eV of energy that seems to be lost on each elementary ET step? In nature, the RC is embedded in the membrane in such a manner that charges of opposite sign arising from the photochemical act are separated between the inside and outside surfaces of the membrane (Fig. 16). These spatially segregated charges thus provide direct sources of the chemical potential used in photophosphorylation.

3.4. Molecular Organization of the Photosynthetic Apparatus

Structural organization of the photosynthetic apparatus has received the greatest emphasis to date in photosynthetic research, for it is its specific

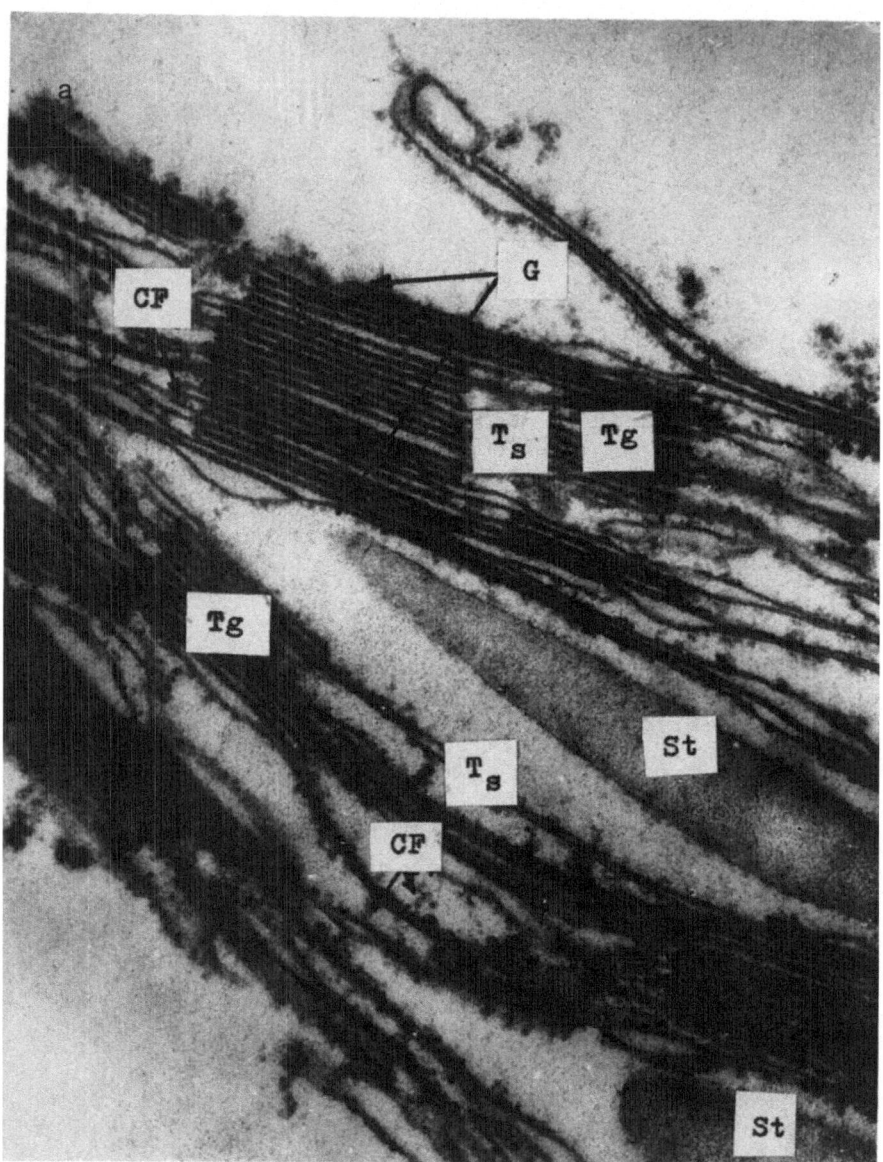

FIGURE 17. Ultrathin section of a pea chloroplast before (a) and after (b) ruby laser irradiation, 50 MW cm^{-2}. G, grana; Tg, thylacoids of grana; T$_s$, thylacoids of stroma; DF, coupling factor particles; St, starch grains. ×97,500.

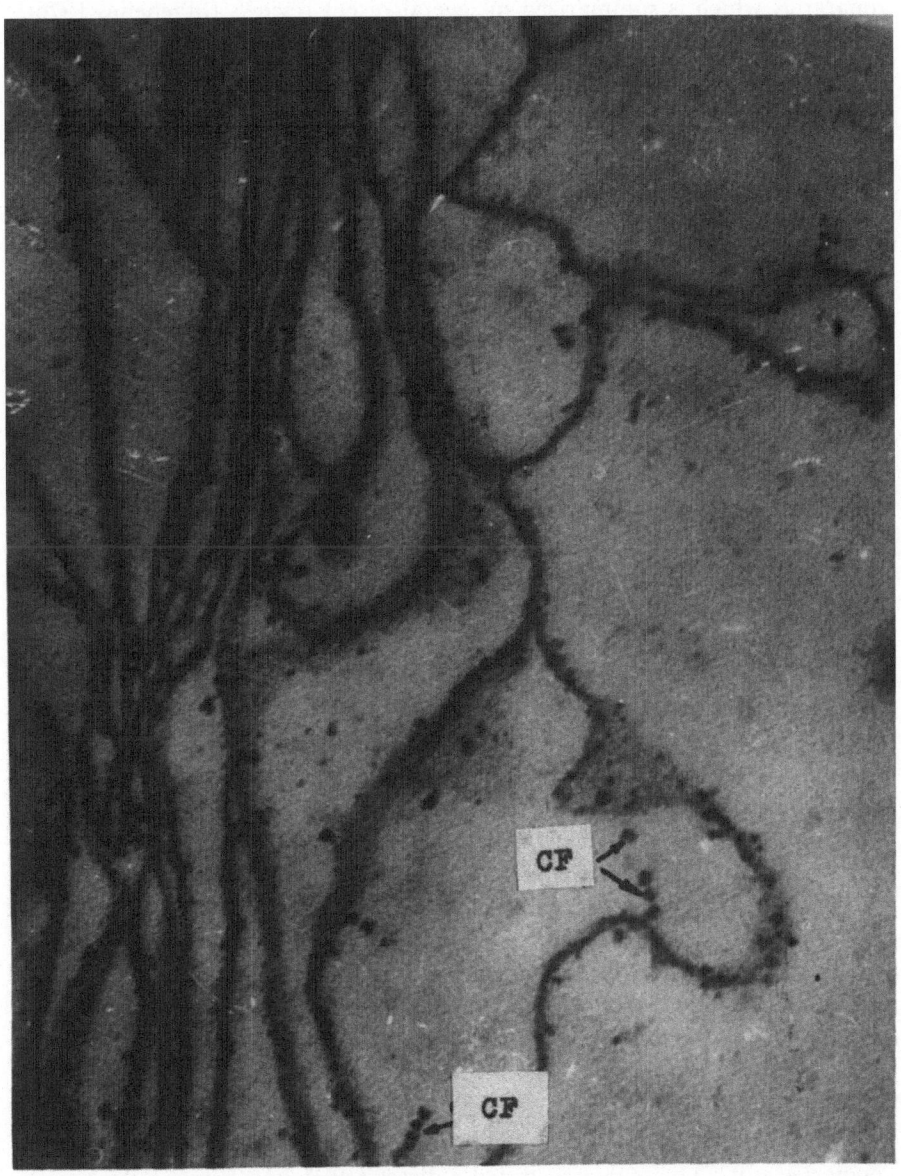

FIGURE 17 (Continued)

molecular architecture that provides the unique functional properties of the entire process. Most of our knowledge on the structure of the photosynthetic apparatus, which is summarized in several recent reviews (Staehelin, 1976; Arntzen, 1978), comes from optical and electron microscopy observations. Laser techniques and laser interactions with biological systems have opened the door for new approaches to solving this problem.

3.4.1. Action of Ruby Laser Radiation

The photosynthetic apparatus is characterized by a broad absorption spectrum in the red region with a maximum at ~680 nm. The overall band width of the spectrum is determined by a nonuniform spectral broadening of different antenna Chls and pigments in PS I. It is evident that ruby laser radiation at 694.3 nm falls within the absorption band of the photosynthetic apparatus and will be absorbed, in the first place, by long-wave Chls associated with PS I. One would expect that high-power laser irradiation will cause, owing to a very fast energy absorption and concomitant heat release, a short-term, nonstationary heating of the chlorophyll–protein complexes. If the use time of the pulse is shorter than the thermal relaxation constants of the photosynthetic matrix (Hayes and Wolbarsht, 1968), the heating temperature may produce acoustic and pressures waves leading to disruption and dissociation of constituent structural components of the photosynthetic apparatus. The essential feature of laser damage is its local effect on excitation absorbing centers of Chl molecules. This makes it particularly suitable for selective damage of structures containing resonantly absorbing Chls.

Experimental investigations of the relation between structure and function based on such an approach were commenced by us in 1969. Laser radiations at powers from 2 to 10 cm^{-2} were found to cause selective damage to PS I, as demonstrated by ESR signal 1, oxidation–reductions of P700, and methylviologen dye reduction (Rubin et al., 1971; Krendeleva et al., 1972a, b). PS I damage was attended by a marked suppression of phosphorylation. More powerful radiations, >10 cm^{-2}, caused damage both to PS I and PS II (Khitrov et al., 1979).

Electron microscopy studies (Keyhani et al., 1971; Floyd et al., 1971; Popov et al., 1977) have demonstrated a direct correlation between the laser-induced inhibition of phosphorylation and the removal of coupling factor (CF$_1$) particles from surfaces of photosynthetic membranes exposed to laser irridiation (Fig. 17). Note that no suppression of the Hill reaction was observed in these experiments.

Electron microscopy techniques, notably freeze-fracturing and freeze-etching, have provided many new and valuable insights into the structure of the photosynthetic membrane (Deamer and Brauton, 1967, Moor and Mühlethaler, 1963). Freeze-fracturing of the chloroplast thylakoid exposed numerous small (110 Å) and a few larger (175 Å) particles on its fracture faces. Attempts to identify them were first made by Arntzen and co-workers (Arntzen et al., 1969). Analysis of freeze-fractured particles of different size classes led to the concept of an asymmetric particle distribution. The small particles found on inner fracture faces were identified as sites of localization of PS I centers; the larger particles on the outer fracture faces appeared to be enriched in PS II activity. It was then established that large particles protrude through two adjacent thylakoids, thereby making them fuse together. (See review articles by Singer and Nicolson, 1972; Arntzen, 1978.)

Further freeze-fracturing studies using ruby laser irradiations (Popov et al., 1977) revealed that laser exposure causes discrete size increments in the particles, occurring simultaneously with changes in their amount and occupied area (Fig. 18). These observations demonstrated the occurrence of particle aggregation on PF faces (P, protoplasmic half-membrane leaflet), and deaggregation on EF faced (E, endoplasmic half-membrane leaflet). Perhaps the difference in behavior of the two size-class particles is the result of the asymmetric localization of Chl molecules, for the latter are responsible for the absorption of electromagnetic ruby laser radiation and its transduction into thermal and mechanical energy. It stands to reason that the laser-induced lesions will occur first in the sites of antenna Chl, since, in fact, this pigment acts as some intrinsic heat or mechanical stress susceptible "label."

Figure 18 represents photographic images of chloroplast membranes after exposure to a laser fluence of 0.5 MW cm^{-2}. The disappearance of particles in the vicinity of the "core" is evidence for identifying this area as a site of Chl localization. This conclusion is in accord with the idea (Anderson, 1975) that Chl molecules are localized around the lipoprotein globule embedded into the outer surface of the membrane.

The investigation has demonstrated lability to mechanical stress of the structure of the photosynthetic apparatus (stacking or unstacking), suggesting that laser irradiation may be employed successively for fractionating intact chloroplasts without detergent treatment. In order to varify this idea, we examined the effects of a giant 1060-nm pulse from a Nd^{3+} laser on a chloroplast suspension. The chloroplasts were found to retain photosynthetic activity at exposures up to 500 MW cm^{-2}; however, structural changes were observed in the presence of carbon particles (~ 500 Å in diameter), showing many similarities with those encountered after detergent or uncoupler treatment. In particular, one observed the removal of coupling factor particles, thylakoid swelling, as well as disintegration of the

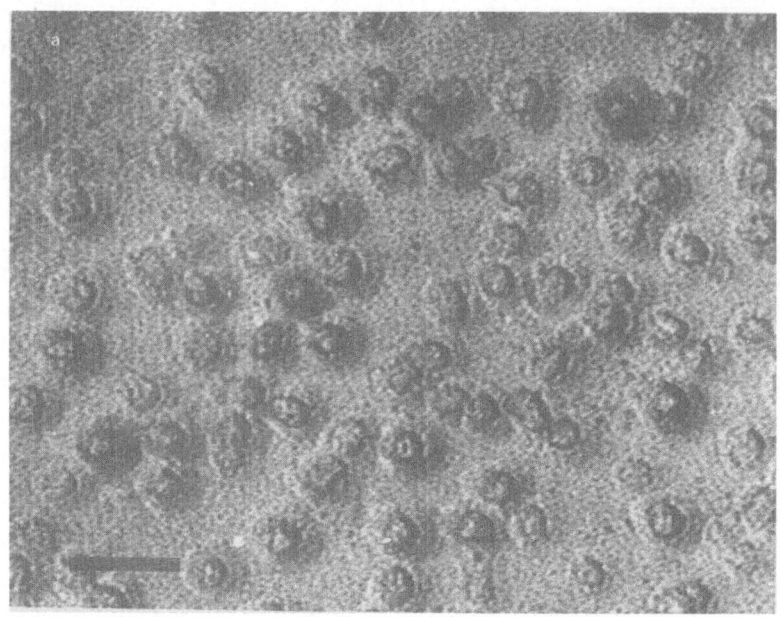

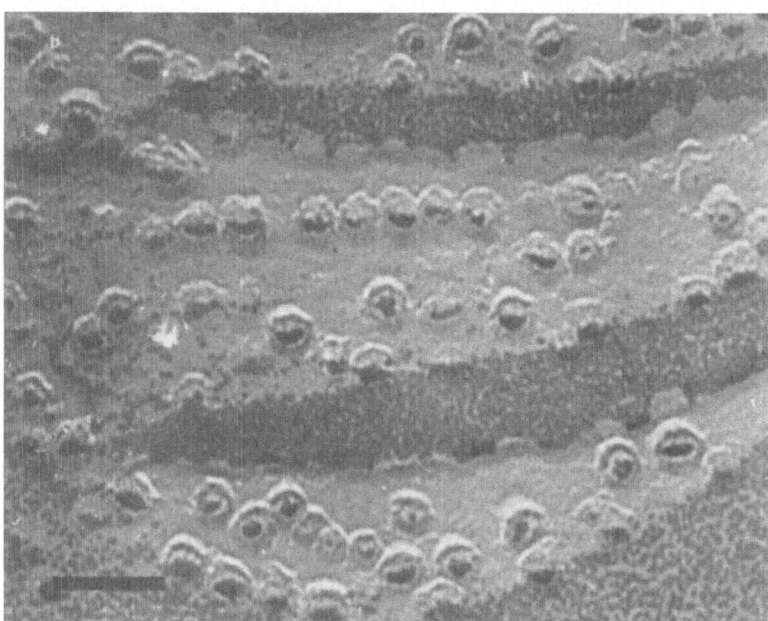

FIGURE 18. The outer fracture face of the membranes of stacked thylakoids of the pea chloroplasts (a) before and (b) after ruby laser irradiation 0.5 MV cm^{-2}. Fracture faces were coated with platinum plus carbon, upon which a layer of carbon was superimposed. Scale bars: 500 Å.

chloroplast membranes. Note that in this case there were no specific changes in a "core" structure.

These data clearly illustrate that strong laser radiation may be profitable in studies of Chl topography by virtue of its specific inaction with the photosynthetic systems. It is felt that use of various wavelengths and laser power levels may, in principle, provide some clue to understanding of localization, functional role, and nature of spectrally different Chls and other accessory pigments.

3.4.2. Raman Scattering Spectroscopy

For a long time, the nature of molecular organization of the LHA complex, and interactions between its pigment and the proteins and lipids of the photosynthetic membrane have been studied by optical absorbance and fluorescence methods. The understanding arising from those studies yielded a consistent view about the existence of a variety of spectral forms of Chl *in vivo* (Litvin and Sineshchekov, 1975; Brown, 1972) with different aggregation capabilities. In other words, the existence of dimerized Chls and its large aggregates—oligomers—have been postulated in a native photosynthetic apparatus, along with monomers.

Meanwhile, because of the insufficient sensitivity of optical methods the explicit answers to the questions as to the nature and patterns of interactions of Chl with its surroundings have not so far been obtained. IR absorption and Raman scattering phenomena, which are a reflection of intermolecular effects, might provide valuable information in this regard.

Katz and co-workers, by their low-temperature absorbance studies of Chl solutions in the visible and IR spectral region, have produced evidence that water and other molecules containing $R-O-H$, $R-S-H$, and $R-N-H$ groups play some role in the formation of spectrally different forms of Chls *in vivo* (Cotton *et al.*, 1978). Instead of aggregates (Chl) these forms were found to be composed as a chain of Chl molecules, linked by hydrogen and/or coordination bond via H_2O molecules or amino acids. Thus, the IR spectroscopic studies made it clear that the origin of the spectral forms of Chls is not the dimerization.

The studies of Lutz in France (Lutz, 1972, 1977; Lutz and Breton, 1973; Lutz and Kleo, 1974) have proven RR spectroscopy to be very fruitful in investigating structural aspects of the photosynthetic apparatus, and the results, in good accord with those obtained by IR spectroscopy, clearly demonstrate the validity of this method.

As is known, the spectral lines in the RR spectrum at about 1700 cm^{-1}, and their position, width, and intensity, which originate from vibrations of groups of Chl molecules when $9 - C = 0$, can serve as a reliable probe of intermolecular interactions of Chl molecules *in vivo*. In a wide variety of photosynthetizing organisms, vibrations at about 1700 cm^{-1} are characterized by a broad band, about 50 cm^{-1} width, resolvable at 35 K into 5–6 spectral lines, thus indicating the presence of 5–6 spectral forms of Chl *a* and two forms of Chl *b*. Failure to observe the forms of this kind in monocrystals of Chl, $(Chl)_n$ oligomers of $(Chl \cdot 2H_2O)_n$ chains led Lutz to the conclusion about the lack of such structures in a native photosynthetic apparatus. Presumably, *in vivo*, Chl *a* and Chl *b* form special complexes linked to extraneous molecules, amino acids most probably, via the $9 - C = 0$ (Chla) or $3 - C = 0$ (Chlb) groups. This conception is in good accord with what has been reported on the basis of IR spectral observations.

It can now be taken as accepted that interactions of *in vivo* Chl molecules with the adjacent proteins are factors that determine the geometrical configuration of the pigment. Pigment–pigment interactions, a source of concentration quenching and wasteful energy dissipation (conversion into heat), are completely eliminated in such a configuration. The advantages of a structure having a protein matrix as a base are evident. Owing to great conformational mobility of the protein matrix the dipole moments of the Chl molecules can readily be adjusted so as to optimize directional flow of excitation energy toward the RCs for the most efficient operation of the process. In the light of what has been said, we are left with the conclusion already stated about an important role of conformational changes in the regulation of photosynthetic activity.

In our laboratory, we have measured RR spectra of chloroplasts and subchloroplast particles (including PS-11-enriched fragments) isolated from different plants. In all variety of preparations, the RR spectra proved to be nearly the same. Meanwhile, in chloroplasts of a barley mutant with blocked Chl *b* synthesis, the RR spectrum appeared to be entirely different. The most marked difference was in the region of 1700 cm^{-1}, a region that is associated with vibrations of the $9 - C = 0$ groups of Chl *a*. The absence of Chl *b* blocks the formation of the LHA (Searle *et al.*, 1979), thereby preventing the appearance of native bonds between the Chl *a* and proteins of the LHA.

Using RS spectroscopy, reaction center preparations from photosynthetizing bacteria also have been investigated (Lutz, 1978). By analysis of their RR spectra it was found that in native reaction centers spirilloxanthyne exists in the form of a 15-*cis* isomer, while in model solution it is in all *trans* form. One would expect that further investigations in this direction may help to advance our understanding of the principles of organization and function of the bacterial reaction centers.

4. CONCLUSION

The results that have been reported to date clearly illustrate that photosynthetic research has benefited greatly from the use of laser techniques. Still, laser spectroscopy and nonlinear optics, and principally Raman scattering, have not yet been used in full scope. Surprisingly, no attempts have been made to apply the Rayleigh and Doppler spectroscopic methods to photosynthetic investigations, although in other fields of biological research their usefulness is well accepted. Even Raman spectroscopy has so far been limited in application, although valuable prospects lie in its application, for instance, in vision research. There are good reasons to expect that dynamic Raman spectroscopy, offering the advantage of a time resolution as high as 10^{-12} s, may provide important clues to fundamental questions regarding the regularities that establish strict correspondence and synchronism of the events involved in excitation transfer, charge separation, and associated conformational rearrangements in the photosynthetic apparatus. The remarkable advantages of the coherent anti-Stokes techniques are most encouraging. It is evident that such well-developed methods as laser microbeams (to study chloroplast topography, excitation distribution in thylakoids, etc.), ultrafast laser cell selection (for preparing homogeneous material, for genetic studies), laser probing (control on crops, crop yields, etc.) promise to throw new light on the matters. One can readily predict that coming years will see new applications of laser techniques and nonlinear optics in this field.

REFERENCES

Adams, M. C., Sibbett, W., and Bradley, D. I., 1978, Linear picosecond electron-optical chronoscopy at a repetition rate of 140 MHz, *Opt. Commun.* **26**:(2): 273–276.

Agranovitch, V. M., 1968, *A Theory of Excitons*, Moscow, Nauka (Russian).

Agranovitch, V. M., and Galanan, M. D., 1978, *Transfer of the Energy of Electronic Excitation in Condensed Media*, Moscow, Nauka (Russian).

Akhmanov, S. A., and Khokhlov, R. V., 1966, Parametrical amplifiers and Generators of Light, *Usp. Fiz. Nauk.* **88**:439–460 (Russian).

Akhmanov, S. A., and Koroteev, N. I., 1977, Spectroscopy of light scattering and nonlinear optics, nonlinear optic methods of active spectroscopy of Raman and Raleigh scattering, *Usp. Fiz. Nauk* **123**(3): 406–471 (Russian).

Alfano, R. R., and Shapiro, S. L., 1970a, Emission in the region 4000–7000 Å via four-photon coupling in glass, *Phys. Rev. Lett.* **24**(11): 584–587.

Alfano, R. R., and Shapiro, S. L., 1970b, Observation of self-phase modulation and small-scale filaments in crystals and glasses, *Phys. Rev. Lett.* **24**(11): 592–594.

Anderson, J. M., 1975, The molecular organization of chloroplast thylacoides, *Biochim. Biophys. Acta* **416**:191–235.

Arntzen, C. J., 1978, Dynamic structural features of chloroplast lamellae, *Current Top. Bioenerg.* **8**:111–160.

Arntzen, C. J., Dilley, R. A., and Crane, F. L., 1969, A comparison of chloroplast membrane surface visualized by freeze-etch and negative-straining techniques; and ultrastructural character ization of membrane fractions obtained from digitonin-treated spinach chloroplasts, *J. Cell Biol.* **43**:16–31.

Arsenjev, V. V., Paschenko, V. Z., Gavanin, V. A., Protasov, S. P., Rubin, A. B., and Rubin, L. B., 1973, Pulse fluorometer with picosecond laser excitation, *J. Appl. Spectrosc.* **18**(3): 1003–1005 (Russian).

Babenko, V. A., Kudinova, T. A., Malishev, V. I., Prohorov, A. M., Cichev, A. A., Tolmachev, A. U., and Shelev, M. Ja., 1977, New fast relaxing passive shutter for the laser on Nd glass, *GETF Lett.* **25**(8): 306–369 (Russian).

Beddard, G. S., Fleming, G. R., Porter, G., Searle, G. F. M., and Synowiec, J. A., 1979, The fluorescence decay kinetics of *in vivo* chlorophyll measured using low intensity excitation, *Biochim. Biophys. Acta* **545**:165–174.

Bradley, D. J., and Sibbett, W., 1975, Subpicosecond chronoscopy, *Appl. Phys. Lett.* **27**(7): 382–384.

Brody, S. S., and Rabinowitch, E., 1957, Excitation lifetime of Photosynthetic Pigments *in vitro* and *in vivo*, *Science* **125**:555.

Brown, J. S., 1972, Forms of chlorophyll *in vivo*, *Ann. Rev. Plant Physiol.* **23**:73–86.

Butler, W. L., 1978, Energy distribution in the photochemical apparatus of photosynthesis, *Ann. Rev. Plant Physiol.* **39**:345–378.

Butler, W. L., and Kitajima, M., 1975, Energy transfer between photosystem II and Photosystem I in chloroplasts, *Biochim. Biophys. Acta* **396**:72–85.

Campillo, A. J., and Shapiro, S. L., 1978, Picosecond fluorescence studies of exciton migration and annihilation in photosynthetic systems, A review, *Photochem. Photobiol.* **28**:975–989.

Campillo, A. J., Shapiro, S. L., Kollman, V. H., Winn, K. R., and Hyer, R. C., 1976a, Picosecond exciton annihilation in photosynthetic systems, *Biophys. J.* **16**:93–98.

Campillo, A. J., Kollman, V. H., and Shapiro, S. L., 1976b, Intensity dependence of the fluorescence lifetime of *in vivo* chlorophyll excited by a picosecond light pulse, *Science* **193**:227–229.

Campillo, A. J., Hyer, R. C., Monger, T. G., Parson, W. W., and Shapiro, S. L., 1977, Light collection and harvesting processes in bacterial photosynthesis investigated on a picosecond time scale, *Proc. Natl. Acad. Sci. U.S.A.* **74**(5): 1997–2001.

Carriera, L. A., Maguire, T. C., and Malloy, T. B., 1977, Excitation profiles of the coherent and anti-Stokes resonance Raman spectrum of β-carotene, *J. Chem. Phys.* **66**:2621–2626.

Carriera, L. A., Goss, L. P., and Malloy, T. G., 1978, Preresonance enhancement of fluorescent compounds, *J. Chem. Phys.* **68**:280–284.

Chamorovsky, S. K., Remennikov, S. M., Kononenko, A. A., Venedictov, P. S., and Rubin, A. B., 1976, New experimental approach to the estimation of rate of electron transfer from the primary to secondary acceptors in the photosynthetic electron transport chain of purple bacteria, *Biochim. Biophys. Acta* **430**:62–70.

Clayton, R. K., 1977, Fluorescence of photosynthetic reaction centers at low temperatures, in *Photosynthetic Organelles*, Special Issue N3 of *Plant and Cell Physiology*, pp. 87–96.

Clayton, R. K., and Yamamoto, T., 1976, Photochemical quantum efficiency and absorption spectra of reaction centers from *Rhodopseudomonas sphaeroides* at low temperature, *Photochem. Photobiol.* **24**:67–70.

Clayton, R. K., Fleming, H., and Szuts, E. Z., 1972, Photochemical electron transport in photosynthetic reaction centers from Rhodopseudomonas spheroides. II. Interaction with external electron donors and acceptors and a reevaluation of some spectroscopic data, *Biophys. J.* **12**:46–63.

Cotton, T. M., Loach, P. A., Katz, J. J., and Ballschmitter, K., 1978, Studies of chlorophyll–chlorophyll and chlorophyll–ligand interactions by visible absorption and infrared spectroscopy at low temperatures, *Photochem. Photobiol.* **27**:735–749.

De Maria, A. J., 1971, Picosecond laser pulses, in: *Progress on Optics* (E. Wolf, ed.), pp. 9, 31–71, North-Holland, Amsterdam.

De Maria, A. J., Stetser, D. A., and Heynau, H., 1966, Self mode locking of lasers with saturable absorbers, *Appl. Phys. Lett.* **8**(7): 174–176.

De Vault, D., 1964, Photochemical activation apparatus with optical laser, in *Rapid Mixing and Sampling Techniques in Biochemistry* (B. Chance, R. H. Eisenhardt, Q. H. Gibson, and K. K. Lonberg-Holm, eds.), pp. 165–174, Academic, New York.

De Vault, D., and Kung, M. C., 1978, Interactions among photosynthetic antenna excited states, *Photochem. Photobiol.* **28**:(6): 1029–1038.

Deamer, D. W., and Brauton, D., 1967, Fracture planes in an ice-bilayer model membrane system, *Science* **158**:655–658.

Duguay, M. A., and Hansen, J. W., 1969, An ultrafast light gate, *Appl. Phys. Lett.* **15**(6): 192–194.

Dutton, P. L., Kaufmann, K. J., Chance, B., and Rentzepis, P. M., 1975, Picosecond kinetics of the 1250 nm band of the *Rps. sphaeroides* reaction reaction center: The nature of the primary photochemical intermediary state, *FEBS Lett.* **60**:275–280.

Eckardt, R. C., Lee, C. H., and Bradford, J. N., 1974, Effect of self-phase modulation on the evolution of picosecond pulses in a Nd:glass laser, *Opto-electronics* **6**(1): 67–85.

Emerson, R., 1958, The quantum yield of photosynthesis, *Ann. Rev. Plant Physiol.* **9**:124.

Emerson, R., and Arnold, W., 1932, The photochemical reaction in photosynthesis, *J. Gen. Physiol.* **16**:191–205.

Fadeev, V. V., 1978a, Lasers in oceanology, *Nature* **9**:54–59.

Fadeev, V. V., 1978, Distance laser monitoring of photosynthetic organisms, *Quant. Electron.* **5**(10): 2221–2226 (Russian).

Fajer, J., Brune, D. C., Davis, M. S., Ferman, A., and Spaulding, L. D., 1975, Primary charge separation in bacterial photosynthesis: Oxidized chlorophylls and reduced pheophytin, *Proc. Natl. Acad. Sci. USA* **72**:4956–4960.

Fenton, J. M., Pellin, M. J., Govindjee, and Kaufmann, K. J., 1979, Primary photochemistry of the reaction center of photosystem I, *FEBS Lett.* **100**(1): 1–4.

Floyd, R. A., Keyhani, E., and Chance, B., 1971, Membrane structure and function. II. Alterations in the photoinduced absorption changes after treatment of isolated chloroplasts with large pulses of the ruby laser, *Arch. Biochem. Biophys.* **146**:627–634.

Friesner, R., Dismukes, G. S., and Sauer, K., 1979. Development of electron spin polarization in photosynthetic electron transfer by the radical pair mechanism, *Biophys. J.* **25**:277–294.

Fujita, I., Davis, M. S., and Fajer, J., 1978, Anion radicals of pheophytin and chlorophyll a: Their role in the primary charge separations of plant photosynthesis, *J. Am. Chem. Soc.* **100**:6280–6282.

Geacintov, N. E., Breton, J., Swenberg, C. E., and Peillotin, G., 1977a, A single pulse picosecond laser study of exciton dynamics in chloroplast, *Photochem. Photobiol.* **26**:629–638.

Geacintov, N. E., Breton, J., Swenberg, C., Campillo, A. J., and Shapiro, S. .L., 1977b, Picosecond and microsecond pulse laser study of exciton quenching and exciton distribution in spinach chloroplasts at low temperatures, *Biochim. Biophys. Acta* **461**:306–312.

Guljaev, B. A., and Tetenkin, B., 1979, Light-harvesting pigment–protein complex of higher plants, *Dokl. Akad. Nauk SSSR* **248**(3):345–347 (Russian).

Halsey, Y. D., and Parson, W. W., 1974, Identification of ubiquinone as the secondary electron acceptor in the photosynthetic apparatus of *chromatium vinosum*, *Biochim. Biophys. Acta* **347**:404–416.

Hayes, J. R., and Wolbarsht, M. L., 1968, Thermal model for retinal damage induced by pulsed lasers, *Aerospace Med.* **39**:474–480.

Hellwarth, R. W., 1961, Control of fluorescent Pulsations, in *Advances of Quantum Electronics* (J. R. Singer, ed.), pp. 334–341, Columbia University Press, New York.

Hirsch, M. D., Marcus, M. A., Lewis, A., Makr, H., and Frigo, N., 1976, A method for measuring picosecond phenomena in photolabile species The Emission lifetime of Bacteriorhodopsin, *Biophys. J.* **7**:189–227.

Holton, D., and Windsor, M. W., 1978, Picosecond flash photolysis in biology and biophysics, *Ann. Rev. Biophys. Bioeng.* **7**:189–227.

Ippen, E. P., and Shank, C. V., 1975, Dynamic spectroscopy and subpicosecond pulse compression, *Appl. Phys. Lett.* **27**(9):488–490.

Ippen, E. P., Shank, S. V., and Dienes, A., 1972, Passive mode-locking of the CW dye laser, *Appl. Phys. Lett.* **21**(8):348–350.

Katz, J. J., Janson, T. R., and Wasielewski, N., 1977, A biomimetic approach to solar energy conversion, in Proceedings of the Karcher Symposium on Energy and Chemical Science, May 7, University of Oklahoma, Norman.

Kaufmann, K. J., Dutton, P. L., Netzel, T. L., Leigh, J. S., and Rentzepis, P. M., 1975, Picosecond kinetics of events leading to reaction center bacteriochlorophyll oxidation, *Science* **188**(4195):1301–1304.

Ke, B., Demeter, S., Zamaraev, K. I., and Khairutdinov, R. F., 1979, Charge recombination in photosystem 1 at low temperatures. Kinetics of Electron Tunneling, *Biochim. Biophys. Acta* **545**:265–284.

Keyhani, E., Floyd, R. A., and Chance, B., 1971, Membrane structure and function I. An electron microscope study of the alteration induced by laser irradiation on the chloroplast lamellar membranes, *Arch. Biochem. Biophys.* **146**:618–626.

Khitrov, Yu. A., Kaurov, B. S., Gavrolov, A. G., and Ribin, L. B., 1979, The influence of powerful ruby laser radiation on the electron transport and conjugated processes in pea chloroplasts (*Pisum sativum*), *Physiol. Rasteniy.* **26**(4):808–814 (Russian).

Klimov, V. V., Allahverdov., S., and Paschenko, V. Z., 1978, Determination of the energy of excitation of the fluorescence lifetime of Chl PSII, *Dokl. Akad. Nauk SSSR* **242**(5):1204–1207 (Russian).

Klyshko, D. N., and Fadeev, V. V., 1978, Distance measurements of impurity concentrations in water by laser spectroscopy method with calibration according to Raman Scattering, *Dokl. Akad. Nauk SSSR* **238**:320–323 (Russian).

Kononenko, A. A., Knox, P. P., Adamova, H. P., Paschenko, V. Z., Timofeev, K. N., Rubin, A. B., and Morita, S., 1976, Spectral and photochemical properties of photosynthetic reaction center preparation from *Rhodopseudomonas spaeroides* strain 1760–1, *Stud. Biophys.* **55**:183–198.

Kononenko, A. A., Korvatovsky, B. N., Rubin, A. B., Rubin, L. B., and Tusov, V. B., 1978, Investigations of picosecond fluorescence kinetics of chromatophores and reaction centers from *Rhodopseudomonas sphaeroides* in the temperature range 295–80°K, *Dokl. Akad. Nauk SSSR* **242**(6):1421–1424 (Russian).

Korvatovsky, B. N., Kukharskyh, G. P., Tusov, V. B., Paschenko, V. Z., and Rubin, L. B., 1979,

Picosecond fluorometry of a pigment–protein complexes enriched by the reaction centers of PSI, *Dokl. Acad. Sci. SSSR* **253**(5):1251–1255.

Krendeleva, T. E., Nizovskaja, N. V., Ivanov, A. V., and Rubin, L. B., 1972a, The influence of ruby laser irradiation on primary photochemical reactions in photosynthesis, *J. Biochem.* **37**:158–162.

Krendeleva, T. E., Nisovskaja, N. V., Ivanov, A. V., and Rubin, L. B., 1972b, On selected damage of photosystem I of isolated pea chloroplasts after treatment with ruby laser radiation, *Nature New Biol.* **240**(13):223–224.

Leigh, J. S., Netzel, T. L., Dutton, P. L., and Rentzepis, P. M., 1974, Primary events in photosynthesis: Picosecond kinetics of carotenoid Bandshifts in *Rhodopseudomonas sphaeroides* chromatophores, *FEBS Lett.* **48**(1):136–140.

Litvin, F. F., and Sineshchekov, V. A., 1975, Molecular organization of chlorophyll and energetics of the initial stages in photosynthesis, in *Bioenergetics of Photosynthesis*, Chap. 12, Academic Press, New York.

Lukashev, E. P., Timofeev, K. H., Kononenko, A. A., Venedictov, P. S., and Rubin, A. B., 1976, Temperature dependence of the reduction kinetics of photooxidized reaction centre bacteriochlorophyll in dark adapted chromatophores of purple bacteria, *Photosynthetica* **10**:423–430.

Lutz, M., 1972, Spectroscopie moleculaire. Spectroscopie Raman de résonance de pigments végétaux en solution et inclus dans des lamelles chloroplastiques, *C. R. Acad. Sci. Paris* **275B**:497–500.

Lutz, M., 1977, Antenna chlorophyll in photosynthetic membranes. A study by Resonance Raman Spectroscopy, *Biochim. Biophys. Acta* **460**:408–430.

Lutz, M., and Breton, J., 1973, Chlorophyll associations in the chloroplast: Resonance Raman Spectroscopy, *Biochem. Biophys. Res. Commun.* **53**(2):413–418.

Lutz, M., and Kleo, J., 1974, Spectroscopy Moléculaire. Diffusion Raman de résonance de la chlorophylle d, *C. R. Acad. Sci. Paris* **279D**:1413–1416.

Lutz, M., Agalidis, I., Hervo, G., Cogdell, R. J., and Reiss-Husson, F., 1978, On the state of carotenoids bound to the reaction centers of photosynthetic bacteria: A resonance Raman Study, *Biochim. Biophys. Acta* **503**:287–303.

Magde, D., and Windsor, M. W., 1974, Picosecond flash-photolysis spectroscopy: 3,3^1-Diethyl-oxadicarbocyanine iodide (DODCI), *Chem. Phys. Lett.* **27**:31–36.

Martin, J. L., Astier, R., Migus, A., and Antonetti, A., 1979, Subpicosecond photodissociation and time-resolved spectroscopy of recombination processes in carbon monoxide hemoglobin, Europhysics Conference, Lasers in Photomedicine and Photobiology, Conference digest, p. 55.

Mathis, P., and Van Best, J. A., 1978, A study of photosystem 2 ractions by flash absorption spectroscopy, in Proceedings of the Fourth International Congress on Photosynthesis, U.K. 4–9 September 1977 (D. O. Hall, J. Coombs, J. W. Geodwood, eds.), London, pp. 387–396.

Mauzerall, D., 1976, Multiple excitations in the photosynthetic systems, *Biophys. J.* **16**:87–92.

Moor, H., and Mühllethaler, K. J., 1963, Fine structure in frozen-etched yest cells, *J. Cell Biol.* **17**:609–627.

Moskowitz, E., and Malley, M. M., 1978, Energy transfer and protooxidation kinetics in reaction centres on the Picosecond time scale, *Photochem. Photobiol.* **27**:55–59.

Netzel, T. L., Struve, W. S., and Rentzepis, P. M., 1973, Picosecond spectroscopy, *Ann. Rev. Phys. Chem.* **24**:473–492.

Nitsh, V., and Kifer, V., 1977, Coherence antistokes resonance Raman spectroscopy, *J. Quant. Electron.* **4**:2555–2559 (Russian).

Paillotin, G., 1977, Organization of the photosynthetic pigments and transfer of excitation energy, in *Proceedings of the Fourth International Congress on Photosynthesis*, U.K. 4–9 September,

1977 (D. O. Hall, J. Coombs, J. W. Goodwin, eds.), The Biochemical Society, London, pp. 33–44.

Parson, W. W., and Cogdell, R. J., 1975, The primary photochemical reaction of bacterial photosynthesis, *Biochim. Biophys. Acta* **416**:105–149.

Parson, W. W., Clayton, R. K., and Cogdell, R. J., 1975, Excited states of photosynthetic reaction centers at low redox potentials, *Biochim. Biophys. Acta.* **387**:265–278.

Paschenko, V. Z., Protasov, S. P., Rubin, A. B., Timoveev, K., Zamazova, L. M., and Rubin, L. B., 1975a, Probing the kinetics of photosystem I and photosystem II fluorescence in pea chloroplasts on a picosecond pulse fluorometer, *Biochim. Biophys. Acta.* **408**:147–153.

Paschenko, V. Z., Rubin, L. B., Rubin, A. B., Tusov, V. B., and Frolov, V. A., 1975b, High time resolution and sensitivity pulse fluorometer with streak camera recording, *J. Tech. Phys.* **45**:1122–1127 (Russian).

Paschenko, V. Z., Kononenko, A. A., Protasov, S. P., Rubin, A. B., Uspenskaya, N. Ya., and Rubin, L. B., 1977, Probing the fluorescence emission kinetics of the photosynthetic of *Rhodopseudomonas sphaeroides* strain 1760–1 on a picosecond pulse fluorometer, *Biochim. Biophys. Acta* **461**:403–412.

Paschenko, V. Z., Kononenko, A. A., Rubin, A. B., and Rubin, L. B., 1978, Picoseond fluorometry of *Rhodopseudomonas sphaeroides*, strain 1760–1, *J. Biophys.* **23**(5):833–838 (Russian).

Pellegrino, F., Yu, W., and Alfano, R. R., 1978, Fluorescence kinetics of spinach chloroplasts measured with a picosecond optical Kerr gate, *Photochem. Photobiol.* **28**(6):1007–1013.

Pellin, M. J., Wraight, C. A., and Kaufmann, K. J., 1978, Modulation of the primary electron transfer rate in photosynthetic reaction centers by reduction of a secondary acceptor, *Biophys. J.* **24**(1):361–369.

Popov, V. I., Tagueeva, S. V., Gavrilov, A. G., Kaurov, B. S., and Rubin, L. B., 1977a, Effect of ruby laser irradiation on the ultrastructure of pea chloroplasts membranes, *Photosynthetica*, *II* **1**:76–80.

Popov, V. A., Kaurov, B. S., Gavrilov, A. G., Tageeva, S. V., and Rubin, L. B., 1977b, Application of the freeze-fracturing method for detecting chloroplast membrane damage induced by the laser irradiation, Thesis of the 2nd Union Symposium Cryogenic methods in electron microscopy, Puschino, pp. 32–34.

Porter, G., Tredwell, C. J., Searle, G. F. W., and Barber, J., 1978, Picosecond time-resolved energy transfer in *Porphyridium Cruentum*, *Biochim. Biophys. Acta* **501**(2):232–245.

Raman, C. V., and Krishnan, K. S., 1928, A new type of secondary radiation, *Nature* **121**:501–502.

Reed, D. W., and Clayton, R. K., 1968, Isolation of a reaction center fraction from *Rhodopseudomonas Sphaeroides*, *Biochim. Biophys. Res. Commun.* **30**:471–475.

Rentzepis, P. M., Topp, M. R., Jones, R. P., and Jortner, J., 1970, Picosecond emission spectroscopy of homogeneously broadened, electronically excited molecular states, *Phys. Rev. Lett.* **25**(26):1742–1744.

Rockley, M. G., Windsor, M. W., Cogdell, R. G., and Parson, W. W., 1975, Picosecond detection of an intermediate in the photochemical reaction of bacterial photosynthesis, *Proc. Natl. Acad. Sci. USA* **72**(6):2251–2255.

Rubin, A. B., and Osnitskaya, L. K., 1963, On the connection between physiological state and average fluorescence lifetime of bacteriochlorophyll in photosynthetic bacterial cells, *J. Microbiol.* **32**:2000–2003 (Russian).

Rubin, L. B., and Rubin, A. B., 1978, Keynote address, Picosecond fluorometry in primary events of photosynthesis, *Biophys. J.* **24**(1):84–92.

Rubin, L. B., Krendeleva, T. E., Shantorenko, N. V., Paschenko, V. Z., Timofeev, K. N., Ivanov, A. V., and Petrov, V. V., 1971, Ruby laser radiation action on the pigment apparatus of photosynthetic organisms, *J. Pricladnoy Spectrosc.* **14**(1):78–81 (Russian).

Rubin, L. B., Paschenko, V. Z., and Rubin, A. B., 1979, Nonlinear quenching of the laser induced fluorescence of photosynthetic organisms, 3rd Conference on Luminescence, Conference Digest, Vol. 1, Szeged, Hungary, September 4–7, pp. 193–196.

Rubin, L. B., Paschenko, V. Z., and Rubin, A. B., 1980, Picosecond fluorometry of the exciton diffusion in green plants antenna chlorophyll, in: *Lasers in Photomedicine and Photobiology* (R. Pratesi and C. A. Sacchi, eds.), Springer-Verlag, Berlin, pp. 221–228.

Ruddock, I. S., and Bradley, D. J., 1976, Bandwidth-limited subpicosecond pulse generation in mode-locked CW dye lasers, *Appl. Phys. Lett.* **29**(5):296–297.

Sauer, K., Mathis, P., Asker, S., and Van Best, J. A., 1979, Electron acceptors associated with p-700 in triton solubilized photosystem I particles from spinach chloroplasts, *Biochim. Biophys. Acta* **503**:120–134.

Searle, G. F. M., Barber, J., Harris, L., Porter, G., and Tredwell, C. J., 1977, Picosecond laser study of fluorescence lifetimes in Spinach chloroplast photosystem I and photosystem II preparations, *Biochim. Biophys. Acta* **459**:390–401.

Searle, G. F. M., Barber, J., Porter, G., and Tredwell, C. J., 1978, Picosecond time-resolved energy transfer in Porphyridium cruentum. Part II. In the isolated light harvesting complex (phycobilisomes). *Biochim. Biophys. Acta* **502**:246–256.

Searle, G. F. M., Tredwell, C. J., Barber, J., and Porter, G., 1979, Picosecond time-resolved fluorescence study of chlorophyll organization and excitation energy distribution in chloroplasts from wild-type barley and a mutant lacking chlorophyll b, *Biochim. Biophys. Acta* **545**:496–507.

Seibert, M., Alfano, R. R., and Shapiro, S. L., 1973, Picosecond fluorescent kinetics of *in vivo* chlorophyll, *Biochim. Biophys. Acta* **292**:493–495.

Shank, C. V., and Ippen, E. P., 1974, Subpicosecond kilowatt pulses from mode-locked CW dye laser, *Appl. Phys. Lett.* **24**(8):373–375.

Shapiro, S. L., Kollman, V. H., and Campillo, A. J., 1975, Energy transfer in photosynthesis: Pigment concentration effects and fluorescent lifetimes, *FEBS Lett.* **54**(3):358–362.

Shuvalov, V. A., Klevanik, A. V., Shartkov, A. V., Matveetz, Ju. A., and Krukov, P. G., 1978, Picosecond detection of Bchl-800 as an intermediate electron carrier between selectively-excited P_{870} and bacertiopheophytin in *Phodospirillum Rubrum* reaction centers, *FEBS Lett.* **91**(1):135–139.

Shuvalov, V. A., Ke, B., and Dolan, E., 1979a, Kinetic and spectral properties of the intermediary electron acceptor A_1 in photosystem I, *FEBS Lett.* **100**:5–8.

Shuvalov, V. A., Dolan, E., and Ke, B., 1979b, Spectral and kinetic evidence for two early electron acceptors in photosystem I, *Proc. Natl. Acad. Sci. USA* **76**(2):770–773.

Singer, S. J., and Nicolson, G. L., 1972, The fluid mosaic model of the structure of cell membrane, *Science* **175**:720–731.

Smith, W. L., Liu, P., and Bloembergen, N., 1977, Superbroadening in H_2O and D_2O by self-focused picosecond pulses from a YAIG: Nd laser, *Phys. Rev. A* **15**(6):2396–2403.

Staehelin, L. A., 1976, Reversible particle movements associated with unstacking and restacking of chloroplast membranes *in vitro*, *J. Cell Biol.* **71**:136–158.

Suna, A., 1970, Kinematics of exciton–exciton annihilation in molecular crystals, *Phys. Rev. B* **1**(4):1716–1739.

Swenberg, C. E., Geacintov, N. E., and Pope, M., 1976, Biomolecular quenching of excitons and fluorescence in the Photosynthetic unit, *Biophys. J.* **16**:1447–1452.

Swenberg, C. E., Geacintov, N. E., and Breton, J., 1978, Laser pulse excitation studies of the fluorescence of chloroplasts, *Photochem. Photobiol.* **28**:999–1006.

Tanaka, Y., Kushida, T., and Shionroya, S., 1978, Broadly tunable, Repetitive, Picosecond Parametric Oscillators, *Opt. Commun.* **25**(2):273–276.

Tredwell, C. J., Synowiec, J. A., Searle, G. F. W., Porter, G., and Barber, J., 1978, Picosecond time resolved fluorescence or chlorophyll *in vivo*, *Photochem. Photobiol.* **28**:1013–1020.

Treifai, M., 1956, The theory of localized excitons diffusion in solid matter, *Czechoslovak J. Phys.* **6**(6):533–550.

Tusov, V. B., Korvatovsky, B. N., Paschenko, V. Z., and Rubin, L. B., 1980, On the nature of the chloroplast 735 nm fluorescence at room and low temperatures. *Dokl. Acad. Sci. USSR* (Russian, in press).

Van Gorkom, H. J., Tannimga, J. J., and Haveman, J., 1974, Primary reaction, plastoquinone and fluorescence yield in subchloroplast fragments prepared with deoxycholate, *Biochim. Biophys. Acta* **347**:416–438.

Witt, H. T., 1979, Energy conversion in the functional membrane of photosynthesis. Analysis by light pulse and electric pulse methods. The central role of the electric field, *Biochim. Biophys. Acta* **505**:355–427.

Wraight, C. A., and Clayton, R. K., 1974, The absolute quantum efficiency of bacterio-chlorophyll photooxidation in reaction centers of *Rhdopseudomonas Sphaeroides*, *Biochim. Biophys. Acta* **333**:246–260.

Yu, W., Ho, P. P., Alfano, R. R., and Seibert, M., 1975, Fluorescent kinetics of Chlorophyll in Photosystems I and II enriched fractions of spinach, *Biochim. Biophys. Acta.* **387**:159–164.

CHAPTER 2

Requirements and Technical Concepts of Biomedical Microprobe Analysis

Franz Hillenkamp

Institute for Medical Physics
University of Münster
D-4400 Münster/FRLY, Federal Republic of Germany

and

R. Kaufmann

Department of Clinical Physiology
University of Düsseldorf
D-4000 Düsseldorf, Federal Republic of Germany

1. PROLOGUE

In the glare of a hot midsummer afternoon in 1968, laboratory work in the department of physiology at Freiburg Medical School had nearly dropped to zero. A siesta is not so uncommon in this southern district of Germany, where the Mediterranean way of life has chronically and equally infected both private and public affairs. On this particular afternoon, one of the present authors (R.K.) had decided to intensify his suntan in a public swimming pool, and quite conscious that he would not return to his laboratory, had carried along some recent papers related to his scientific concern, the cellular control of muscular contraction. When he had refreshed himself by swimming across the pool several times, he retired to a silent corner of the park and returned to the problem of intracellular Ca kinetics. This, he realized was the key to the Ca-mediated control of muscular contraction.

The idea that a highly efficient cellular Ca release and reuptake system must exist in striated muscle cells had gained much attraction among muscle physiologists but suffered from the fact that experimental proofs were totally indirect. Was there really no way of thinking how to localize Ca at subcellular sites with a spatial resolution and a sensitivity good enough to discriminate between, say, myofibrils and sarcoplasmic reticulum? Another check of his calculations showed that even under the most optimistic assumptions, the result was indeed discouraging: How could one hope to discriminate 10^{-6}–10^{-3} mol liter^{-1} Ca against a background of 10^{-1} mol liter^{-1} K at a spatial resolution of at least 0.5 μm? In other words, 10^{-17}–10^{-20} g Ca had to be detected in a sample volume of 10^{-13} cm^3. Bad luck! All the experts in histo- and cytochemistry previously consulted had flatly denied that this matter could be discussed in realistic terms. The faint hope of using autoradiography had vanished after checking the properties of the only available radioisotope, ^{44}Ca.

However, there was still another approach to be kept in mind, but that had not been systematically explored for its potentialities: microprobe analysis. The physical backgrounds of this technique were rather obscure to him. Only recently he had found some remarks on a funny technique called "laser microprobe analysis" in a small book by Leon Goldman dedicated to biomedical laser applications. Would this be a realistic way? Since this book was among the papers he had carried along, he read the relevant chapter. When he left the pool late in the afternoon, he had acquired both a sunburn and a confused idea of a new project. It took two months for him to discard what had initially seemed like an essential part of the idea: emission spectroscopy. This turned out to be much too insensitive for the given purpose. Could mass spectrometry do better, perhaps? It took two weeks to get a favorable response, although embedded in it was the skepticism of the relevant experts.

Ten years followed. Proposals for funding got unfavorable comments by referees, yet some funds were approved. The work progressed by quantum steps, some requiring long detours and periods of stagnation before any achievement was noted. At the end, the vague idea that had once emerged from a young scientist's daydream had transformed itself into a proven

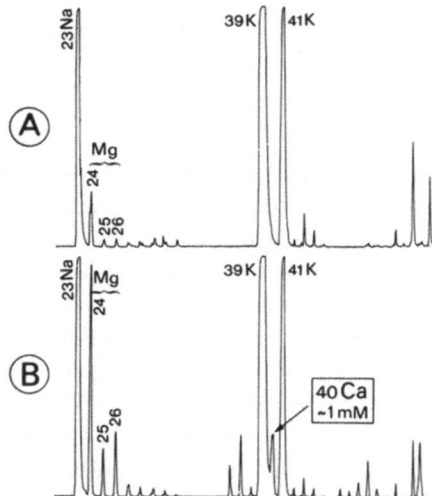

FIGURE 1. Microprobe analysis in biomedical samples (rat skeletal muscle fiber): Detection of small amounts of Ca ($\sim 10^{-3}$ mol liter^{-1}) in the presence of large K concentrations (10^{-1} mol liter^{-1}) at a spatial resolution of 1 μm by laser microprobe mass analysis. A myofibrillar, B interfibrillar space (sarcoplasmic reticulum).

technique with a usable machine. As Fig. 1 demonstrates, the technique of laser-induced microprobe mass analysis, now called LAMMA, has reached its original goal, but—as this review will show—it has also opened a wide field of applications that were neither anticipated nor even imagined.

1.1. Introduction

Since the introduction of light microscopy and, even more, electron microscopy, our insight into the cellular and subcellular structure of living matter has been very much increased. At the same time, biochemistry has contributed a large amount of knowledge about the molecular principles of life. The latter information, however, is usually derived from isolated and purified constituents extracted from living systems. Therefore, a gap in our information exists concerning the control and interdependence of reactions occurring—*in situ*—and at the level of integrated living entities such as tissues, cells, and cellular structures. This has sharpened the search for methods that can perform microchemical analyses in integrated biological specimens with high spatial resolution, that is, on a micrometer or submicrometer scale. Although the rapidly evolving techniques of histo-chemistry and microautoradiography are moving in this direction, they have inherent physical or technical limitations, which still leave major fields of interest (for instance, the subcellular distribution of physiological cations) inaccessible.

Therefore, in the past 15 years, many life scientists have looked for alternative approaches to overcome this limitation. However, the only instruments presently available are ones primarily developed for materials research that can execute microchemical analyses on solid state matter with a submicrometer resolution. These are commonly referred to as microprobe analysis devices. These instruments usually consist of a probing beam of primary energy (such as an electron, ion, or laser beam), which can be focused and directed onto the microarea or the microvolume to be analyzed. In the analytical part of the instruments not only changes in the primary beam after interaction with the probed material (electron energy loss, absorption) are used to gain the desired analytical information but also secondary effects such as emission of secondary electrons, photons, x-rays, or secondary ions.

Despite recent successes, these techniques (which will be briefly dis-cussed below) still leave unfulfilled many of the particular requirements for biomedical microprobe analysis. In attempts to set down the desirable features of an ideal instrument, the following list of items emerges:

- Spatial resolution (imaging as well as analytical), at least to 1 μm and better to 0.1 μm or less.
- Analytical capability to analyze all elements of the Periodic Table with the same, or at least similar, sensitivity.
- Discrimination between isotopes and determination of isotopic ratios of the highest possible precision.
- Sensitivity enough to analyze trace elements or impurities down to the sub-ppm range.
- Reproducibility and analytical precision good enough to provide for quantitative, or at least semiquantitative, data.
- Analytical access not only to the atomic but also to the molecular composition (inorganic, and even more desirable, organic) of the sample.
- Measuring time short enough to allow for spatially resolved profiling in extended and complicated biological structures.
- Correlation of analytical information easy and unambiguous in the same specimen.

2. ACTUAL MICROPROBE INSTRUMENTS

An evaluation of the capabilities and performances of actual microprobe instruments with respect to the above requirements shows clearly that the ideal instrument does not yet exist and, most probably, will never exist in this form. However, from the viewpoint of the evolution of the techniques, it is most important to consider whether the limitations of a microprobe system are given primarily by the inherent laws of physics, or whether there is room for a "large-scale" technical optimization in the desired direction. The following discussion of the actual microprobe systems that are alternatives to laser microprobe mass analysis will have this orientation.

The most commonly used microprobe systems are based on two principles:

1. Electron probe x-ray microanalysis (for review see Hall, 1971; Echlin and Kaufmann, 1978; Lauchli, 1972).
2. Ion probe secondary ion mass spectroscopy (for review see Andersen and Hinthorne, 1971; Burns-Bellhorn, 1978; Truchet, 1975).

2.1. Electron Probe X-Ray Microanalysis (EPXM)

In electron probe x-ray microanalysis (EPXM), electrons are focused in an electron-optical system onto the spot of interest. Analytical information is obtained by recording the spectrum of the characteristic x-rays excited in the K, L, or M shells of atoms of the analyzed sample volume (typically 10^{-13}–10^{-15} cm^3) by either an energy-dispersive or wavelength-dispersive spectrometer. So far, most of the microprobe investigations in living matter have been carried out with this type of instrument with an imaging electron microprobe (either transmission or scanning) extended for electron microprobe analysis simply by attaching a suitable spectrometer. Thus, the correlation of structural to analytical information seemed in theory to be ideal. On the other side, however, actual usage made it increasingly obvious that electron microprobe instruments, because of their physical limitations, are of rather restricted use if compared to the "ideal" requirements listed above. These restrictions are as follows:

- Detectability: most current instruments can analyze only elements with Z numbers in the range of 9 to 35.

- Sensitivity: under realistic conditions it is in the 100-ppm range at best.
- Discrimination between isotopes, impossible; which, for instance, prevents the study of the exchange and transport kinetics of cations.
- Inorganic or organic molecular constituents not accessible for analysis.
- Further problems arise from the degradation of the biological matter by the electron beam and from electrostatic charging effects in the dielectric sample. Since most of these limitations are of physical rather than of technical nature, a substantial step ahead in the performance of this type of instrument is not to be expected unless completely new concepts, such as highly efficient x-ray focusing devices, emerge.

2.2. Ion Microprobe

Ion microprobe instruments are less restricted than EPXM instruments. They are based on the well-known physical principle of ion sputtering, i.e., the phenomenon that accelerated primary ions (for instance O^+ or Ar^+) impinging on the surface of a specimen detach (sputter) and partly ionize the superficial layer of molecules and atoms. These secondary ions in turn are analyzed by an appropriate mass spectrometer (secondary ion mass spectrometry, SIMS). In order to use this principle for microprobe analysis, the beam of primary ions can be focused onto a selected spot of interest with a spatial resolution about 1–2 μm, and secondary ions can be selected for imaging the specimen with a spatial resolution also in the 1- to 2-μm range. In the latter case, an "image" is obtained that represents the spatial distribution of one ion, say Na or K, over the whole field of view. Both types of instruments have proven to be much more sensitive than EPXM. Furthermore, they are not restricted to elements within a certain range of Z, nor do they exclude the possibility of studying polyatomic ions. Also not to be forgotten is that these machines can discriminate isotopes.

The potential advantages of using ion microprobe instruments for biomedical purposes have not yet been fully exploited. Only a few systematic studies have been performed to date (see Burns-Bellhorn and Lewis, 1976; Truchet, 1975). The main problems of ion microprobe systems as applied to biomedical samples are the strong dependence of ion yield on the composition, and the chemical nature of the matrix, which makes quantitative analyses very difficult, and the rather poor optical imaging system with the inherent difficulty of correlating analytical data with structural details. Last but not least, ion microprobe instruments are neither inexpensive nor easy to operate.

Other microprobe approaches using focused x-rays (Nicholson, 1974; Long and Rockert, 1963) or protons (Bosch *et al.*, 1978) for the primary beams have been proposed and occasionally built as laboratory setups. From the viewpoint of biological microprobe analysis, these instruments have no obvious general advantages on the basis of the actual state of the art. Microprobe systems using Auger spectroscopy or electron energy loss spectroscopy as analytical devices are interesting special developments in surface analysis, useful for very thin sections, layers, or films, but seem of no importance for biomedical samples.

2.3. Raman Scattering Microprobes

Microprobe analysis by Raman scattering has more recently been demonstrated to be a promising tool especially for the microchemical analysis of organic and inorganic constituents in, e.g., airborne particles (Etz *et al.*, 1978) and mineral samples (Delhaye *et al.*, 1977). The inherent potentialities of this microprobe technique are obvious. However, even with the considerable technological improvements that can be expected in the future, the high radiation and thermal load imposed on the specimen during long measurement times due to the very low quantum yield, as well as the complexity of the spectra obtained from nonhomogeneous biological samples, will remain as serious limitations for biomedical microprobe analysis. Laser-stimulated microfluorometry, strangely enough, has not been systematically exploited so far. Some advances can certainly be expected in this method in the future, possibly through better time resolution after excitation with subnanosecond or picosecond laser pulses and fluorescence–phosphorescence transition measurements at low temperatures.

2.4. Laser Microprobe Emission Spectroscopy

Laser microprobe emission spectroscopy was introduced as early as 1962 (Brech, 1962), and a long list of successful applications, particularly in material analysis and forensic medicine, has been published since. A review of this work with particular reference to biomedical applications has been published by Harding-Barlow (1974). Despite considerable efforts by several groups, most of the expectations for further improvements of the technique predicted by the author at that time have not been fulfilled for theoretical rather than technological reasons. In particular, the spatial resolution could not be improved below several micrometers without a prohibitive loss in detection sensitivity and reproducibility of results. Though this straightforward and comparatively cheap technique will continue to be routinely used in a variety of different applications, it cannot be considered a real biological microprobe with the capability of meeting most of the requirements listed above.

2.5. Laser Microprobe Mass Analysis (LAMMA)

Attempts to overcome these limitations and replace the emission spectrograph by a mass spectrometer, which, at least in principle, is much more sensitive, date back to the mid-1960s. The results of these early investigations are reviewed by Harding-Barlow (1974). Though very promising, these experiments were not being pursued toward a useful biomedical microprobe by any of the early investigators. Around 1970, several groups (Vastola and Pirone, 1968; Knox, 1971; Eloy, 1978) took up the idea. But only the LAMMA instrument, developed by our own group as described in this contribution, seems to have come close to achieving a real biomedical microprobe. More recently, another instrument, LIMA, based on the same principle of ions generated by an intense focal UV-laser beam has been marketed (Cambridge Mass Spectrometry Ltd., Milton Road, Cambridge CB4 4BH, England).

3. THE LAMMA INSTRUMENT AND ITS PERFORMANCE

3.1. General Concept

The concept of the laser microprobe mass analyzer, LAMMA, as it stands now is governed by the mutual requirements described above: spatial resolution for analysis and structure recognition of better than $1 \mu m$; absolute detection sensitivity down to 10^{-20} g, corresponding to mass fractions of the detected constituents in the low ppm to ppb range; and the possibility of obtaining information on organic molecules as well as on atomic species and their isotopes. As explained by Harding-Barlow (1974), 10-ns pulses of Q-switched lasers are best suited for producing partially ionized plumes from organic material. As laser-induced emission spectroscopy could clearly not be improved to meet the sensitivity requirements, a suitable mass spectrometer was the obvious choice on the detection side.

The instrument that evolved from these general considerations after several years of experimental investigations (as now commercially available from Leybold-Heraeus, Bonnerstrasse, D-5000 Cologne, Germany) is shown schematically in Fig. 2. A Q-switched solid state laser with optional frequency multipliers and a time-of-flight mass spectrometer (TOF) are suitably arranged around an optical microscope of high magnification ($\times 1000$) and correspondingly high spatial resolution (better than 0.5 μm). This requires the use of high numerical aperture microscopic objectives (usually a Zeiss 100/1.32 Ultrafluar) with short working distances (200–300 μm). This excludes the possibility for laser irradiation and ion extraction from the same side of the sample, as classically used in all the emission instruments. The instrument is therefore designed for the analysis of thin samples, usually thin sections of about 0.5 μm thickness ($0.05 \leqslant d \leqslant 3$ μm), monocellular layers, or small particles of up to 1 μm in diameter, with focusing and observation optics on one side of the sample and the collecting ion optics on the other (Fig. 3). The inset in Fig. 3 shows an enlarged view of the sample stage. Samples are mounted onto grids as used in electron microscopy, either directly or after coating the grids with organic films (Formvar, colloidin, etc.). The grids are placed directly underneath a quartz cover slip that serves as the vacuum window in an x, y-movable vacuum flange used for sample positioning relative to the laser beam focus. The 10- to 20-μm distance between the cover slip and the sample prevents ion emission from the cover slip that would otherwise interfere with the sample

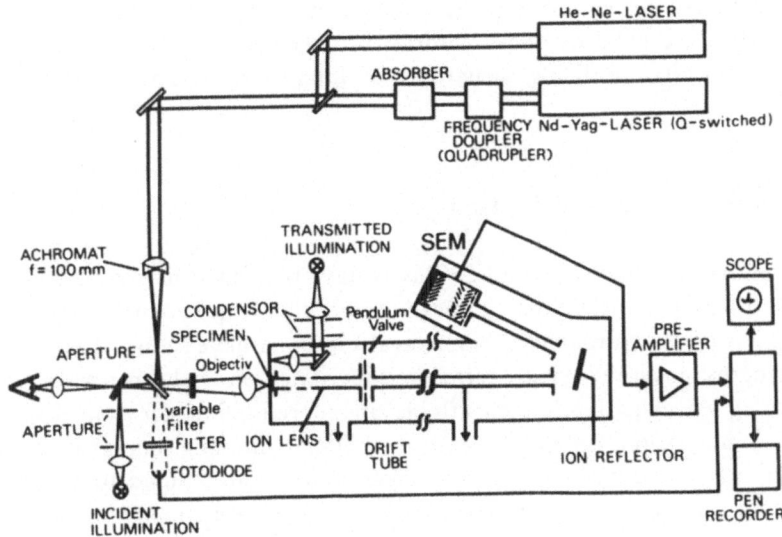

FIGURE 2. Schematic diagram of the LAMMA instrument.

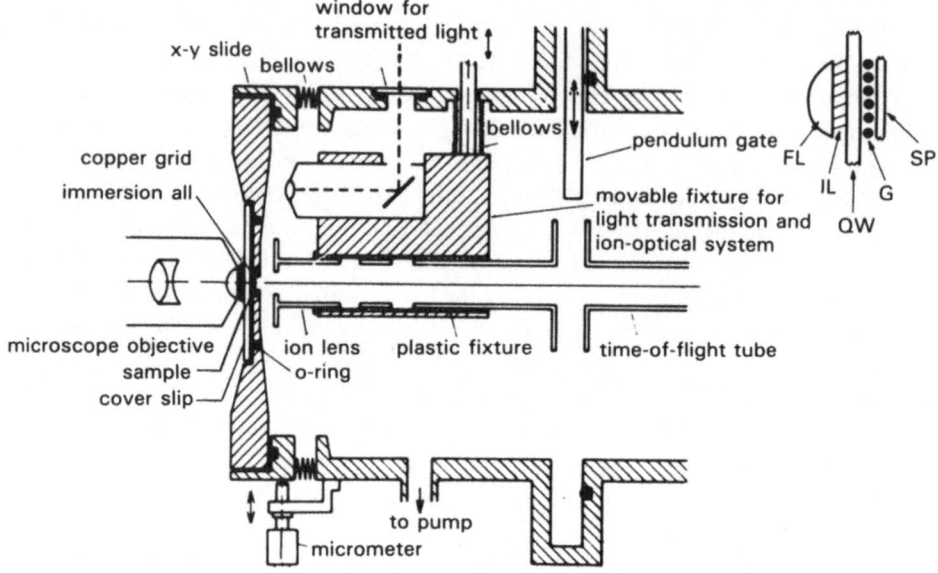

FIGURE 3. Sample stage of the LAMMA instrument. (Inset: Enlarged view of the specimen arrangement; FL, front lens of the mirroscope objective; IL, immersion liquid; QW, quartz window; G, supporting grid; SP, specimen.)

spectrum. The usable numerical aperture is thereby limited to 1.0. The samples can be visualized with top illumination, or transmission-illumination in light field, dark field, or phase contrast to overcome most of the problems encountered in identifying structures of interest in unstained biological samples.

3.2. Laser and Laser Optics

Both Q-switched ruby and Nd lasers have been used mostly at doubled or quadrupled wavelengths; a large number of experiments on a considerable variety of organic or biomedical specimen have shown that shorter wavelengths generally yield better results. The ion yield increases with decreasing wavelength and seems to become more uniform for elements with different ionization potentials and/or different binding energies to the organic matrix. In many cases the background signals, caused by fragment ions from the organic matrix, are concentrated at fewer mass numbers in a lower mass range. In contrast to early expectation (see, e.g., Harding-Barlow, 1974), the natural absorption bands of biomedical specimens do not play a dominating role in the energy transfer and ion formation process. The frequency quadrupled line of the Nd laser at 265 nm has therefore proven to be the best choice for such applications. The optimal irradiance in the focus depends on the mode of operation (desorption or volume evaporation; see below), and the type of sample. For volume analysis of the dielectric biological samples, irradiances are in the range 10^7–10^{10} W/cm^2. In an actual analysis, the threshold irradiance for sample perforation and/or ion detection must be determined experimentally by inserting filters of different optical densities. The best overall analytical results are usually obtained at irradiances at a factor of 3–10 above that threshold. As the corresponding pulse energies are below 1 μJ, the laser damage of optical components frequently encountered in emission probe instruments has therefore not been a problem. UV-transmitting optics focus the laser beam onto a 50- to 70-μm pinhole, which in turn is imaged onto the specimen by the microscope objective. The laser should be operated in the TEM$_{00}$ mode to assure reproducible energy from shot to shot. Part of the laser pulse is monitored by a fast photodiode whose signal serves as a zero time marker in the TOF spectrum and also as a monitor of the pulse energy. A 1-mW He–Ne laser beam aligned collinearly with the power laser serves for target designation and alignment of the system.

3.3. The TOF Spectrometer and Signal Detection

Once the area to be analyzed has been chosen, the transmission illumination condenser is replaced by a three-element immersion ion lens mounted next to it on a pneumatically driven movable stage. Generated ions are accelerated by the first lens element to an energy of typically 3 keV and imaged down a straight section of the flight tube of about 1 m length. An electrode at suitable potential reflects the ions back onto the first dynode of a 17-stage secondary electron multiplier (SEM), typically at a potential of about 5 kV. With a properly chosen electrode potential, the ion reflector acts as a time focusing element, i.e., it compensates for the flight time dispersion caused by the spread of initial energies of the ions, which under the operation parameters given above typically ranges between 5 and 100 eV. The SEM signal is preamplified and fed into a fast transient recorder for analog–digital conversion and storage in memory. Typical features are a sampling rate of 10 ns/channel, an amplitude resolution of 8 bits (254 points) and 2048–32000 channels. Spectra stored in the memory can be read out repetitively and displayed on a screen, written out on a strip chart recorder, or transferred directly to an on-line computer (e.g., HP 21MX or PC-AT) for further data processing.

Figure 4 shows a typical LAMMA spectrum of positive ions obtained from a 0.3-μm-thick section of an Epoxy resin, doped with alkaline and metal ions (Li, Na, K, Sr, Pb) at concentration of 10^{-2} mol/liter. The ordinate gives relative intensities of the ion signals; the abscissa is the flight time of the ions that goes with the square root of the relative molecular mass over the charge of the ions (m/e). As expected, some background signals of fragment ions from the resin matrix occur besides the clearly identifiable peaks due to the various isotopes of the doping elements. At a maximum sampling rate of 10 ns/channel and 2048 channels, all singly charged ions with masses up to a relative mass of about 30 are registered. Higher masses are displayed either by introducing a time delay into the triggering of the transient recorder, by using a recorder with more channels, or by choosing slower sampling rates with a corresponding loss in resolution. Ions with relative masses (M_r) up to 100,000 daltons and above have been measured routinely.

A reversal of all potentials allows the detection of negative ions. In contrast to most of the other ionization methods, e.g., electron impact, signals of negative ions have appeared at intensities comparable to those of positive ions for the large variety of different samples so far investigated. Information in the literature on ion–electron conversion rates at the first SEM dynode in dependence on ion polarity are somewhat conflicting, but it can safely be assumed that with the irradiation parameters given above, ions of both polarities are produced at yields of comparable orders of magnitude. This certainly is a distinct advantage of this method, as spectra of opposite polarity usually contain useful complementary information on the sample investigated. For ion energies of 5 keV at the first dynode, the influence of the ion mass of atomic ions on the conversion rate should also be only slight; for complex molecular ions of large mass, a dependence of the conversion rate is expected and needs further investigation. A spectrum from a NBS standard glass sample with signals from a large number of identified atomic

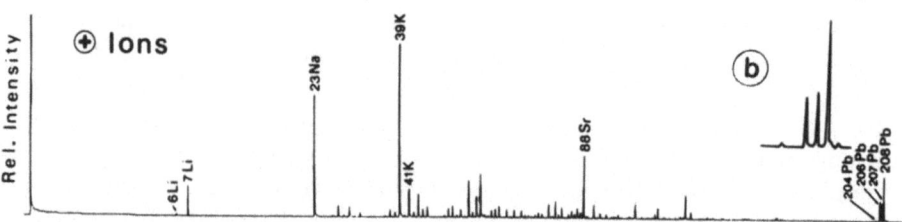

FIGURE 4. Typical LAMMA spectrum of positive ions obtained from a 0.3-μm-thick Epoxy resin standard doped with organometallic complexes of Li, Na, K, Sr, and Pb at equimolar concentrations of 10^{-2} mol/liter.

ions throughout the Periodic Table of elements is shown in Fig. 6 for demonstration. Doubly charged ions have never been identified at moderate irradiances in the focus. Even at settings as large as 100 times ion detection threshold, they are very scarce. For molecular ions, there is a very strong preference for even electron ions; radical ions have been observed only in exceptional cases from condensed aromatic compounds. This is in agreement with the frequent observation of parent ions from organic molecules with alkaline attachment (cationization).

In the volume analysis mode, specimens below about 0.5 μm thickness allow a spatial resolution of 0.5 μm to be reproducibly achieved. This is demonstrated in Fig. 5. With some care, the resolution limit of 0.25 μm can be reached at irradiances near the perforation threshold.

The mass resolution of the system is limited by the flight time dispersion in the initial acceleration section due to the initial energy spread of the ions, the total flight time spread due to slight variations of the path lengths of the ions as imaged down the system, and the time resolution of the transient recorder of 10 ns. In systems with the ion reflector attachment, a resolution of $M/\Delta M$ of 600 at an M of 200 is routinely achieved as demonstrated by the signals of the Pb isotopes shown in the inset of Fig. 4. This is sufficient to discriminate adjacent peaks of + integral mass numbers (M) up to relative masses of about 1000. Under special circumstances, the resolution has been increased to the point that the signal due to an organic fragment ion of nominal mass number 41 could be clearly separated from that of the [41]K isotope (Heinen *et al.*, 1980).

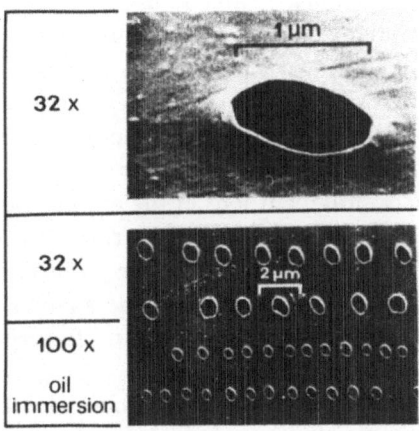

FIGURE 5. Perforations in a thin (0.3-μm) Epoxy resin foil produced with the LAMMA instrument.

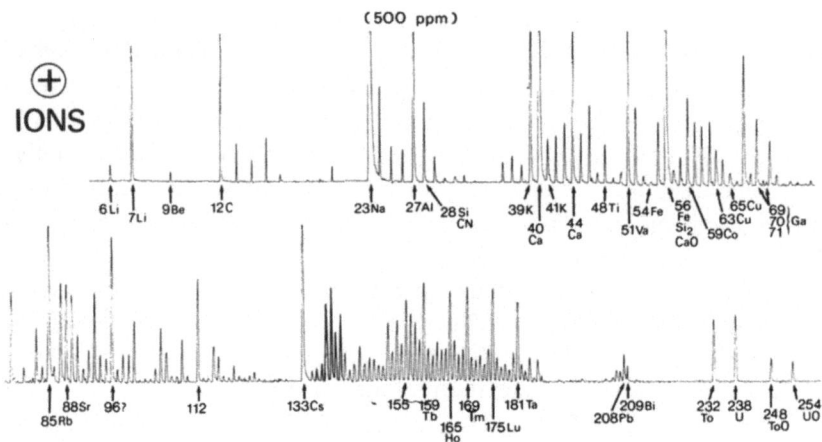

FIGURE 6. LAMMA spectrum of a standard glass sample (NBS 611), volume analyzed 10^{-13} cm^3.

3.4. Sensitivity

The detection sensitivity depends on a number of parameters such as ionization potential for positive and electron affinity for negative ions, binding state of atomic ions to the surrounding organic matrix, bond strength in organic molecules, stability of ions generated, and—most important—the laser wavelength and irradiance in the focus. It can, therefore, be specified for a given analytical problem and well-defined instrumental parameters only. Table I lists the absolute and relative sensitivity for the detection of a number of atomic constituents dissolved as organometallic complexes in an Epoxy resin, commonly used as standard specimen in the LAMMA instrument. These numbers are convincing evidence of the outstanding performance of the method for microanalysis of atomic ions in organic matrices. With the instrument set for highest sensitivity, signals due to fragment ions of the organic matrix may, at times, obscure the signal of atomic ions of interest, particularly in the mass range of between 25 and 100. The choice of a suitable embedding material can, in many instances, circumvent this problem in the analysis of biological tissue, because the matrix signals are reproducible, and fingerprint patterns of the matrix can be discounted in data analysis.

The reproducibility of the ion signals from shot to shot depends not only on the homogeneity of the sample but also, for the analysis of organic molecules originally in solution (see Section 6.1), strongly on the choice of the preparation method. For the particularly interesting case of

TABLE I. Detection Limits of the LAMMA Instrument for Atomic Ions Dissolved as Organometallic Complexes in Epoxy Resin Standards

	Absolute (g)	Relative (ppmw) .
Li	2×10^{-20}	0.2
Na	2×10^{-20}	0.2
Mg	4×10^{-20}	0.4
Al	2×10^{-20}	0.2
K	1×10^{-20}	0.1
Ca	1×10^{-19}	1.0
Cu	2×10^{-18} (10^{-17})	20.0 (200)
Rb	5×10^{-20}	0.5
Cs	3×10^{-20}	0.3
Sr	5×10^{-20} (10^{-19})	0.5 (5.0)
Ag	1×10^{-19}	1.0
Ba	5×10^{-20}	0.5
Pb	3×10^{-20}	0.3
U	2×10^{-19}	2.0

physiological or toxic elements in organic material, reproducibility is typically better than $\pm 20\%$. As successive spectra can be recorded at a rate determined by the selection of a new site for analysis with the microscopic stage, i.e., at a rate of several spectra per minute, averaging over a suitable number of spectra is easy and usually advisable for increased reliability of results.

3.5. Quantitative Analysis

Relative quantification of a given ion species in comparable samples is a straightforward procedure, because the ion signal has been shown to increase linearly with the concentration over several orders of magnitude (Fig. 7). This solves a substantial number of problems frequently encountered in biological microanalysis, such as the determination of ion concentration profiles within a cell or across biological membranes, or in kinetic studies of metabolic processes. LAMMA lends itself particularly to the latter set of problems, as isotope dilution techniques with stable isotopes can be applied. As is true for most actual analyses with other microanalytical methods, absolute quantification still poses considerable difficulties. The details of the physical process of the laser-induced ion formation and its

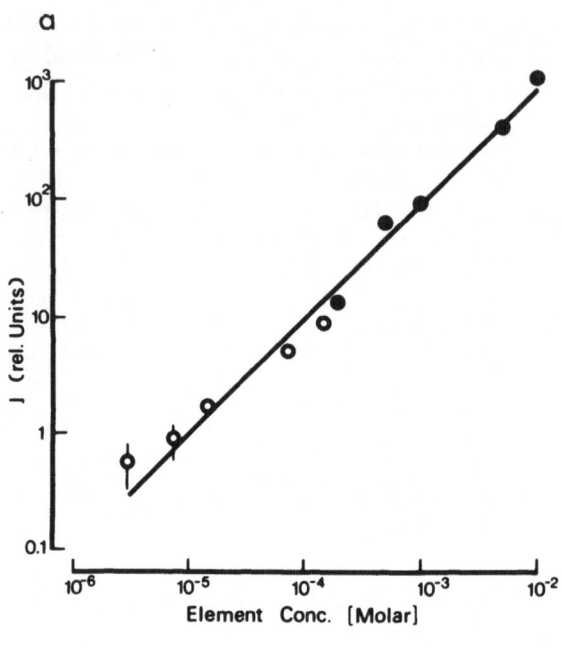

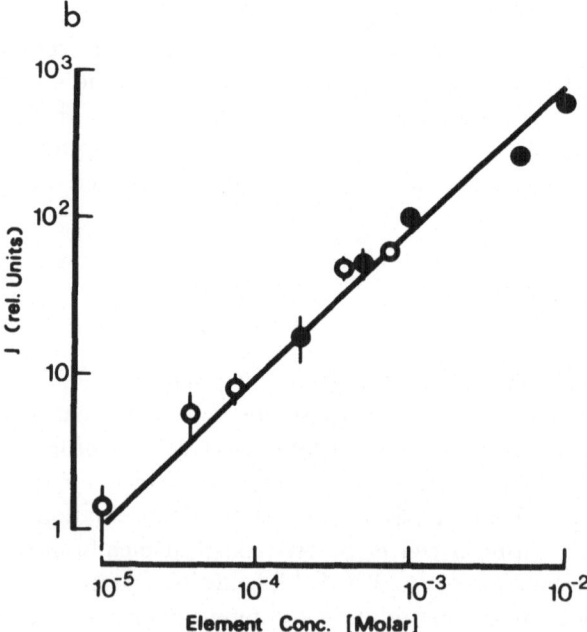

FIGURE 7.   Dependence of the ion signal intensity on element concentration of (a) Pb and (b) Li in Epoxy resin standard specimen.

dependence on the irradiation parameters, the properties of the atomic or molecular ions of interest, and the influence of the surrounding matrix, or solid state structure of the sample are as yet far from being completely understood. No theoretical model can, therefore, be devised at the moment that would allow a prediction of the yield and the properties of the different ions of interest in any given measurement situation. In several applications, the use of suitable external standards such as the Epoxy resin ones mentioned above or of internal standards, e.g., for the assessment of the Ca^{2+} concentration in photoreceptors of crayfish (Schroeder et al., 1980), has been successfully demonstrated. Great care must, however, be taken to assure the equivalence of the standard with the sample investigated. This includes comparability of the interaction between the element of interest and the matrix both in the solid phase and after evaporation, comparable physical parameters of standard and specimen, as well as comparable geometry and instrumental parameters during analysis. In favorable cases, the instrumental parameters, particularly the laser irradiance on the sample, can be chosen such that the ion signal saturates. Further increase of laser irradiance does not, in this case, yield any further increase of the ion signal in saturation. For such cases, some of the above strict requirements for quantification can be relaxed, and sensitivity factors can be given that allow for a direct intercomparison of the concentration of different ion species (see Tables I, II, and III).

In summary, it can be stated that the laser-micro-mass-analyzer

TABLE II. Composition of K 309 and K 491 Glasses (Weight Percentage), as Specified by the U.S. National Bureau of Standards

	K 309	K 491
Li_2O		0.001
B_2O		0.11
Al_2O_3	15	0.16
SiO_2	40	0.19
CaO	15	
TiO_2		0.26
Fe_2O_3	15	0.26
CeO_2		37.98
ZrO_2		0.40
CeO_2		
BaO	15	
PbO		59.35

TABLE III Sensitivity Factors E_x (with Reference to Fe) for Various Metallic Elements Contained in Glass Fibers or Glass Microspheres Produced from NBS Standard Reference-Glass Materials K 491 and K 309

	K 491		K 309	
	E_x	E_x^{13}		E_x
Li	140			
B	0.8			
Al	2.9		Al	2
Si	0.4	0.7	Si	0.49
Ti	3.1	4.5	Ca	4.2
Fe	1	1	Fe	1
Ge	0.01		Ba	0.64
Zr	0.8	2.1		
Ce	1.38			
Ta	0.29			

^{a} E_x^{13} values as reported by Kovalev *et al.* (1978).

LAMMA offers a performance that meets, if not all, certainly a great number of the requirements for an ideal microanalytical instrument for biomedical investigations as listed above:

- The spatial resolution is sufficient for the analysis of many subcellular compartments.

- All elements of the Periodic Table can be detected, with sensitivities differing by at most two orders of magnitude.

- Absolute and relative detection sensitivity is, in most cases, high enough to identify elements present at physiological or trace concentration. Isotopic ratios of elements can be determined and used for kinetic studies.

- Information on the molecular constituents in the analyzed volume can be, at least in principle, extracted from the spectra by fingerprinting methods. Parent ions even of large, nonvolatile, and/or thermally labile organic molecules are frequently detected. Information on a large number of elements or molecular constituents is obtained for each—destructive—single shot. Many spectra can be taken in short time to facilitate a statistical evaluation of the results.

4. LAMMA APPLICATIONS IN BIOMEDICAL MICROPROBE ANALYSIS

Since the advent of the LAMMA technique in 1976, a large number of biomedical applications have already been made. They were related to a rather wide spectrum of scientific problems, which are much too broad to be reviewed in detail. However, a selection of LAMMA applications will be probed that cover the most significant cases to indicate the broad field of usages. A more extensive and detailed collection of LAMMA work in general and biomedical applications in particular is contained in the proceedings of the first LAMMA symposium (Hillenkamp and Kaufmann, 1981).

4.1. Specimen Preparation for LAMMA Analysis

Unlike in material sciences, where the majority of samples to be microprobed consist of solid state matter, most biomedical samples are highly hydrated and, with a few exceptions, are not directly accessible to any kind of microprobe analysis, laser microfluorometry and Raman microprobe being the only techniques principally applicable to living hydrated systems. Therefore, specimen preparation for microprobe analysis basically means a transformation of the specimen into a solid state phase from which, for analysis in the LAMMA 500 instrument, semithin (0.3–1 μm) histological sections can be cut. Conventional fixation and dehydration procedures are not usable if one is interested in the analysis of highly diffusible water soluble compounds such as physiological cations. To prevent unacceptable leaching or redistribution of such compounds, cryotechniques are most commonly employed. These all start with shock freezing the specimen either by immersion in an appropriate liquid coolant (liquid N_2, freon, propane), or by contact with a polished metal surface cooled to a liquid N_2 or He temperatures. This technique gives satisfactory structural preservation over a limited specimen depth where the cooling rate is fast enough to prevent the formation of larger ice crystals. While from the viewpoint of an electron microscopist, a good ultrastructural preservation of shock frozen specimen is found over a specimen depth of 10–15 μm only, the situation is not so restricted for analytical purposes, since ice crystal damages can be tolerated as long as they stay below the spatial resolution of the microprobe system. In the case of the LAMMA analysis, a 100-μm-thick layer of the specimen is usually of acceptable quality.

After shock freezing, the specimen can be processed further in three ways:

1. It can be sectioned in a cryoultramicrotome and the section can be analyzed either in the hydrated state (cryostage in the analytical instrument) or after freeze drying at room temperature.
2. The bulk specimen can be freeze dried *in toto* in a special freeze-drying device at temperatures below $-80°C$ followed by Epoxy resin embedding under vacuum with Spurr's low-viscosity medium and dry sections cut on conventional ultra microtome.
3. Freeze substitution by an apolar medium can be employed before embedding and sectioning instead of freeze drying.

Most of the biological samples used in LAMMA studies were prepared by freeze drying as in procedure 2. In special cases (analysis of Ca), precipitation techniques appear to prevent dislocation of the element to be analyzed even if wet fixation, dehydration, and staining procedures are employed. A more extensive discussion of preparation techniques for biomedical microprobe analysis is given in Echlin and Kaufmann (1978).

4.2. Distribution of Physiological Cations and Trace Elements in Soft Biological Tissues

4.2.1. Ca in Retinal Tissues and Photoreceptors

The retina, because of its functional and structural complexity, is an attractive object on which to perform microprobe studies, which eventually may contribute to a better understanding of the physiological principles of phototransduction and related events. The role of Ca in photoreception is still a rather exciting and real question (see Yoshikami and Hagins, 1973; Liebmann, 1978; Fleming and Brown, 1979). In a recent study, Schroeder *et al.* (1980) used LAMMA for elucidating the role of pigment granules in the Ca turnover of the isolated crayfish retina (*Astacus leptodactylus*). The goal aimed at was to get information on possible Ca stores related to the Ca turnover of photoreceptor cells. Besides the analytical results summarized below, this study is also remarkable from the technical point of view, since for the first time in the history of LAMMA analysis, the authors introduced a stable isotope (^{44}Ca) for measuring the kinetics of Ca turnover in a sub-cellular compartment and made use of an internal standard (evaporated inorganic salts of Pt, Ca, Mg, K, etc.) deposited on the surface of the section as a homogeneous layer of known thickness. The material that appeared most suitable for creating such an internal standard by vacuum deposition was MgF_2 (see Fig. 8). With layer thicknesses of a few nanometers, the homogeneity of the deposited material was better than $\pm 10\%$ SD (^{26}Mg isotope). It appears that this standardization method not only allows for normalization of the spectra with respect to variations of the sample volume but lends itself to absolute quantification provided that sensitivity factors for standard and analyzed elements can be determined under the chosen experimental conditions. This does not appear to be too difficult according to Kaufmann *et al.* (1979, 1980). However, the authors also emphasize that at high spatial resolution (laser perforation 2–3 μm), the linear relationship between signal intensity and the apparent sampled volume (given by the cross-section area of the laser perforation times the section thickness) is lost. This is most probably as the result of border effects, i.e., contributions of standard element ions desorbed from areas outside the visible perforation area. For more details of this interesting but still controversial subject, see the original paper of Schroeder *et al.* (1980).

Some results from a histological section of a crayfish (*Astacus leptodactylus*) retina are shown in Fig. 9. Two sorts of pigment granules, a black pigment (DP) and a reddish proximal pigment (PP), can be distinguished. The hypothesis was to be tested whether these pigments may play a role as Ca stores in the Ca turnover of the crayfish photoreceptor. As shown by the LAMMA spectra in Fig. 10, the DP granules contained the bulk amount of Ca present in these photoreceptors. For the absolute Ca concentration in the DP an estimation of 120–140 mmol/liter was found in contrast to a Ca

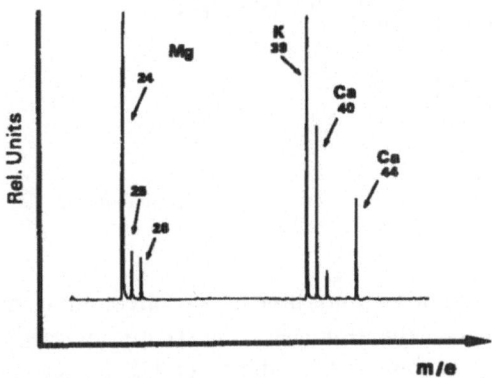

FIGURE 8. LAMMA spectrum of an *Astacus* retina specimen (0.3 μm section of Epoxy-resin-embedded material) after incubation with the stable ^{44}Ca isotope. For quantification, MgF_2 was vacuum deposited onto the specimen as an internal standard. The signal of the ^{26}Mg isotope (11.71 % abundance) was taken (Schroeder *et al.*, 1980) for calibration.

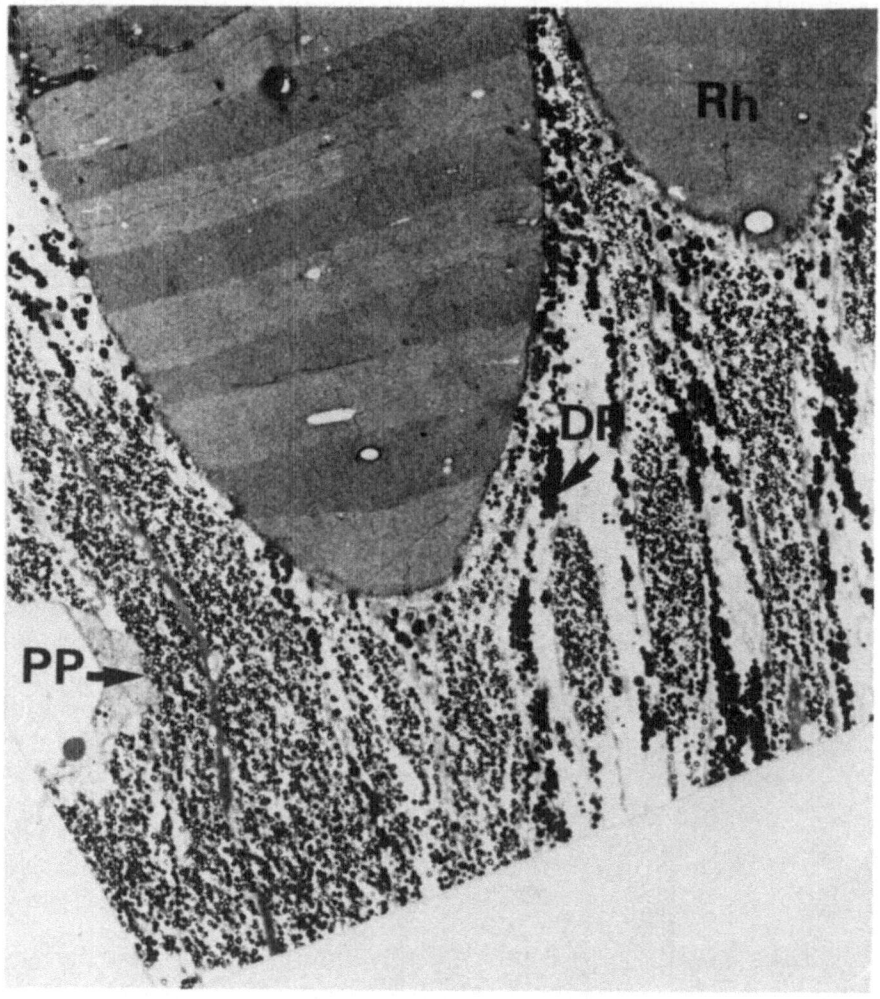

FIGURE 9. Astacus retina (EM micrograph). Rh, rhabdome; DP, distal pigment; PP, proximal pigment (Schroeder *et al.*, 1980).

concentration of less than 40 μmol/liter in the surrounding material. In the microvilli of the so-called rhabdome (see Fig. 9), an intermediate Ca content of 3–10 mmol/liter was found. The ability of DP for Ca uptake and Ca release after incubation in varying extracellular Ca concentrations has been shown by subsequent LAMMA analyses to be high in Ca-rich media, while Ca is lost in low-Ca solutions. To prove that the Ca accumulation really comes from external sources, the stable [44]Ca isotope was added to the per-

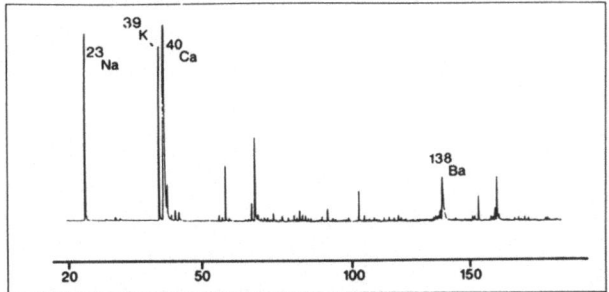

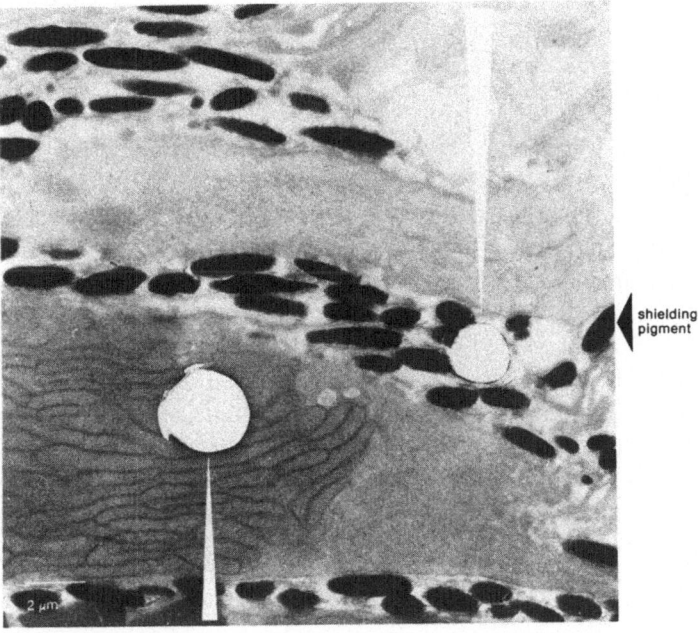

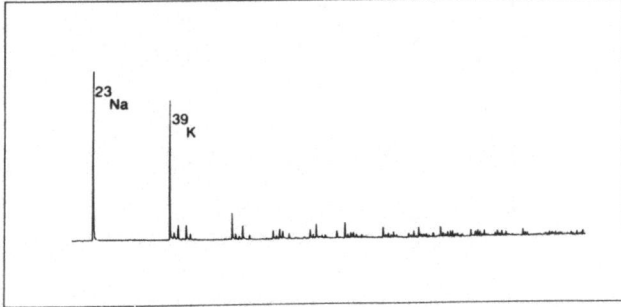

FIGURE 10.   Two LAMMA spectra obtained in the distal pigment (upper panel) and the surrounding material (lower panel) of an *Astacus* retina (Schroeder *et al.*, 1980).

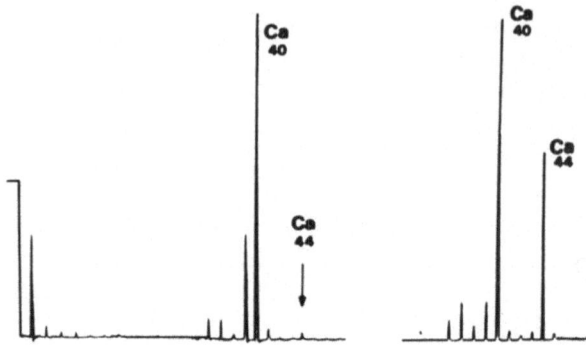

FIGURE 11. LAMMA spectra of the distal pigment granules of an *Astacus* retina before (left) and after (right) incubation with ^{44}Ca (Schroeder *et al.*, 1980).

fusion medium, and the ^{44}Ca uptake in the DP was measured by LAMMA analysis after a previous Ca depletion. A typical LAMMA spectrum after 120 min incubation time in ^{44}Ca medium (10 mmol/Ca) and the ^{44}Ca uptake measured in the DP by LAMMA are shown in Fig. 11.

The Ca concentration profile along single rods has been sought in frog retina by Schroeder *et al.* (1980, 1984) and also Kaufmann (1980). As shown in Fig. 12, a bimodal Ca-distribution profile is found along single rod receptors. A first maximum (4–5 mmol/liter) was found along the area supposed to be filled with disk material. Surprisingly enough, practically no Ca could be detected in that part of the rod where mitochondria are densely packed in the inner segment. A second Ca maximum (up to 15 mmol/liter) could be

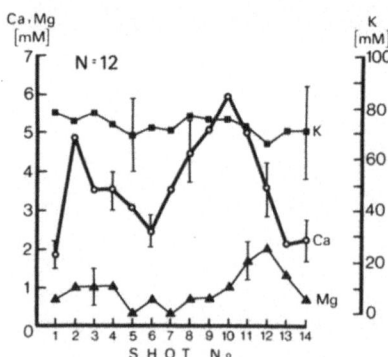

FIGURE 12. Mass concentration profile of K, Ca, and Mg along 12 frog photoreceptors. The profile proceeds from the edge of the rod outer segment (shot No. 1) to the edge of the rod inner segment (shot No. 14).

located in the proximal part of the inner segment. A nearly identical bimodal concentration profile was determined for Mg reaching 1 mmol/liter in the disk area and 2 mmol/liter in the proximal part of the inner segment. This obvious nonhomogeneous distribution of Ca and Mg in frog photoreceptor cells is in contrast to the rather homogeneous K and Na concentration profiles.

4.2.2. Ba in Melanin Granules of the Retinal Pigment Epithelium

Some years ago Burns-Bellhorn and Lewis (1976) in an ion miroprobe study detected rather large amounts of Ba in the retinas of cat and cow. In collaboration with the first author, Kaufmann (1979) reinvestigated retinal tissues of cat, frog, and man by LAMMA analysis and found Ba highly concentrated and strongly bound to the melanin granules of the pigment epithelium and the choroid. An example of such a measurement is given in Fig. 13. Here, a chemically (glutaraldehyde) fixed human retina was investigated after conventional dehydration and embedding procedures. The local preservation of Ba in such chemically fixed and dehydrated specimen led us to suggest that Ba is rather tightly bound to a chemical structure related, most probably, to the melanin.

The biochemical features of the melanin are still not fully explored. It is highly insoluble and chemically very inert. It contains a large quantity of (protected) free radicals. From this basic information, it is quite conceivable that melanin, either in the course of some sort of free radical polymerization during its biosynthesis or by specific cation trapping, can bind or incorporate divalent cations. However, there must be an extremely high preference for Ba^{2+} over other physiological divalent cations such as Mg^{2+}, Ca^{2+} present at concentration levels several orders of magnitudes higher than that of the (toxic!) trace contaminant Ba^{2+}. It is interesting to note, however, that Ca and Mg (equally tightly bound) do coexist with Ba in the melanin granule at about equal molar concentrations (10–30 mmol/liter) as determined by use of a reference standard specimen.

An attempt was made to check whether or not the high affinity for Ba is a peculiarity of the ocular melanin or rather a general property of other melanins as well. To this end, tissue specimens taken from a freshly dissected human melano-blastoma (courtesy of Professor Pau and Dr. R. Hennekes, Department of Ophthalmology, Düsseldorf University, Medical School) were analyzed after glutaraldehyde fixation and conventional preparation. As demonstrated in Fig. 14, the melanin granules of this tumor also contained considerable amounts of Ba (accompanied again by Ca and Mg), although at a concentration level presumably half an order of magnitude

(a)

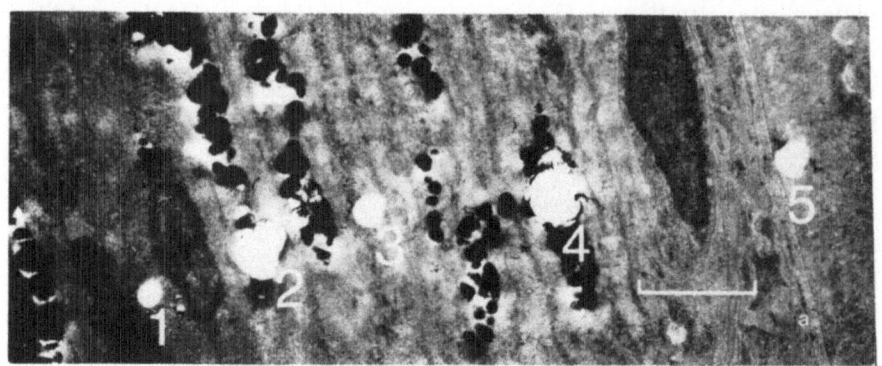

(b)

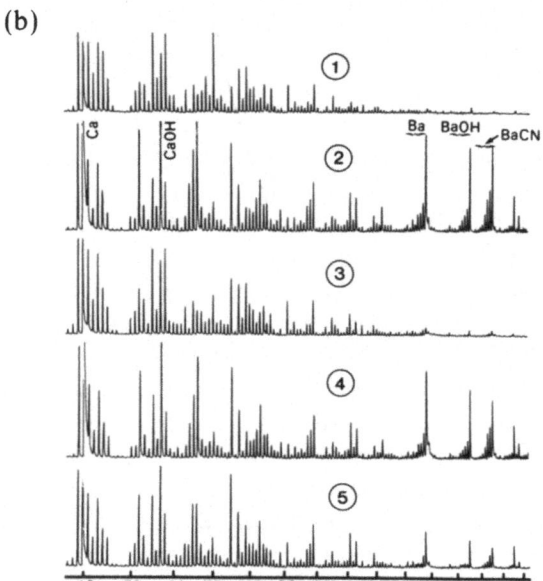

FIGURE 13. (a) LAMMA spectra of five areas of interest taken in a human retina specimen. (b) Note the strong Ba-related signals in spectra 2, 4, and 5 recorded in the melanin granules. No Ba was detected in the surrounding material (shot and spectrum Nos. 1 and 3).

below the one found in the retinal pigment. Whether or not the lower Ca content of the melanin granules in this growing tumor tissue has something to do with the fast *de novo* synthesis of melanin (and, hence, the shorter time elapsed for the melanin to act as ion trap) deserves further systematic studies. It is interesting to note, however, that in newborn kittens, the retinal melanin granules do not contain measurable amounts of Ba.

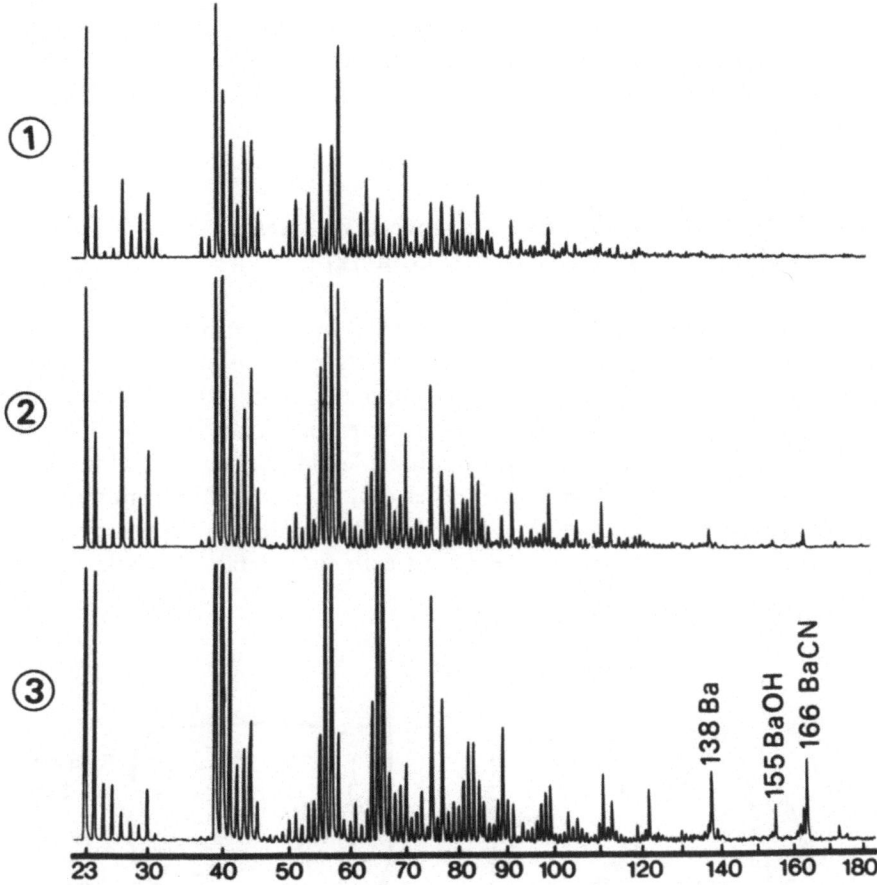

FIGURE 14.   Detection of Ba in the melanin granules of a human melanoblastoma tumor.

Despite the fact that Ba accumulation does not appear to be a very specific property of ocular melanin, the question of a possible physiological role for Ba in the physiology of photoreceptors remains to be evaluated. Recent experimental evidence (Brown and Fleming, 1979) suggests that the intensity–response curve of dark adapted toad rods is steepened and shifted to the left (i.e., to higher sensitivities) by intracellular Ba^{2+}. Similar observations were made in photoreceptors of bullfrog (Bolnick and Sillman, 1979) where Ba^{2+} was found to increase the receptor potential of rods but decreased the responses of cones (in which it also delayed the onset of rapid dark adaptation). The concentration ranges for Ba^{2+} found to be effective in these experiments were 0.4–2.0 mmol/liter. It remains to be shown whether and eventually to what extent Ba is employed as a physiological modulator of photoreceptor responses. LAMMA analysis performed in photoreceptors of the non-dark-adapted frog retina demonstrated the presence of only trace amounts of Ba close to the detection limit of the instrument for Ba (10^{-5} mol/liter).

4.3. K/Na Ratio in Transport Active Epithelia of the Inner Ear (Stria Vascularis)

The labyrinth of the inner ear is divided into a perilymphatic and an endolymphatic space, both filled with electrolyte solutions of very different composition. Whereas in the perilymphatic space the K/Na ratio corresponds to the usual composition of the extracellular fluids (1:30), the K/Na ratio is reversed in the endolymphatic liquid to about 30:1. Since the two spaces are separated from each other only by rather thin membraneous structures, such as Reissner's and basilar membranes (for orientation see Fig. 15), an active transport system must exist to maintain the rather large

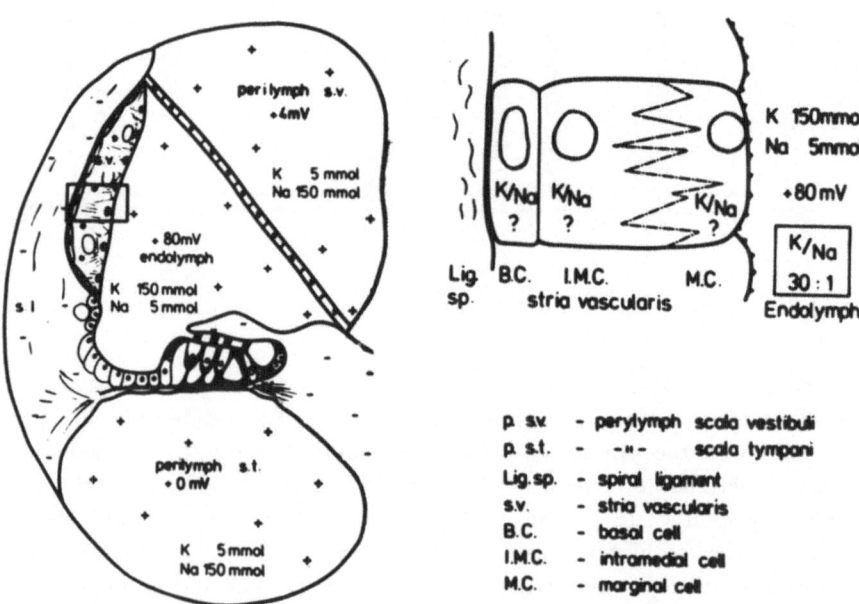

FIGURE 15. Schematic diagram illustrating the electrolyte concentration within the endolymphatic and perilymphatic stria vascularis and the cellular layers of the stria vascularis (right side).

concentration gradients for K and Na. It has been suspected that a thin strip of cells located on the lateral wall of the triangular endolymphatic space, the so-called stria vascularis, might be the location of the active ion transport system involved. Thus, disturbances at this site may induce changes in the endolymphatic ionic composition, which in turn appears to be the physiological reason behind Meniere's disease. Figure 15 also illustrates schematically the histological features of the stria vascularis.

The topographical inaccessibility of the inner ear structures in general, as well as the minute dimensions of the stria vascularis in particular, have so far excluded the application of conventional approaches usually employed

FIGURE 16.   Schematic representation of analytical profiles taken with LAMMA through the three cell layers of the stria vascularis of the guinea pig (for orientation see Fig. 15). The numbers in each area indicate the K/Na ratio measured in this spot (Orsulakova et al., 1981).

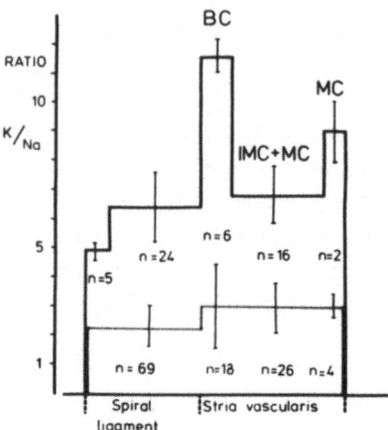

FIGURE 17. Averaged data of the K/Na ratio obtained by profiling across the stria vascular as indicated in Fig. 16 before (solid line) and after (dashed line) 2–3 min anoxia (Orsulakova *et al.*, 1981).

for the study of transport by active epithelia. Therefore, we do not have direct evidence for a transport active system located in the stria vascularis nor do we know anything about the transport mechanisms involved. In this situation, LAMMA appeared to be a promising approach, provided a preparation technique could be devised for an appropriate cryofixation of the inner ear structures. Such a procedure was elaborated by Orsulakova *et al.* (1980). Figure 16 gives an impression of the analytical LAMMA procedure in an attempt to obtain concentration profiles of the physiological cations across the three cell layers of the stria vascularis. In the schematic drawing of Fig. 16, the numbers given for each analytical spot indicate the K/Na ratio found there. Averaged data of this kind are plotted in Fig. 17 and demonstrate clearcut concentration profiles and concentration gradients both obviously correlated to some of the cell layers of the stria vascularis. That these gradients must be the result of energy-dependent active transport mechanism(s) can be concluded from the results obtained after 2–3 min anoxia. The K/Na ratios measured by LAMMA along the various cell layers of the stria vascularis generally declined, and the concentration profiles or gradients flattened drastically (see Fig. 17). These results, although rather preliminary, clearly favor the concept of an active ion transport system located in the stria vascularis and, thus, call for further systematic analyses as to the nature and site of this ion pump. (For further information see Orsulakova *et al.*, 1981, and Meyer zum Gottesberge-Orsulakova *et al.*, 1985.)

4.4. K/Na/Ca Ratio in Single Isolated Skeletal Muscle Fibers

One of the earliest LAMMA studies (see Kaufmann *et al.*, 1978) concerning physiological cation distributions was performed on single isolated fibers of frog skeletal muscle. As such fibers are comparatively easy to isolate, they are rather ideal objects for cryofixation (see Section 4.1); their functional state prior to fixation can largely be controlled; and last, but not least, their overall content of intracellar cations is well known. Therefore, such fibers can also serve as some sort of standard to check for problems that may have occurred during dissection, fixation, or specimen preparation. The results shown in Fig. 18 demonstrate to what extent intracellular ion concentration may change during dissection (but also during inappropriate cryofixation) even before any changes in structures or functions are visible. For example, in the case of muscle fiber, the striation pattern, excitability, and contractile response are not notable altered in spite of different Na/K/Ca ratios. These values were determined in isolated single frog skeletal muscle fiber, and although rather reproducible in each individual fiber, were found to vary rather widely from one fiber to another. The data of five fibers (A–E in Fig. 18) are shown. Only in fiber A was a "healthy" K/Na ratio of 24:1 found with a Ca-concentration much too low to be detected at the instrument setting chosen. Fibers B–E exhibited nonphysiological K/Na ratios declining from 7.2:1 (B) to 2:1 (E) with a Ca signal increasing and (fiber E) finally exceeding the K signal. It is assumed that, with the exception of fiber A, the fibers were damaged during dissection and had lost K and taken up Ca via membrane leakage, the latter ion being rapidly taken up by the structures of the sarcoplasmic reticulum. It was interesting to note that not until the late stage of Ca accumulation, as represented by fiber E, had been reached did the fibers show visible signs of damages (contractures) or loss of contractile activity.

4.5. Fe/Mg Ratio in Single Human Red Blood Cells

The determination of element concentrations and concentration ratios in single red blood cells as a function of various experimental, physiological, or pathophysiological conditions is a typical goal for microprobe analysis, especially if distribution histograms are sought rather than global analyses of the whole cell population. The limited sensitivity of electron probe microanalysis initially restricted such investigations to Na and K and was extended to Fe (for instance, see Kirk and Lee, 1978), but elements like Mg or Ca are still inaccessible for such kind of analyses.

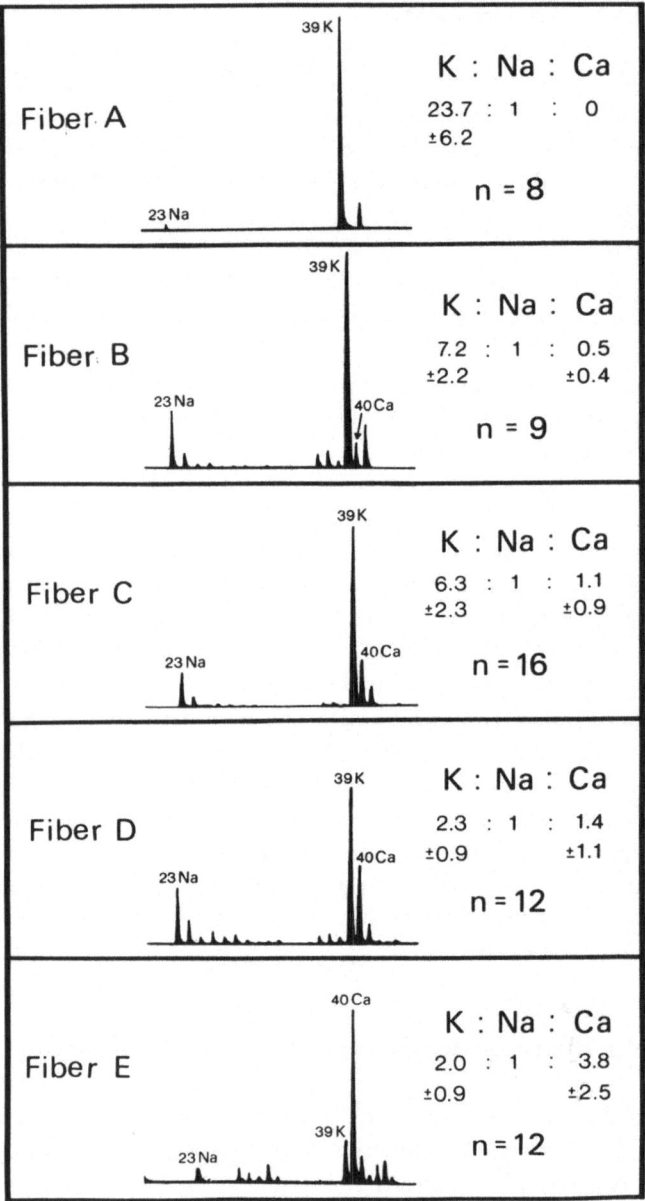

FIGURE 18. LAMMA spectra and K/Na/Ca ratios (averaged) in individual frog skeletal muscle fibers (isolated *in vitro* shock frozen, freeze-dried, and plastic-embedded specimen). See text for further explanation.

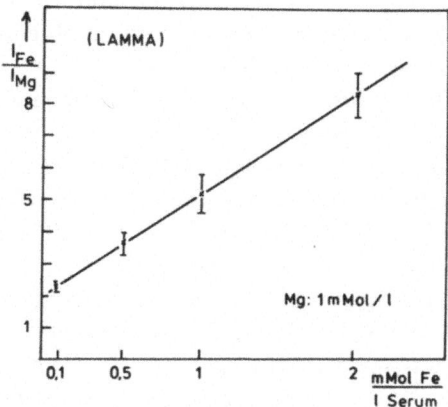

FIGURE 19. Signal intensity ratio versus concentration plot for the |Fe|/|Mg| concentration ratio in air-dried standard specimen of human blood serum containing known concentrations of FeCl and Mg (Schmidt *et al.*, 1980).

The much higher sensitivity of LAMMA was used by Schmidt *et al.* (1980) to determine the Mg/Fe ratio and the Mg/Fe ratio histograms in red blood cell populations of dialysed patients suffering from renal insufficiency. By using appropriate standards (air-dried sera) of known Fe/Mg ratios, a linear relationship between the measured signal intensity ratios I_{Fe}/I_{Mg}, and the concentration ratios I_{Fe}/I_{Mg} in the standard was established (see Fig. 19).

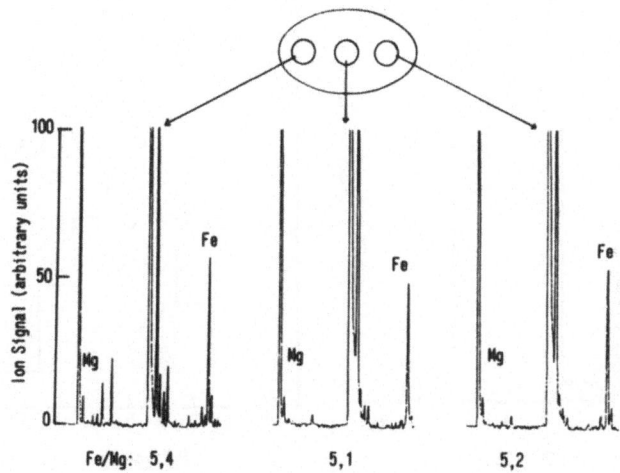

FIGURE 20. Three LAMMA spectra obtained in the same single human erythrocyte. The spectra are scaled for detection of Mg and Fe. Na and K signals are largely overloading the dynamic range (Schmidt *et al.*, 1980).

In addition, it was documented whether or not the signal variation in individual erythrocytes is small enough so as not to obscure the histogram of the distribution of inter-individual differences.

As shown in Fig. 20, the spectra obtained from single erythrocytes are fairly reproducible. Under similar conditions, reproducibility of the Na/K ratio in single human erythrocytes is of $\pm 10\%$ SD. Therefore, one can be confident that the distribution histogram of the Fe/Mg ratio (shown in Fig. 21) determined by Schmidt *et al.* (1980) in two different patients reflects the true interindividual variability for this parameter. Also, the differences in the distribution histograms of the two patients showed a remarkable reproducibility, as checked by repetitive measurements in different samples from the same patient (see Table IV).

In the same work of Schmidt *et al.* (1980), the potentialities given by LAMMA for individual blood cell analysis are nicely demonstrated by another interesting detail. The LAMMA analysis of single echinocytes (a pathological form of red blood cells characterized by a spiculated appearance) revealed that these cells had lost not only the Fe content but their Mg as well.

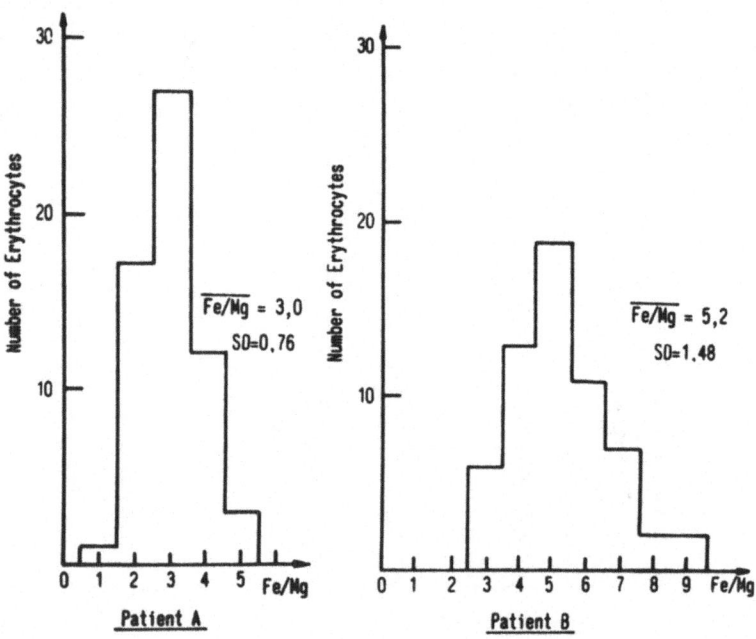

FIGURE 21. Histogram of the Fe/Mg ratio determined in the red blood cells of two different patients.

TABLE IV. Reproducibility of the Averaged Fe/Mg Ratio ($N = 30$) in Red Blood Cells Determined in Three Samples of the Same Patients[a]

Measurement in sample	I_{Fe}/I_{Mg}	$\pm$ SD
1	3.22	0.87
2	3.00	0.78
3	3.40	0.80

[a] Reproducibility of the averaged Fe/Mg ratio ($n = 30$) in red blood cells determined in three samples of the same patients.

4.6. Detection and Localization of Trace Elements in Biological Tissues

LAMMA is superior to electron probe microanalysis in detectability and sensitivity by allowing analysis and localization of trace elements down to concentration levels hitherto inaccessible. This may give analytical access to the subcellular distribution or transport kinetics of toxic heavy metals such as Pb and to the detection and localization of essential trace elements such as F, Cu, or Fe and offer the possibility of using atomic labels on pharmacological agents to follow their distribution in the cellular or subcellular level. The following collection of examples may illustrate the present state of the art. It should be noted, however, that up to now the subjects taken for LAMMA analysis have been selected randomly rather than as part of systematic programs to delineate the full extent of the field.

The detection of Ba in the pigment granules of the retina has already been given as an example of the capabilities of the LAMMA instrument in this respect. Some species of bacteria (*Pedomicrobium*) or algae (*Dunaliella*) are known (or suspected) to concentrate particular divalent cations such as manganese, copper, or uranium. The LAMMA instrument could detect Mn or U in cellular individuals scattered within a nonhomogeneous population (Sprey and Bochem, 1981).

LiCl is known to have favorable effects in certain types of mental disease, but nothing is known so far about either its cellular site of action or its distribution in the brain. We, therefore, tested whether the LAMMA instrument could detect the small amounts of Li present in brain cells of a rat after administration of a single therapeutic dose of LiCl (200 mg/kg). The spectrum shown in Fig. 22 demonstrates that this is possible. So, for the first time, we have the possibility of studying the cellular and subcellular distribution of Li in the various regions of the central nervous system.

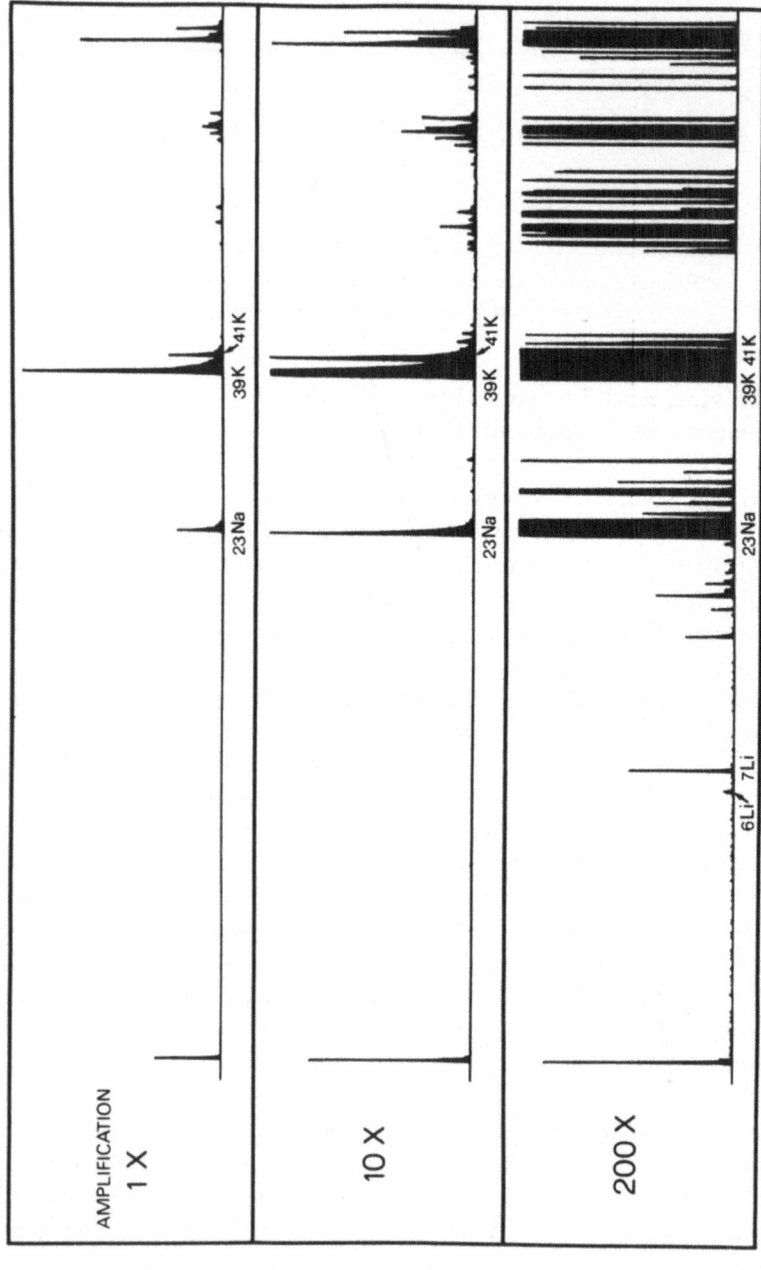

FIGURE 22. Detection of Li in a rat cerebellum specimen 12 h after intravenous administration of 200 mg/kg LiCl; cryosectioned, freeze-dried, unembedded, 1-μm section.

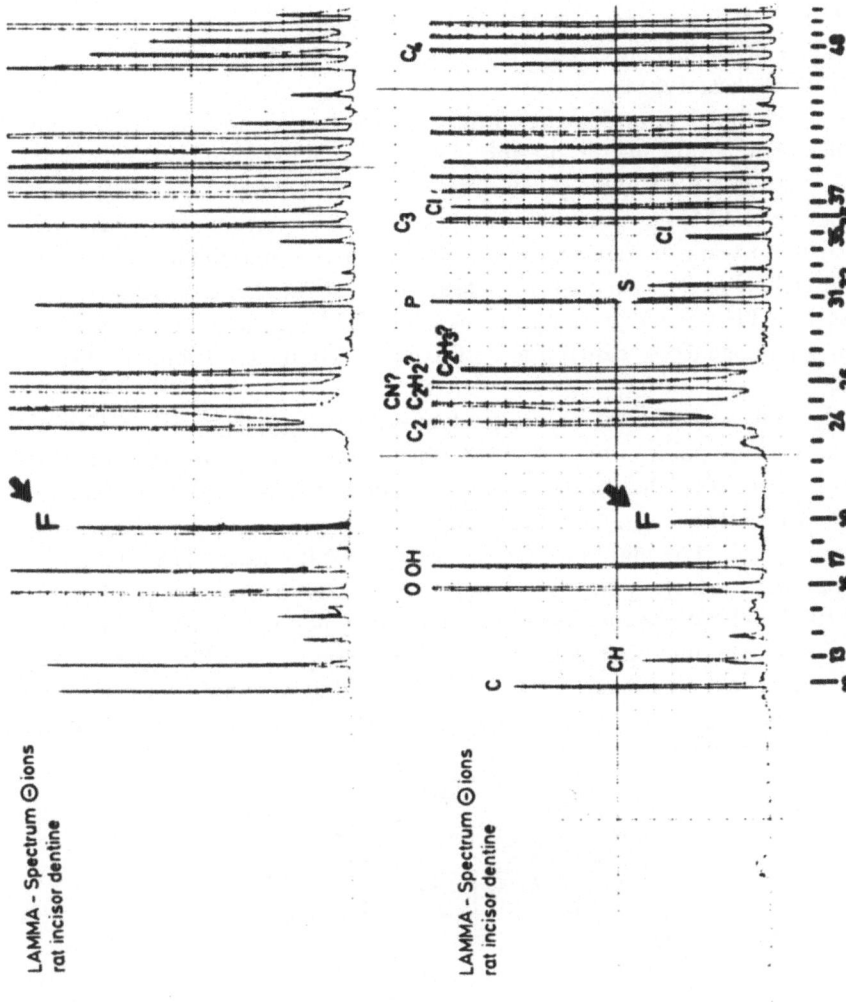

LAMMA - Spectrum ⊖ ions
rat incisor dentine

LAMMA - Spectrum ⊖ ions
rat incisor dentine

FIGURE 23. LAMMA spectra of negative ions obtained in the dentine material of a rat incisor specimen. Note the clearcut presence of a ¹⁹F signal in both analyses (Gabriel *et al.*, 1981).

Fluorine can protect teeth from dental caries, but we need to know how it reaches the sites of action if administered externally (toothpaste), or internally (drinking water), and further how it is incorporated into the crystalline matrix of enamel and dentine. The LAMMA instrument can detect the small amount of F (10–100 ppm) present in human or animal teeth (Fig. 23) and, hence, may help to solve some of these questions (see also Gabriel *et al.*, 1981).

By histochemical approaches, iron was found to accumulate within the granules of macrophages or uterine glandular cells (decidua) of pregnant *Tupaia* (Professor Kuhn, Department of Pathology, University of Göttingen). LAMMA analyses were performed in order to study the Fe-distribution and Fe-concentration profiles across glandular and other uterine cells (Fig. 24). Kaduck *et al.* (1980) investigated a case of Wilson's disease in a 14-year-old boy. By means of LAMMA analysis, they found a rather inhomogeneous distribution of copper within hard muscle tissue specimen.

The detection of trace amounts of heavy metals such as Pb or U is a particular strong point of the LAMMA instrument. Here, its detection limits are even better than that of the ion microprobe in which sensitivity tends to decline with increasing Z number. This is demonstrated by the example shown in Fig. 25. The material investigated was a NBS bovine liver standard material with an (averaged) nominal Pb mass content of 0.35 ppm (1.5×10^{-6} mol). We found (expectedly) a rather nonhomogeneous Pb concentration in the microparticulate material. Averaging the Pb signal of 48 measurements led to a mean signal intensity for that nominal concentration of 0.35 ppmw which corresponds to the spectrum shown in the Fig. 25. Thus,

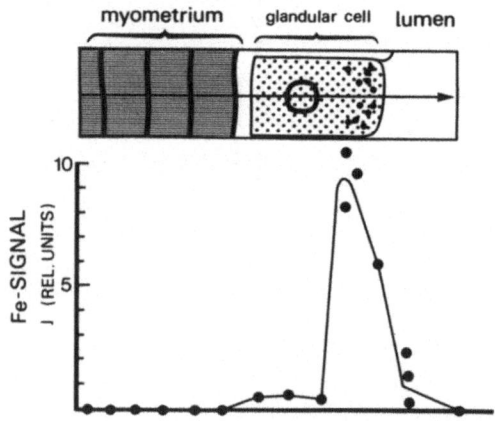

FIGURE 24. Profile of Fe content across a glandular cell of a pregnant *Tupaia* uterus.

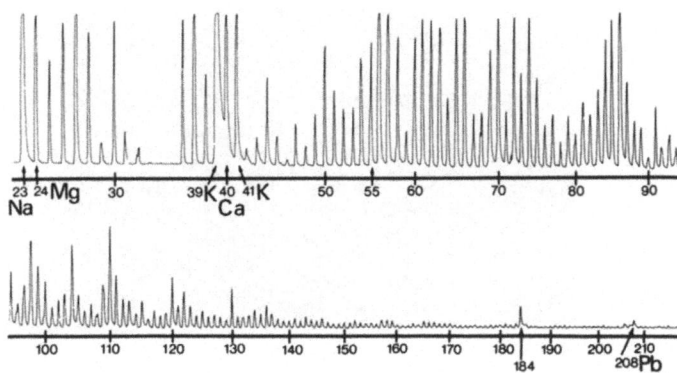

FIGURE 25. Detection of Pb in an NBS bovine liver standard containing a nominal mass of 0.35 ppm of Pb.

in an incidental case of human Pb intoxication during early stage of pregnancy, the Pb content of various fetal tissues was sufficient to be measured and to perform cellular and subcellular distribution studies (Schmidt *et al.*, 1980). LAMMA analysis has been used further not only to study microdistribution of trace amounts of Pb in plant materials and algae (*Phymatodocis nordstedtiana*, Lorch and Schafer, 1981; see also Fig. 26 and Table V) but also to estimate the Pb contents of individual airborne microparticles (Wieser *et al.*, 1980).

TABLE V. Lead (Pb) Signal Intensity in Various Regions of the Algae *Phymatadocis nordstedtiana* after Preincubation with a Medium Containing Pb at a Concentration of 1 mmol/liter

Sample area	Relative lead concentration
Cell wall	82; 116[a]; 21; 14; 20; >140[a]; t; t; 62; t; 50; 133; t; 21; 62[b]; 96; 33; 75; 130; 84; 138; 28; 60[b]; 36; 14; 55; 76
Chloroplast and cytoplasm	70; 29; 120; >140
Chloroplast	>140; >140; >140
Pyrenoid	140
Nucleus	137; >140
Cytoplasm	26; 14; t[c]; 14; 20
Intracellular space	13; 18; t[c]; 9; t[c]
Epon	n.d.[d]; n.d.[d]

[a] Tangential section of cell wall; mainly outer layer included in analysis.
[b] Sample area in the isthmus region, where semicells overlap.
[c] t, traces of lead.
[d] n.d., Lead not detectable in the LAMMA spectra.

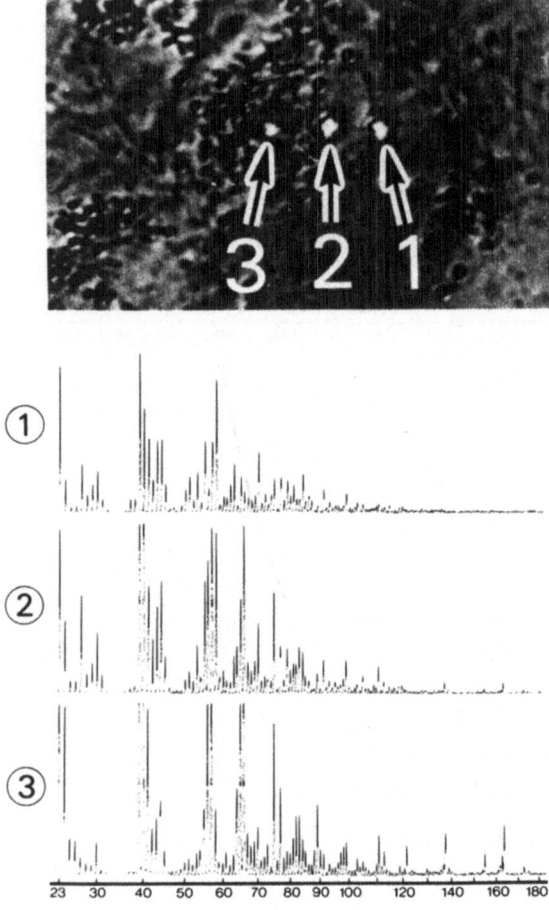

FIGURE 26. (Upper panel) EM micrograph of an Epoxy-resin-embedded specimen of the algae *Phymatodocis nordstedtiana (chlorophyta)* after (lower panel) LAMMA analysis at various areas of interest (Lorch and Schafer, 1981).

Other examples of the LAMMA technique capabilities are given by the detection of Sr in particular areas of fresh water gastropod shells, the role of Ca and Sr in the formation of the outer layer of the cell wall in certain desmid species (Baumgarten, personal communication). The micro-distribution in pine wood of chromium, copper, and fluorine induced by a pretreatment with an inorganic preservative has also been found (Klein and Bauch, 1979). For further information see also Schmidt (1984).

5. LAMMA APPLICATIONS IN PARTICLE ANALYSIS

5.1. Microanalysis of Small Particles: Some General Considerations

In the chemical analysis of microparticulate material, all the conventional methods are applicable as long as one is interested only in the average composition of many particles. However, in many current problems of particle analysis—for instance, aerosol research (forming, aging, and removal of airborne particles), or particle toxicology (induction of pneumomosis)—a single particle approach is mandatory. For this task, a microprobe technique must be capable of obtaining analytical information on the inorganic as well as on the organic composition of individual particles in order to get chemical histograms of heterogeneous particle populations rather than data on the average composition. One of the most stringent prerequisites of an appropriate instrument (besides detection limits and spatial resolution) is its capability to perform a sufficient number of analyses in a reasonable period of time if the results are to be both meaningful and statistically significant, particularly where complex chemical histograms are involved.

5.2. Performance and Limitations of the LAMMA Technique as Applied to Particle Research

With respect to the special requirements for particle analysis, laser microprobe mass analysis offers a number of advantages when compared with alternative techniques (see also Kaufmann and Wieser, 1980):

• As far as atomic composition is concerned, all elements of the Periodic Table can be detected, many of them at sensitivities in the ppm or even sub-ppm range. This usually allows a high sensitivity for detecting many trace contaminants in single particles.

• Since the LAMMA technique separates isotopes, isotopic ratio studies can be performed. For instance, stable isotopes may be used in isotope dilution studies of kinetic processes in aerosol systems, or changes in natural isotopic abundances during the period of particle formation may be detected.

• Elemental analysis is not usually sufficient for the chemical characterization of aerosol particles if the aim is to establish their origin, history, and ecological significance. In this regard, one needs as much information as possible about both organic and inorganic molecular constituents. Surely, an ideal case would be the possibility of detecting the presence of carcinogenic or toxic compounds. One of the unique features of LAMMA is that not only are the classical organic fingerprint spectra obtained, but also under favorable conditions and appropriate instrument settings, very characteristic and simple "desorption" spectra of major organic constituents (see also Section 6 below). This offers the hope that this technique will prove to be a considerable step forward in the organic microprobe analysis of particulate materials.

• If microprobe techniques are applied to single-particle analysis, the problem of distribution statistics is usually raised. If the measuring procedure itself or the time required to perform a single analysis excludes the possibility of analyzing large populations of individual particles, important statistical information is missing. Since with the LAMMA instrument the time needed for the analysis is limited only by data processing and repositioning of the specimen holder to the location of the next particle, a large number of analyses can be executed in a rather short period of time.

• Many studies also require the particle morphology (size and shape), the distribution of particles over the support, and the exact position of individual particles chosen for analysis, either as separate data or to correlate with the analytical data. This requires some imaging capability. The LAMMA instrument itself can give this structural information, at least within the physical limitations of optical microscopy, including UV absorption, phase contrast, polarization contrast, and dark field illumination. Most important, all techniques of optical pattern analysis can be easily fitted to the optics of the LAMMA instrument (see McCrone and Delly, 1973; Morton and McCarthey, 1975). If the structural elements of the particles or the whole particle itself are below the resolution of the optical microscope, there is still the possibility of using size dispersive devices for controlled particle deposition (Cadle, 1975; Mercer, 1973). It is preferable, however, to make a scanning electron microscope picture of the particle(s) to be analyzed prior to the LAMMA analysis.

There are also some limitations and disadvantages of the LAMMA technique imposed by the physical and procedural conditions:

• LAMMA analysis is essentially destructive. This, for instance, excludes the possibility of recording both positive and negative ion spectra from the same particle. Further, optimizing the operational conditions of the instrument, e.g., selecting the best power densities, requires some trial shots with the loss of a corresponding number of particles.

• The fact that the whole particle is lost at once (up to ~ 2 μm particle size) may sometimes turn into a disadvantage, particularly if surface analysis

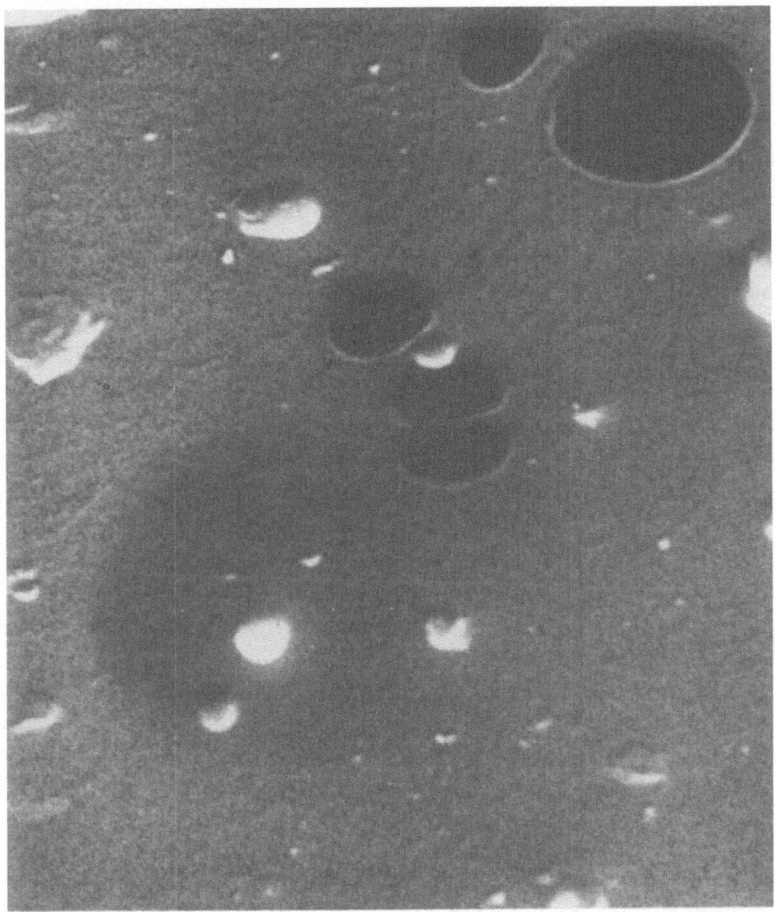

FIGURE 27. SEM picture of aerosol particles deposited on a thin (50-µm) Formvar foil. The laser perforation (diameter 2 µm) indicates the former location of an individual particle that has been evaporated for LAMMA analysis.

is desired. With larger particles, however, it has been found that LAMMA allows for some modest kind of depth profiling, at least under favorable conditions.

• As in most other microprobe techniques, quantitative analysis imposes some problems. In the absence of an established model for the laser-induced plasma formation, an empirical approach is so far the only practical way. However, as studies of Haas, Wieser, and Wurster (1981) indicate, the properties of the observed ions from single particles may be compatible with the assumption of a local thermodynamic equilibrium (LTE) model, as developed in plasma physics, at least, under conditions present in the pyrolytic mode of LAMMA analysis. Apart from this approach, semiquantitative analyses are possible on the basis that (1) matrix effects on ion yield are much less important in LAMMA than they are in *SIMS*; (2) signal-concentration plots for elements are linear over several orders of magnitude, at least, in standard specimens; and (3) reference to standards is possible whenever appropriate reference materials are available.

• With every laser-induced evaporation of an individual particle, part of the supporting film is also evaporated (see Fig. 27). This material, of course, contributes to the recorded LAMMA spectrum and inevitably creates some kind of a "background" (see also Section 3.4). The ratio of these background signals to those from the particle itself depends on the size of the particle, the thickness of the supporting foil, and the chemical nature of both materials. Since Formvar, collodion, or carbon foils can be made as thin as 10 nm, there is usually no background problem encountered with particles in the range down to about 0.5 μm in size. With smaller particles, or if trace constituents are to be analyzed, the foil-induced background can no longer be neglected. However, since the LAMMA spectra of the supporting foil are usually well defined and highly reproducible, methods of background subtraction can be successfully applied.

• The need to bring the specimen into vacuum will create problems when the analysis of liquid or volatile particles is of interest. However, it should not be too difficult to design a suitable cold stage for LAMMA analysis of particles partially or fully in the liquid phase.

5.3. LAMMA Analysis of Reference Particles

For the sake of reference and the optimum setting of instrumental conditions, particles of known chemical composition such as $(NH_4)_2SO_4$, $KMnO_4$, ammonium tartrate $(NH_4C_4H_4O_5)$, kaolinite $[Al_2(OH)_4(Si_2O_5)]$, talcum $(Mg_3OH_2Si_4O_{10})$, as well as particulate glass made from various NBS glass materials were analyzed (Kaufmann and Wieser, 1980; Kaufmann, Wieser, and Wurster, 1980). As demonstrated by Fig. 28, the LAMMA spectra obtained from these reference particles are rather simple and are easily related to the respective chemical composition of the particle under investigation. It is obvious that here an immediate and unambiguous characterization of the particles' chemical nature is possibly based on the complementary information contained in the spectra of positive and negative ions. This is particularly true for the spectra of $(NH_4)_2SO_4$ and $NH_4C_4H_4O_5$ particles. Although in the former, the appearance of the

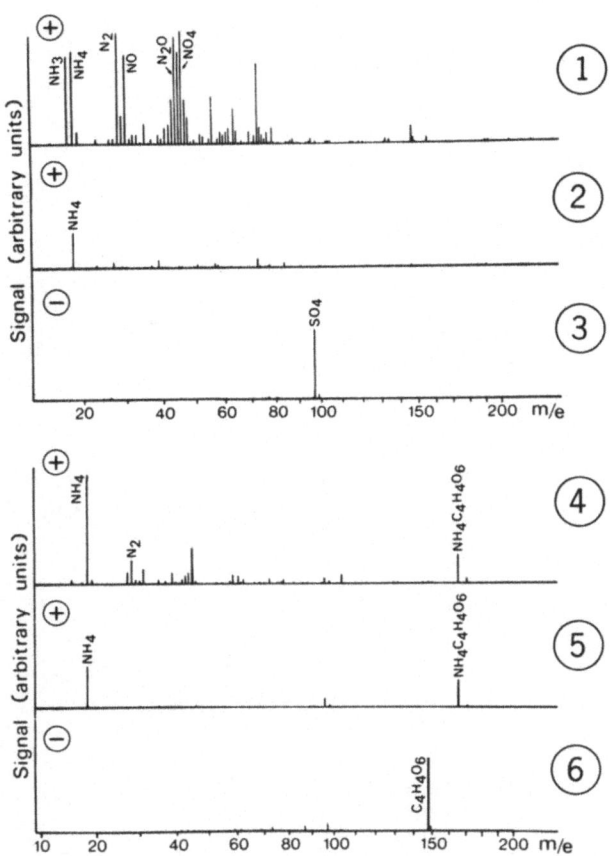

FIGURE 28. LAMMA spectra of positive and negative ions obtained from $(NH_4)_2SO_4$ (panels 1–3) and $(NH_4)_2C_4H_4O_6$ (panels 4–6) reference particles. The spectra in panel 1 and 4 were recorded at about 10 times the threshold for ion formation, whereas all other spectra were taken at 3 times threshold laser intensity (from Wieser *et al.*, 1980).

$96:SO_4^-$ ion is a little bit surprising according to the rules of classical ionization mechanisms, its identification in the laser-induced mass spectrum is not questionable as cross-checked by LAMMA analysis of Na_2SO_4. These particles with a leading peak at mass number 96 can unambiguously be attributed to SO_4, because in this case, the NH_2SO_3 fragment possibly present in the $(NH_4)_2SO_4$ particles is excluded.

The LAMMA spectra of silicate reference particles are more complex. Although fairly reproducible, fingerprint spectra have been obtained in both $Al_2(OH)_4(Si_2O_5)$ and $Mg_2(OH)_2(Si_4O_{10})$ particles; nevertheless, most of the prominent peaks can be unequivocally assigned in the spectra of the positive as well as the negative ions. The signal intensity of the Si^+ ion is

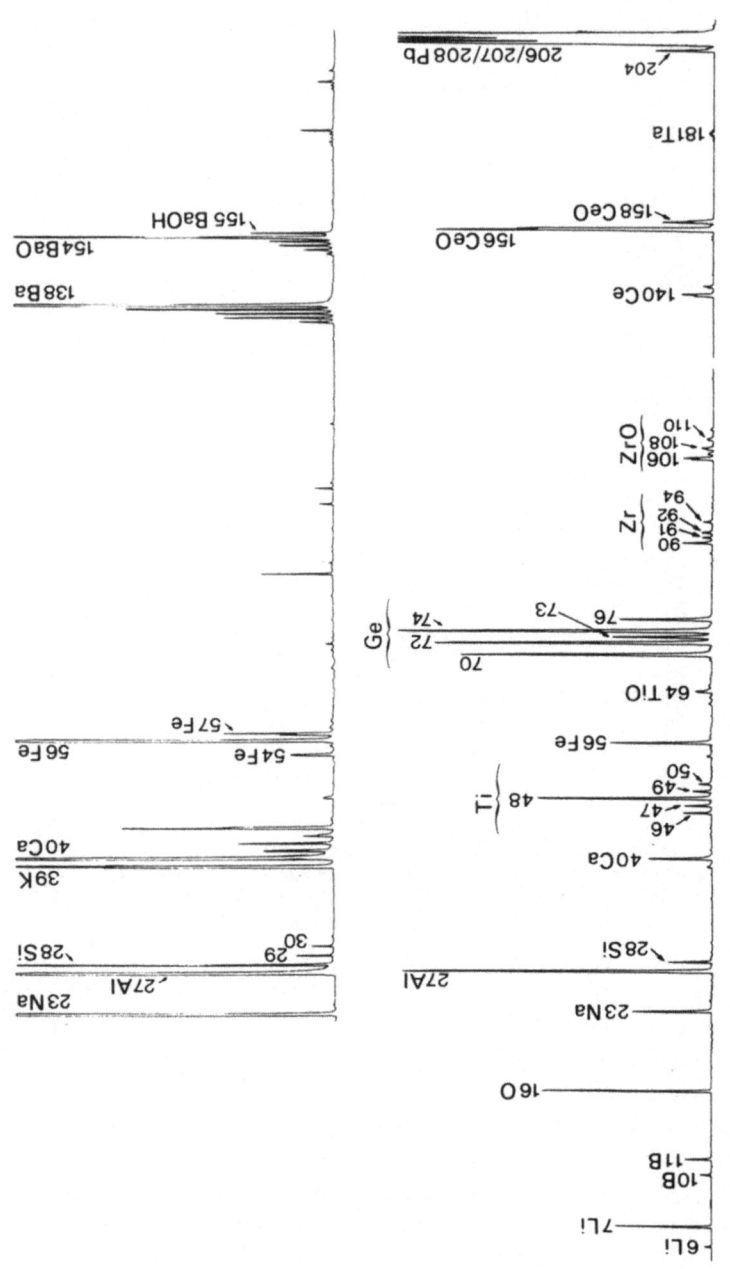

FIGURE 29. LAMMA spectra of positive ions recorded from glass particles (diameter 1 μm) made of NBS K 309 (upper panel) and K 491 (lower panel) standard reference glass materials. Nominal chemical compositions of these materials are given in Table II.

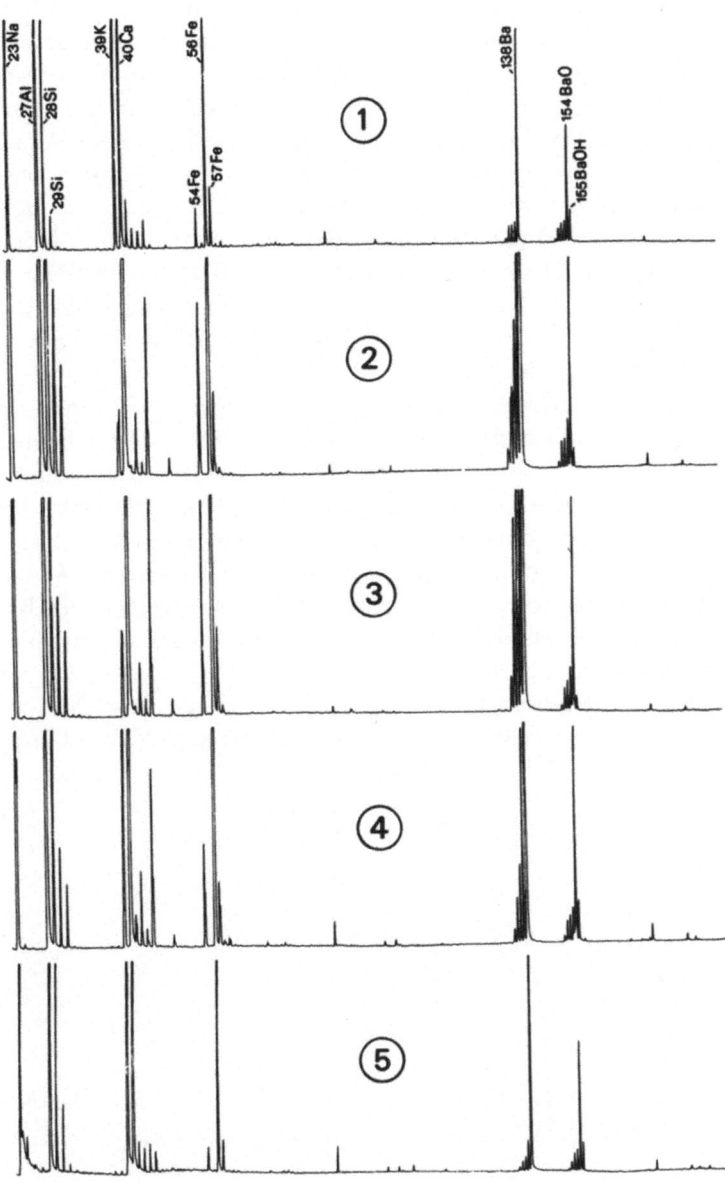

FIGURE 30. Reproducibility of LAMMA analysis in five different glass particles of NBS K 309 glass material.

rather low, at least at the laser irradiance of 3 times threshold chosen. At higher irradiances (10–15 times threshold), the yield of positive $28:Si^+$ ions increases in relation to the negative silicon containing molecular ions.

It is interesting to compare the $(NH_4)_2SO_4$ LAMMA spectra with those obtained by SIMS from equal reference particles (Cambridge Instrument study, 1978) of upper-atmospheric fly-ash particles. A prominent $18:NH_4^+$ signal was recorded in the SIMS spectrum of positive ions similar to the LAMMA spectrum together with strong $1:H^+$ and $17:NH_3^+$ signals not seen in the LAMMA spectrum. Since negative secondary ions are not easily available in SIMS, the information related to the $96:SO_4^-$ group is contained only in a number of very small signals assignable to a large variety of positively charged molecules consisting of S in any possible combination with H, N, and O.

In Fig. 28, LAMMA spectra obtained at either low (3 times threshold for evaporation) and high irradiance (10 times threshold for intensity) are compared. The latter are more complicated and much less suited for a straightforward chemical identification. However, although here the primary information contained in the low irradiance spectra is lost, there is the advantage that, in the analysis of particles with unknown chemical composition, atomic trace contaminants (for instance, heavy metals) can be found at optimal detection limits. These are in the ppm to sub-ppm ranges in the favorable cases such as Pb or U.

LAMMA measurements in glass particles made from NBS standard reference glass materials (see Fig. 29) revealed that in particles of presumably

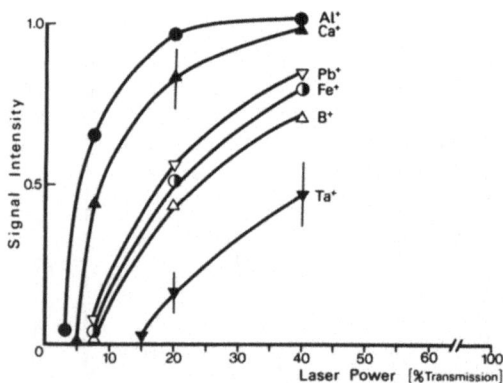

FIGURE 31. Signal intensity for various elements present in particles of NBS standard reference glass (K 491) depending on irradiance in the laser focus. For most of the elements, signal intensity ratios become independent of laser irradiance at about 10–12 times the threshold for ion formation. Sensitivity factors (see Table III) can thus be derived for materials of similar composition.

both homogeneous intraindividual and interindividual chemical (inorganic) composition:

- A fairly satisfactory analytical reproducibility can be achieved (see Fig. 30).
- Conditions can be found under which signal intensity ratios (ion yield) become independent of laser intensity (see Fig. 31) such that elemental sensitivity factors can be worked out (see Table III) which apply at least for materials of similar composition.
- Absolute quantitation might be possible if the particle volume can be determined and the particle size does not exceed 2/3 of the diameter of the irradiated area (see Fig. 31).

5.4. LAMMA Analysis of True Aerosols

The first investigations of true aerosols by means of LAMMA analysis were done by Kaufmann and Wieser (1980) and Wieser *et al.* (1980). The particles were collected from the airborne state near the ground by means of a "spectral impactor" first introduced by Zebel and Hochrainer (1972). This selects particle subsets according to aerodynamic size. Two distinct populations were investigated: large particles (diameter 0.5–1 μm) and giant particles (diameter 1–10 μm) according to a classification introduced by Whitby (1978).

Figure 32 shows examples of both positive and negative ion LAMMA spectra from large aerosol particles, whereas Fig. 33 demonstrates such spectra obtained from giant particles of the same specimen. It was found (and expected) that the population of giant particles (most probably representing that part of the aerosol which has been mechanically generated) contains constituents and/or trace contaminants of industrial origin such as Cd, Ni, Cu, Zn, Fe, or molecular fragments related to soil material (SiO_2, SiO_3, SiO_3H, Ca, CaOH, Ca_2O, Ca_2OOH).

On the other hand, large·aerosol particles had much more prominent S- and NH_4-related signals than giant particles. In about 30% of both large and giant particles, lead was detected. Obviously, the exhaust following the combustion of leaded fuel initially produces small lead particles by nucleation, which subsequently agglomerate with the various types of particles already present in the atmosphere.

Tables V and VI summarize the most frequent signals obtained in LAMMA analysis of aerosol particles together with a tentative peak assignment and their frequency of incidence.

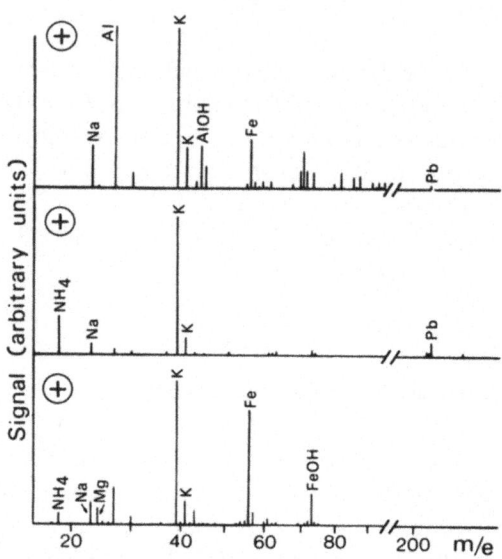

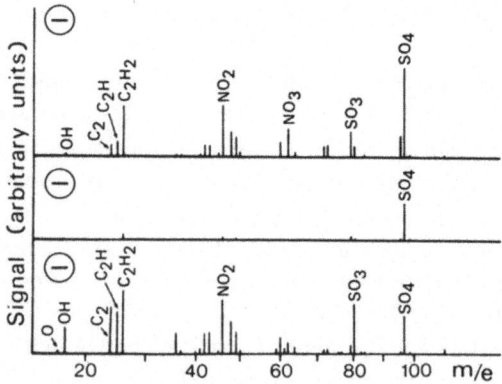

FIGURE 32. LAMMA spectra of positive (panels 1–3) and negative (panels 4–6) ions recorded from large atmospheric aerosol particles (Wieser *et al.*, 1980).

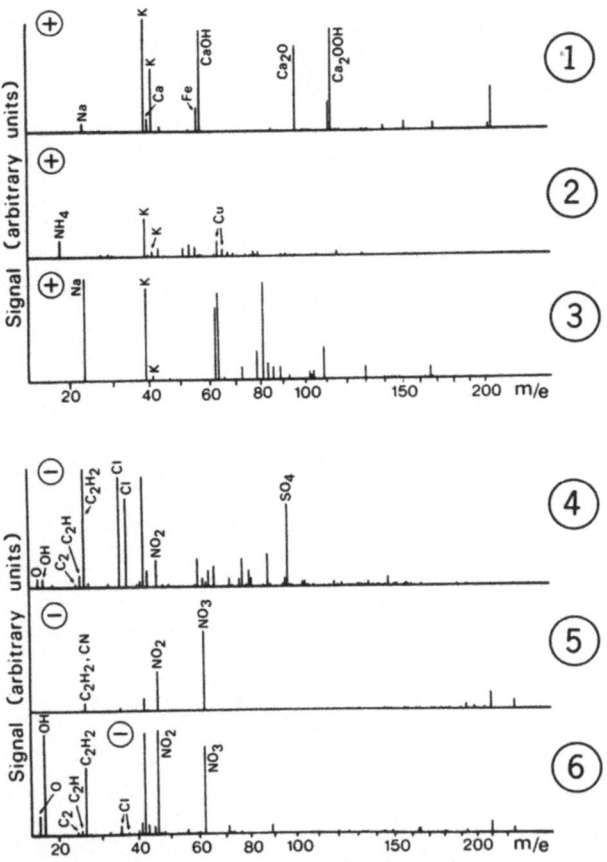

FIGURE 33. LAMMA spectra of positive (panels 1–3) and negative (panels 4–6) ions recorded from giant atmospheric aerosol particles (Wieser *et al.*, 1980).

5.5. LAMMA Analysis of Asbestos and Man-Made Mineral Fibers

Some preliminary studies demonstrating the capability of LAMMA to characterize asbestos fibers according to their main constituents, as well as their trace contaminants, are shown in Figs. 34 and 35. Spurny *et al.* (1981) have performed a more systematic study in both asbestos fibers (AF) and man-made mineral fibers (MMMF) in order to get information on the chemical stability or instability of such fibers under various conditions. The chemical instability of chrysotile in either acidic or alkaline solutions has already been demonstrated (Harrington *et al.*, 1975; Palekar *et al.*, 1979). In

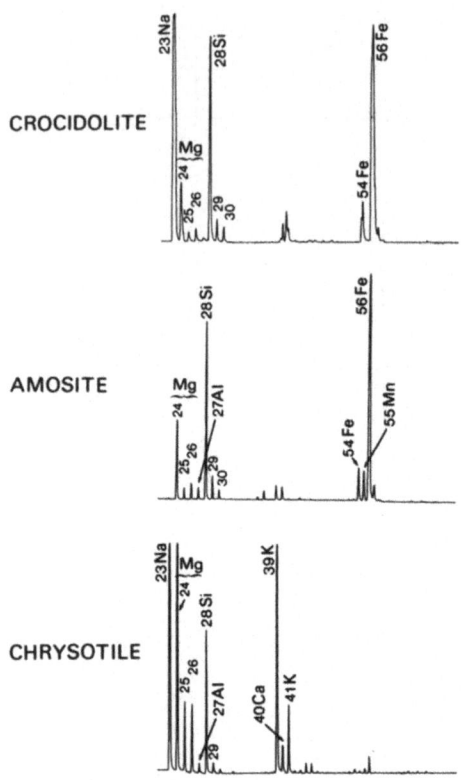

FIGURE 34. LAMMA spectra of positive ions obtained from UICC asbestos fibers (Spurny *et al.*, 1981).

various solutions or media, including living tissue, chrysotile decomposes by an almost complete removal of MgO. It was not known if in MMMF similar alterations occur. For this purpose, UICC standard asbestos and IM100, IM104, and IM106 glass fibers (Johns Manville, Inc.) were investigated by LAMMA prior to and after incubation in various inorganic and organic media (Spurny, 1983). As demonstrated by Figs. 36 and 37, a preferential leaching of some elements also occurs in glass fibers after exposure to either acidic or basic media. The extent of which elements such as Na, K, Ca, and Zn are lost depends on exposure time and fiber diameter. The smallest fibers exhibit the fastest loss. LAMMA analyses further revealed that both the original chemical composition of the investigated fibers (glass fibers as well as asbestos fibers) as well as the degree of leaching were highly inhomogeneous along individual fibers. The leaching effect varied also from fiber to fiber within a fiber population of comparable

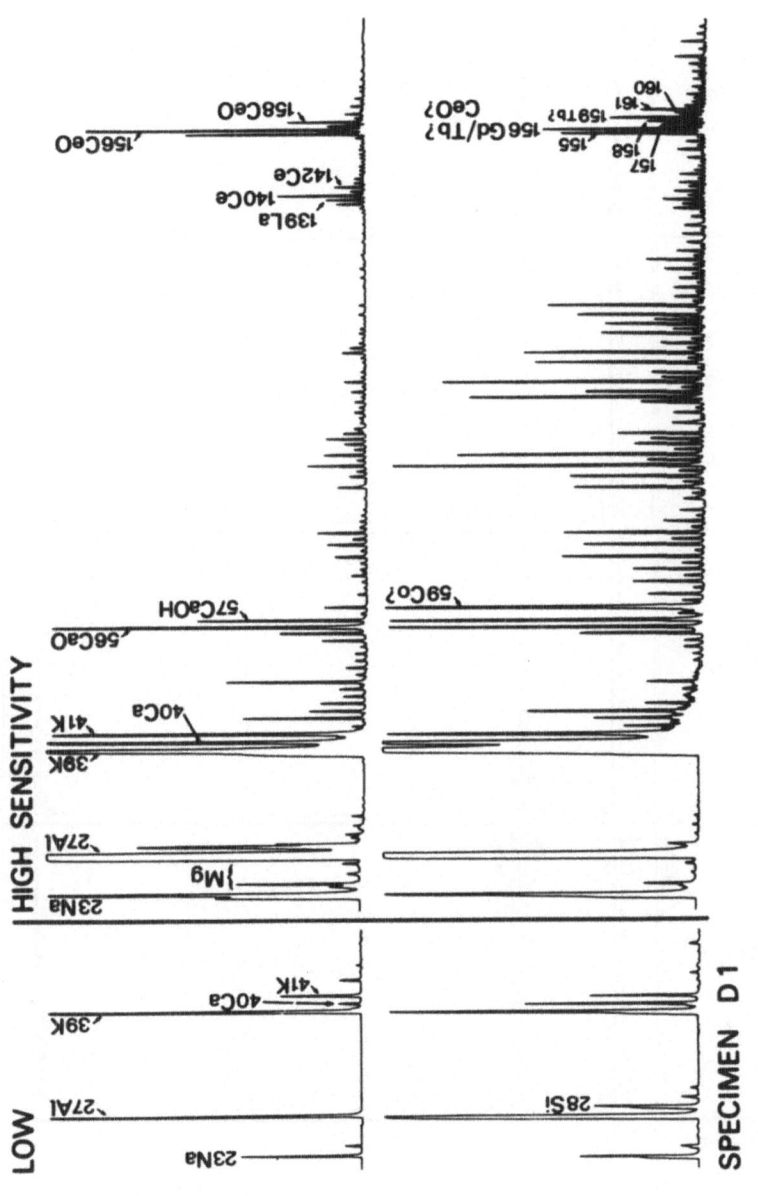

FIGURE 35. LAMMA analysis (positive ions) of two erionite fibers (diameter: 1–2 μm, length: 10–15 μm) at both low and high sensitivity. Note the various trace contaminants tentatively assigned to rare earth elements (Spurny *et al.*, 1981).

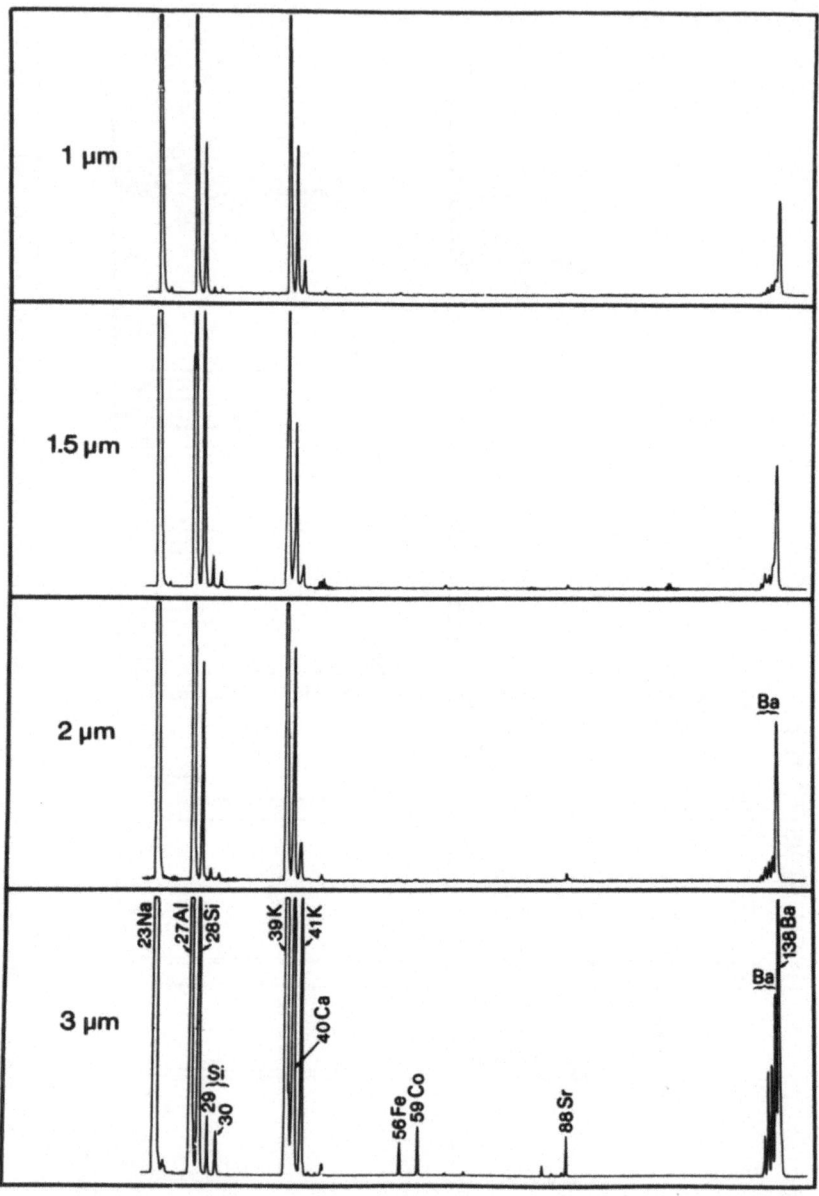

FIGURE 36.   LAMMA spectra obtained from man-made glass fibers of various fiber diameter (Spurny *et al.*, 1981).

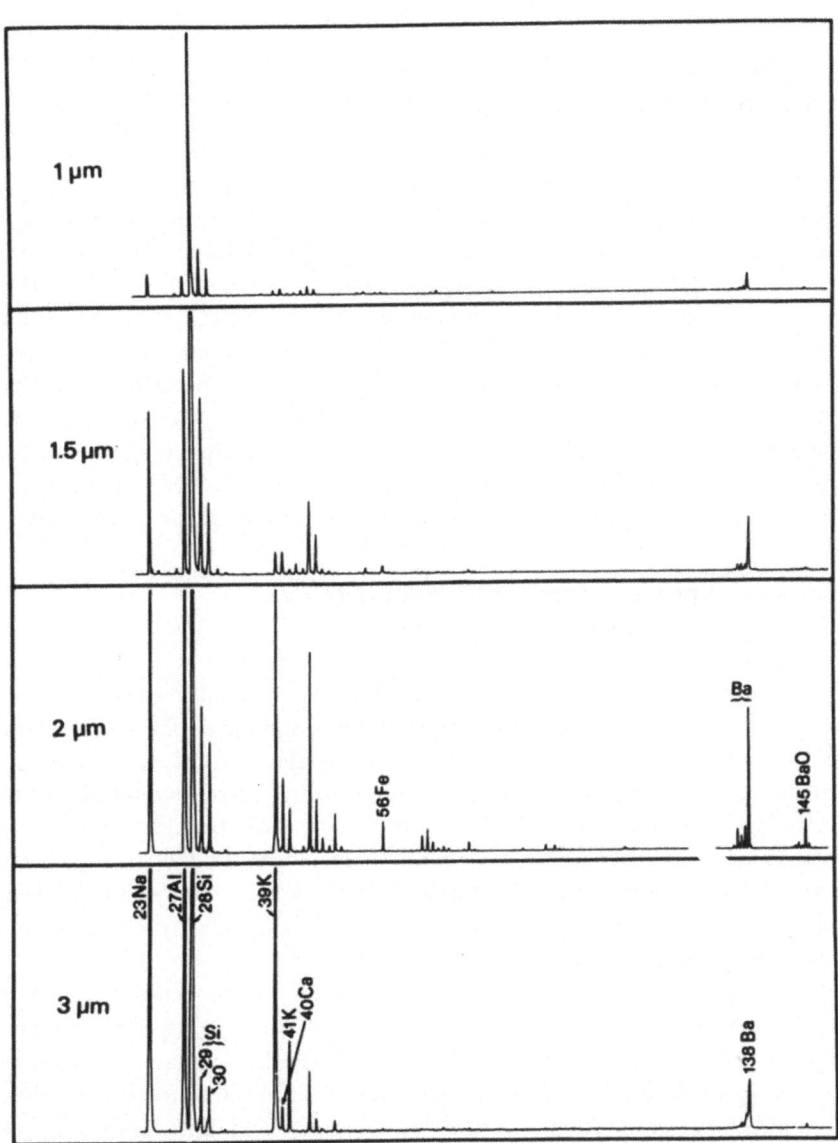

FIGURE 37. LAMMA spectra obtained from the same type of glass fibers as in Fig. 36 after 1-week exposure to a 1% H_2SO_4 solution. Note the diameter-dependent leaching effect associated with all elements except Si.

diameter. In general, glass fibers turned out to be more resistant to exposure to acidic solutions and living tissue than chrysotile fibers.

6. LAMMA APPLICATIONS IN ORGANIC MASS SPECTROMETRY*

It has been pointed out above that LAMMA spectra or organic samples contain useful information on the molecular constituents in addition to that on the atomic species. The extraction of the pertinent data from the spectra and their interpretation is, however, far more difficult than that from atomic constituents. For a number of synthetic organic polymers, it has already been demonstrated (Gardella *et al.*, 1980; Graham and Hercules, 1982) that important information on the chemical composition, in particular, in relation to the side groups to the backdone polymer structure, can be elicited from the LAMMA spectra. In another attempt to systematically assess this problem for biological matrices, a series of LAMMA investigations of suitably prepared, pure bioorganic substances has been initiated.

6.1. Mass Spectroscopy of Nonvolatile or Thermally Labile Compounds

Though the identification of at least some molecular constituents in complex biomedical samples remains as the ultimate goal (for a first example, see Seydel and Lindner, 1983), the application of the LAMMA instrument to organic mass spectrometry has during these investigations turned out to be an interesting field of its own. The basic problem in the mass spectrometry of all but a few bioorganic molecules is the fact that they are essentially nonvolatile and/or thermally labile, whereas most of the conventional techniques such as electron impact (EI) mass spectrometry require ionization in the gas phase.

Several analytical techniques routinely used in biochemistry, such as HPLC and HV electrophoresis, could be substantially improved in sensitivity would a more sensitive detection method be available that needs only a very small amount of the sample. High spatial resolution is, therefore, not a primary requirement in these applications and may eventually be relaxed to some extent, the extremely small sample mass per test shot being a distinct advantage that LAMMA can offer for such application.

---

* *Note added in proof.* Several important developments have taken place since the original submission of this manuscript. For latest developments in this field see, e.g., Hillenkamp *et al.* (1987), Holm *et al.* (1987), Karas *et al.* (1985, 1987, 1988), and Spengler *et al.* (1987).

The investigation of such important groups of substances as amino acids and small peptides, nucleic acids and their building blocks, glycosides, various hormones and enzymes, and a selection of other pharmacological agents have shown that the analytical result depends very strongly on the sample preparation employed. Sample preparation will not only strongly affect the irradiance threshold at which ions begin to appear as well as the ion yield above threshold, it also has a decisive influence on the fragmentation pattern of the molecular ions. Two techniques have so far proven successful. Both start from highly diluted (10^{-4} mol/liter^{-1}) solutions of the substances in ethanol or water. Small droplets (1 μl) of such solutions are put onto a coated EM grid and transferred into the vacuum chamber for a very fast evaporation of the solvent. They yield rather homogeneous layers with no or very little substructure visible in the microscope. Electrospraying these solutions onto the substrate results in even more homogeneous layers but needs conductive substrates that are difficult to make at the thickness of only about 10 nm required for use in the LAMMA 500 instrument. Another more substance-consuming method requiring a larger sample quantity uses a standard vibrating orifice aerosol generator to form small particles of controllable size in the submicrometer range, which are then impacted onto a coated EM grid (see also Section 5). For good and reproducible results, it seems mandatory to avoid crystal formation of a size comparable to or even larger than the focal spot of the laser.

6.2. Laser Desorption Mass Spectrometry

With suitably prepared samples two modes of operation can be distinguished, at least in principle. At low laser irradiances, ions begin to appear with no visible alteration or damage of the sample. In this mode, many successive spectra can be recorded with great reproductibility from the same sample site. These so-called laser desorption (LD) spectra show striking similarities with spectra obtained by field desorption (FD) (Krueger and Schueler, 1980) and to a lesser extent secondary ion mass spectrometry (SIMS) (Eicke *et al.*, 1980), the most important other methods that have been successfully applied in the mass spectrometry of nonvolatile thermally labile bioorganic compounds. The characteristics of these spectra are as follows:

- Substantial, even high yields of parent molecule ions.
- Almost exclusively even electron ions (no radicals). In the positive spectrum, the cation species $(M + Na)^+$ and $(M + K)^+$ more often than not dominate the protonated $(M + 1)^+$ ones. In the negative

spectra, the deprotonated $(M - 1)^-$ ion often appears as the base peak. Anionized molecules are rare.

- The spectra show specific and reproducible fragmentation patterns that depend on the molecular structure in a complex way not yet fully understood. Negative spectra · are often simpler as they show fewer lines than positive ones.
- The more polar the molecules are, the easier it is to obtain desorption spectra.

Similar spectra have also been obtained by several groups that have used CW and pulsed CO_2 lasers at 10.6 μm wavelength (Posthumus *et al.*, 1978; Kistemaker *et al.*, 1980; Stoll and Rollgen, 1979) or free running and Q-switched Nd lasers at 1.06 μm wavelength (Posthumus *et al.*, 1979; Heresch *et al.*, 1980), all with laser irradiated sample areas of 0.1–1 mm in diameter. Apparently, all the laser and nonlaser desorption methods induce transitions of molecules from the solid state into the gas phase, which, at least, locally have a strong nonequilibrium aspect. This leads to the formation of large molecules or ions that would not be thermally stable at temperatures at which their vapor pressure is large enough for them to be generated in significant amounts by thermal evaporation alone. The details of these processes as well as the common and/or different features for the various methods are not yet well understood, with the exception of FFID (Krueger, 1979). These problems will certainly be subject to intensive investigation in the future. For a review of the various laser desorption techniques, see Hillenkamp (1983). This technique is also well adapted for the analysis of the products of laser ablation of tissue as used in surgery. Spengler *et al.* (1987) and Estler and Nogar (1986) have analyzed the plume of various excimer laser ablation of biomolecules such as PMMA to determine what role thermal and other processes play in the molecular breakdown.

6.3. Laser Pyrolysis

If the focal laser irradiance is raised to a level at which damage or even perforation of the sample is visible in the microscope, the spectrum will generally change significantly. The signals of parent molecules will disappear, leaving lines of more and smaller fragment ions to dominate the spectrum. These "laser-pyrolysis" spectra will in their general appearance be similar to those obtained with electron impact ion spectrometry.

6.4. Laser Desorption Spectra of Amino Acids and Peptides

A systematic investigation of all biologically important amino acids has been undertaken by Schiller (1980). Figure 38 shows the positive and negative spectra of valine as an example. The results of this investigation can be summarized as follows:

- In the positive ion spectra of all monocarbonic amino acids, the decarboxylated ion $(M - 45)^+$ appears as the base peak. At least one of the parent molecule ions $(M + 1)^+$, $(M + Na)^+$, and $(M + K)^+$ is always present, although the relative magnitude varies, most probably as a result of various degrees of Na and K contamination present in the samples.
- Dicarbonic amino acids never yield the $(M - 45)^+$ ion, possibly because of hydrogen bonding between the two acidic groups. In the spectra of these amino acids, the parent molecule ion usually forms the base peaks.
- Most positive spectra show a substantial number of other signals that often cannot easily be associated with simple fragments of the amino acid under investigation.
- The deprotonated $(M - 1)^-$ peak usually appears as the base peak in the negative spectra with only a few fragment ion signals present.
- All positive spectra of aromatic amino acids contain rather strong signals of the residue R^+ of the molecule. Occasionally this signal even forms the base peak rather than the $(M - 45)^+$ signal. Figure 39 shows the spectra of tyrosine as an example. To the contrary, ions of the residue have never been observed in spectra of aliphatic amino acids.

The results obtained for the amino acids have been extended to di- and tripeptides. Figure 40 shows the spectra of the two dipeptides, glycyl-tyrosine and tyryl-glycine. As expected, they are significantly different from the spectra obtained from single equimolar mixtures of the amino acids, which yielded only their summed spectra. The parent molecule ions appear in both the positive and negative spectrum. The signals due to both decarboxylated amino acids are present as an indication of a cleavage of the peptide bond. The $(M - 45)^+$ signal of the terminal amino acid is always stronger than that of the initial one. This pattern has also been found in tripeptides and opens the possibility for sequencing the smaller polypeptides, at least. Surprisingly enough, the decarboxylated peptide ion $(M - 45)^+$ has never been observed. Even more surprising is the appearance of the deprotonated

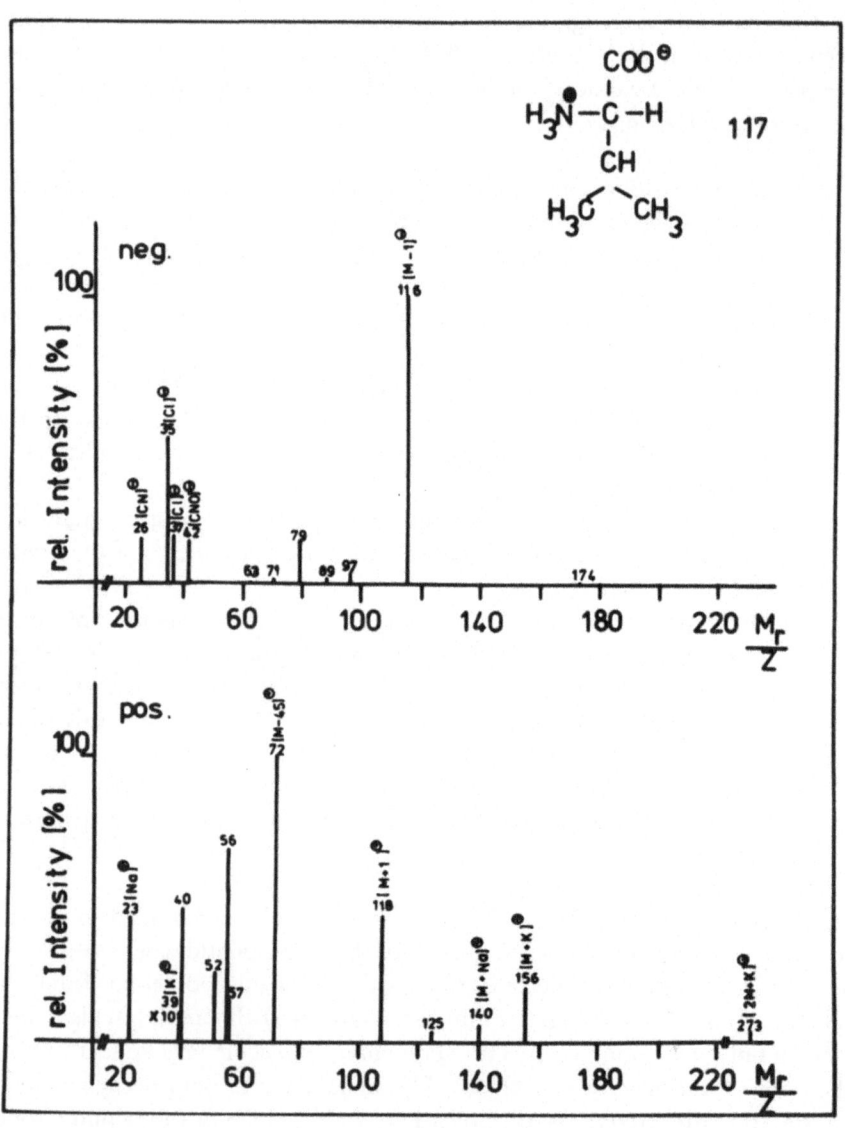

FIGURE 38. Laser desorption spectra of the aliphatic amino acid L-valine.

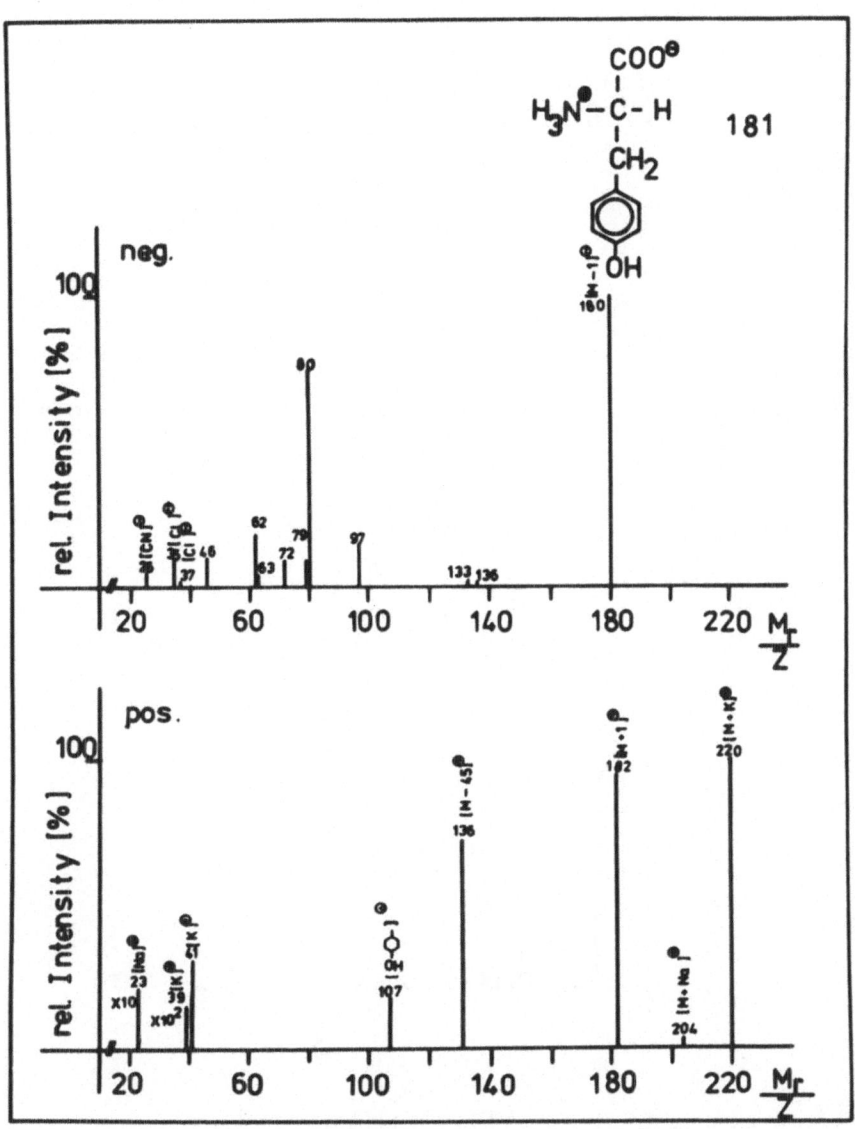

FIGURE 39. Laser desorption spectra of the aromatic amino acid tyrosine.

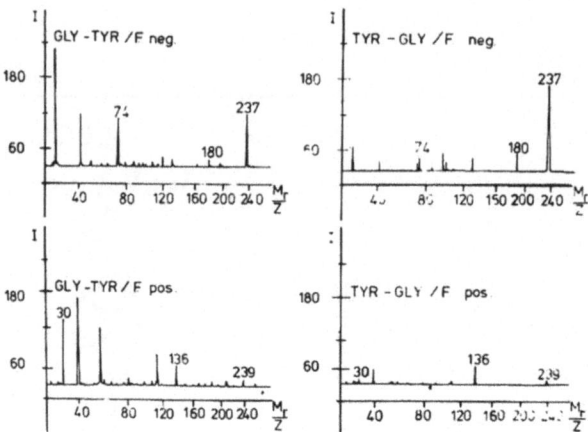

FIGURE 40. Laser desorption spectra of the dipeptides tyrylglycine and glycyl-tyrosine.

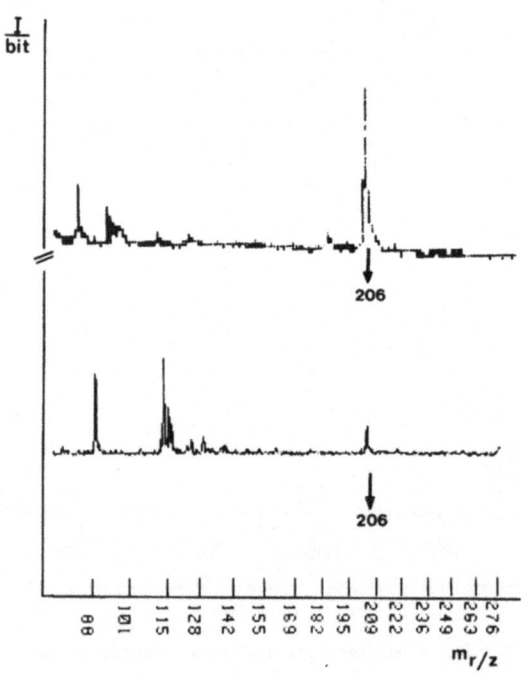

FIGURE 41. Laser desorption spectra of aminocitric acid from a biological RNP (top) and a synthetic sample (bottom).

parent molecule ion $(M_{AA} - 1)^-$ of *both* amino acids in the negative spectrum. This cannot be explained by simple cleavage of the peptide bond and indicates that a substantial degree of chemical reaction occurs during the evaporation and ion formation process.

The experience gained from this investigation of amino acids has been the basis for the identification of an amino citric acid as an essential part of a peptide that in turn had been derived from a calf thymus, ribonucleic protein (Wilhelm and Kupka, 1981). Such ribonucleic proteins are thought to act as highly species- and organ-specific antigens. Their structural and

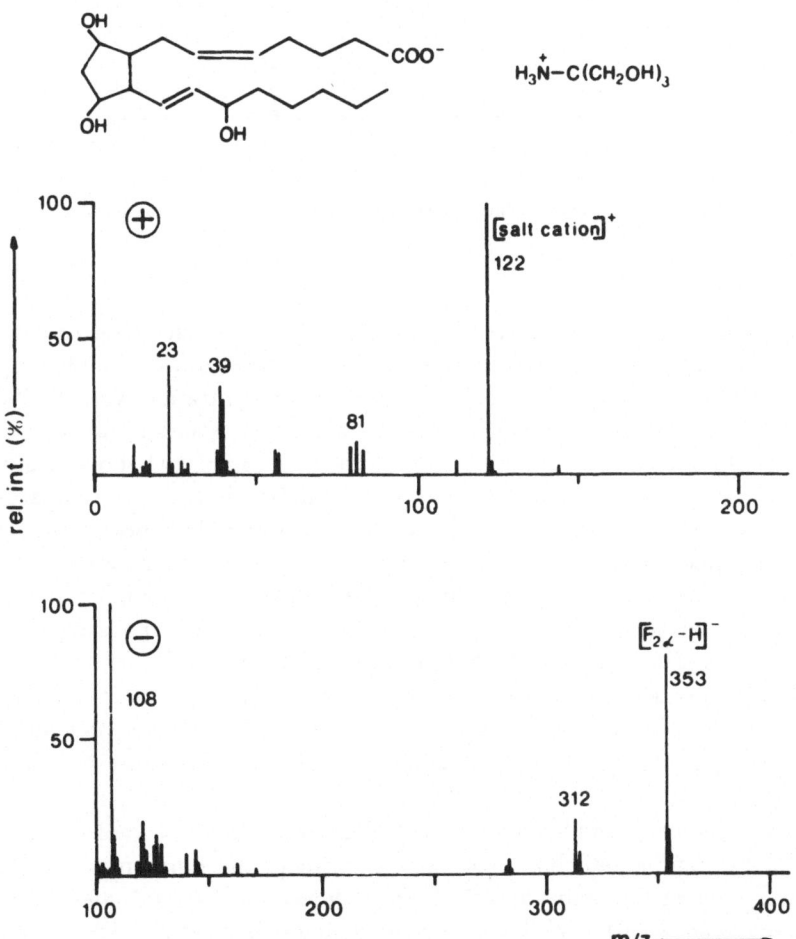

FIGURE 42. Laser desorption spectra of the tromethanine salt of prostaglandin $F_2\alpha$.

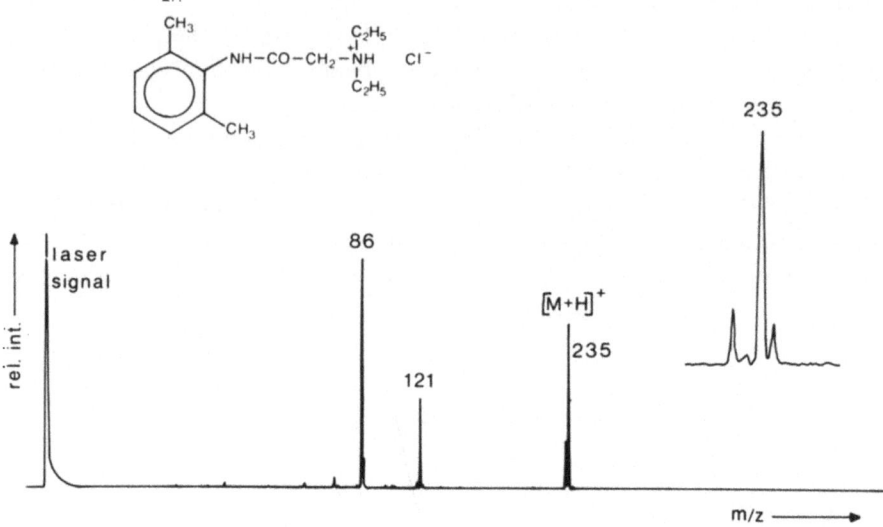

FIGURE 43.   Laser desorption spectrum of a lidocaine-containing ointment.

functional properties are of great interest in current immunological research. In tissues, they are present at extremely low concentrations, and samples containing the amino-citric acid at detectable concentration were available in only minute quantities after a very elaborate isolation procedure of several chromatographic and electrophoretic steps. Figure 41 shows negative ion spectrum for a sample and synthetic amino-citric acid for comparison. This together with other control measurements that had been conducted after the unknown had been identified is convincing evidence for the natural occurrence of this amino acid which so far had not been found to occur in other biological systems. The positive spectra of untreated samples not shown here exhibit a peak at $M = 247$ rather than $M = 208$. This indicates a high calcium affinity of the acid that possibly is part of its biological function. Treatment of the samples with oxalic acid results in precipitation of calcium oxalate and appearance of the $(M + 1)^+$ signal at $M = 208$ as expected.

As further examples for the identification of organic molecules with LAMMA Figs. 42–46 show spectra of the salt of the hormone prostaglandine F_2 and of the pharmaceutical agents lidocaine, barbital, and atropine (Heinen et al., 1980) as well as of several galactose derivatives. All substances yield high signals of parent molecule ions in the negative and/or the positive ion spectrum.

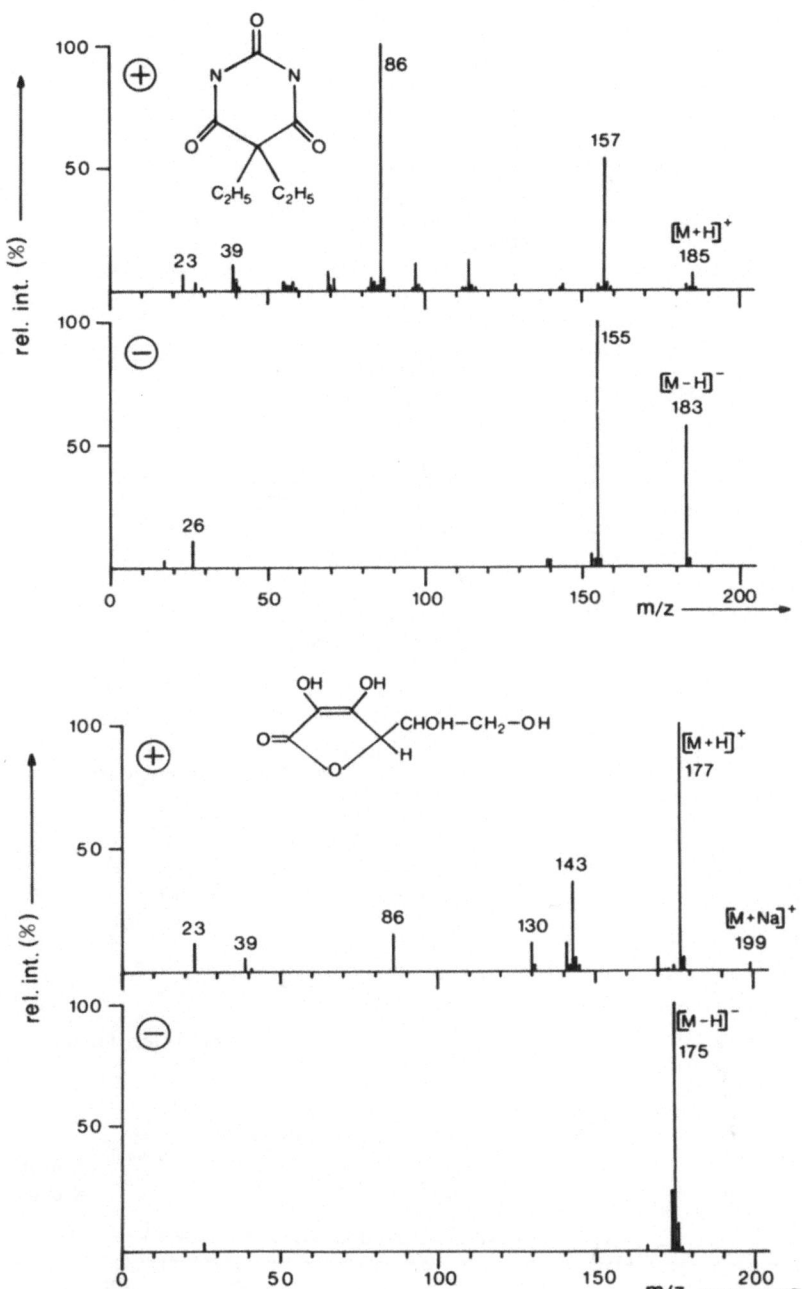

FIGURE 44. Laser desorption spectra of barbital (upper part) and ascorbic acid (lower part).

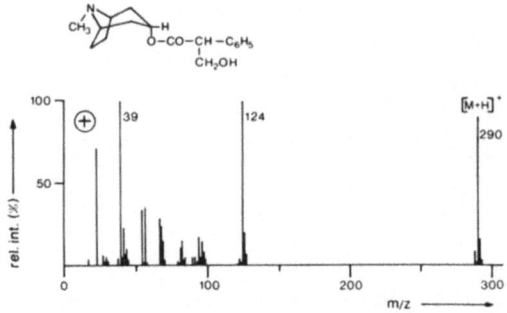

FIGURE 45.   Laser desorption spectrum of atropine.

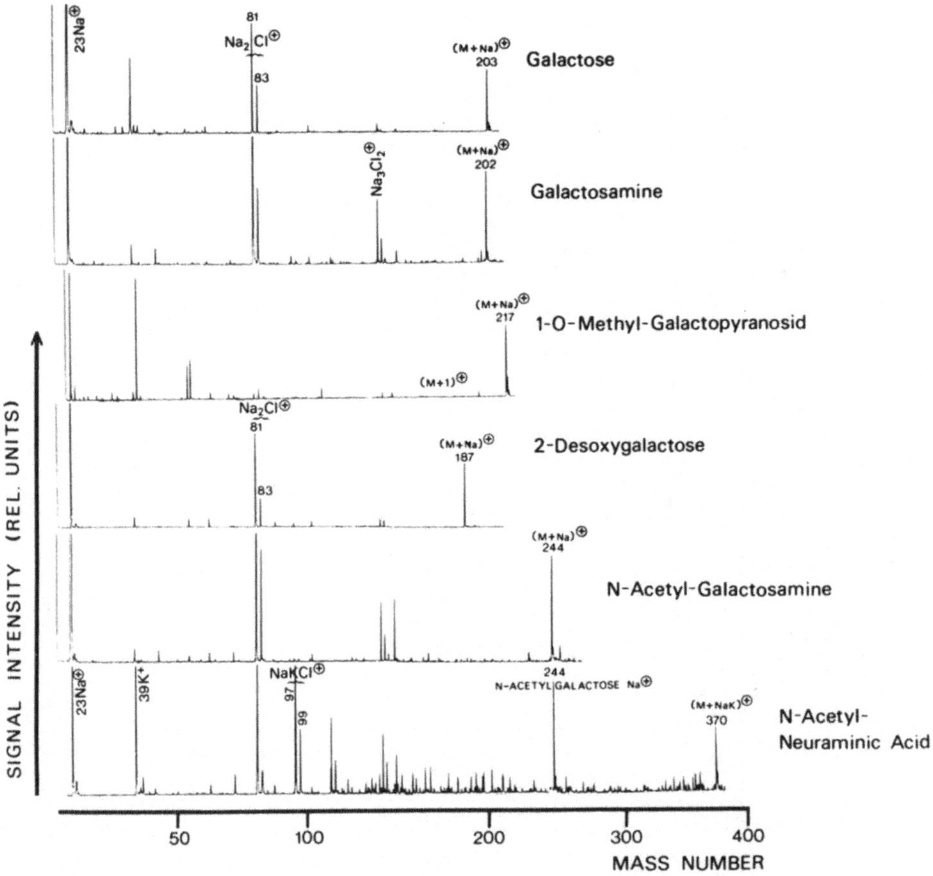

FIGURE 46.   Laser desorption spectra of some derivatives of the monosaccharide galactose.

6.5. LAMMA Investigation of Human Skin

Kupka *et al.* (1980) have published the first attempt to identify a selected organic compound by its contribution to the complex spectrum obtained from sections of skin tissue. They had been interested in the deposition of phthalates in the skin of chronic hemodialysis patients. The phthalate plasticizers are contained in blood-line tubing used in dialyses and are known to dissolve out of the tube walls slowly. Spectra of thin sections of the polyvinylchloride tubing material, of a standard perfusate, pumped through the tubing in a closed circulation for an extended period of time, and of the dry mass of plasma of chronic hemodialysis patients had to be taken and compared to the spectra obtained from skin in order to produce reasonably conclusive evidence that the phthalate plasticizers do indeed accumulate in patient tissue. This work demonstrates that the identification of organic molecules in the spectra of complex biological matrices is by no means simple, particularly if the molecules of interest are rather unpolar and, consequently, must be expected to have a low ion yield. Such investigations require a great deal of experience and carefully designed control measurements of suitably chosen standard or reference systems.

6.6. Mass Spectrometric Fingerprinting of Mycobacteria

A somewhat similar investigation has been undertaken by Seydel and Heinen (1979). Several groups have demonstrated some years ago that different strains of mycobacteria can unambiguously be identified by fingerprint techniques of their mass spectra taken by either Curie-point pyrolysis, laser evaporation, or field desorption. All these methods have the disadvantage that a large homogeneous population of the bacteria must be maintained for single analysis. The growth of such large populations would, at least, require an undesirable amount of effort and time delay in a routine clinical analysis, while in some cases they can not even be grown. Figure 47 shows the pyrolysis spectra of four different strains of mycobacteria. Each of these spectra is the average over approximately 30 shots each pyrolyzing a whole bacterium. The four spectra show enough significant differences for the strains to be unambiguously identified by them. This demonstrates that mass spectrometric identification of mycobacteria is possible even on the basis of an average of a few single cell spectra. Most surprising is the finding, reported by the authors, that for each of the peaks shown, the standard error was less than 10%. This indicates that the complexity of organic matrices and the resulting complexity of the mass spectra does not

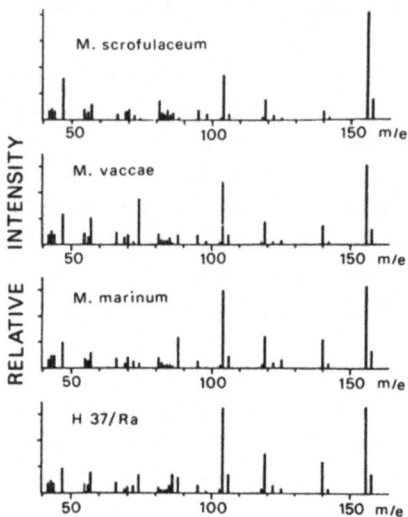

FIGURE 47. Laser-pyrolysis spectra of four different strains of mycobacteria averaged over approximately 30 shots each.

necessarily imply a lack of reproducibility or exclude possibilities for identification of structures of interest by fingerprint techniques, even if the peaks in the spectra cannot in most cases be identified as originating from a known fragmentation reaction of an identified molecular species. For further information, see also Seydel and Lindner (1983).

7. CONCLUSIONS AND OUTLOOK

In view of the many questions raised at the beginning of the LAMMA development, the skepticism expressed by both the authors and their critical colleagues and referees (which, in times of difficulties and seemingly unsurmountable problems during the development, sometimes came close to overcoming the enthusiasm that eventually carried through) LAMMA may be considered a success. Technologically, the instrument can certainly still be improved, and some of these improvements can even be expected in the near future. At the moment, the transient recorder for the fast A/D conversion of the spectra appears to impose the most stringent limit upon the system. An increased time resolution of 5 ns, possibly even 2 or 1 ns per channel, would not only improve the mass resolution in the range of relative molecular masses above about $M = 200$, it should also lead to substantial improvements and simplification of the ion optics and the mass spectrometer. Even more progress would result from an extension of the dynamic range of the recorder from the present 8-bit to, at least, 10-bit (1024 points). Many applications that in principle could be performed with LAMMA are severely hampered by this rather limited dynamic range. There is good reason to believe that the rapid progress in electronic instrumentation during the last decade will continue and eventually provide a recorder better suited for this task. For further information see, e.g., Simons (1984).

A cryo- (or at least a cold) stage may be desirable for a number of applications in biomedicine. The first steps towards its realization have already been taken. The possibility of analyzing bulk surfaces rather than thin specimen is of great interest, not only for many industrial applications but also for biomedical specimens such as hard tissues like bone and teeth. An instrument adapted for such bulk surface analyses has also been developed (for details see, e.g., Feigl *et al.*, 1983, and Heinen *et al.*, 1983) and marketed as LAMMA 1000®. Although the spatial resolution is somewhat less (1–3 μm) than that obtained in the transmission mode, detection limits and other performance features are very similar to those of the LAMMA 500 instrument.

The number and variety of applications that have already been attempted with the LAMMA technique are surprising considering the short time since the realization and availability of the instrument. It can be expected that with wider dissemination of the information on the performance of the method and with more instruments being used by research groups with different backgrounds and interests, many anticipated as well as many unexpected applications will be reported in the years to come. For the developers of a new technique, it is one of the most exciting and rewarding experiences to see how the range of applications widens both quantitatively and qualitatively. Rather than speculate more about the major lines of future applications, the authors leave it up to the needs and ingenuity of potential users to surprise all.

The main problem still unsolved is that of quantification. One—and at the moment, the most straightforward and promising—line of development is to use suitable internal and external standards with reference materials tailored to specific problems by the individual investigators. More work along this line can be expected for the future. A more general solution of this problem would require a much better understanding of the physical processes involved in the laser-induced ion formation and a more rigid control of the instrumental performance. The initial steps have been taken towards the first goal, but the problems are very complex indeed. The investigation is difficult, particularly because of the small sample volumes involved and the limited number of ions generated per shot. The only aspect of the process in which the major points seem reasonably well understood at the moment is that of the energy transfer from the laser radiation field to the solid, dielectric sample (Fürstenau, 1981a). In the first phase of this transfer, nonlinear optical processes, possibly initiated or supplemented by classical absorption, populate the conduction band to a density of ca. 10^{18} electrons/ cm^3. Only after this first phase will intraband transitions of electrons with the conduction band lead to substantial energy transfer into the solid. A series of experiments on the cluster ion distribution from homogeneous inorganic foils has led to crude models about the processes of disintegration of the solid and the ion formation, but much more work needs to be done in this direction (Fürstenau *et al.*, 1981b). Another important problem to be investigated, which so far has been touched only superficially, is that of laser desorption of organic molecules from solid surfaces or possibly even induced phase boundaries within the sample. It can only be hoped that all these open questions will eventually be answered, and an overall reliable model based on a combination of empiricism and true understanding will make quantification possible on a wide scale.

ACKNOWLEDGMENTS.   The authors wish to take the opportunity to thank all their collaborators, funders, and supporters. Financial support was obtained from the Stiftung Volkswagenwerk, the Federal Ministry of Science and Technology (BMFT), and the Deutsche Forschungsgemeinschaft.

REFERENCES

Anderson, C. A., and Hinthorne, J. R., 1972, Ion microprobe mass analyzer, *Science* **175**:853–860.

Bochem, H. P., and Sprey, B., 1979, Laser microprobe analysis of inclusions in *dunaliella salina*, *Z. Planzenphysiol.* **95**:179–182.

Bolnick, D. A., and Sillmann, A. J., 1979, Barium delays the onset of rapid dark adaptation in bullfrog cones, *Invest. Ophthalmol. Vis. Sci.* **23**:875–878.

Bosch, F., Martin, B., Nobiling, R., Povh, B., and Traxel, K., 1978, The Heidelberg proton microprobe, *Microscop. Acta Suppl.* **2**:331–340; *Microprobe Analysis in Biology and Medicine* (P. Echlin and A. Kaufmann, eds.), Hirzel Verlag, Stuttgart.

Brown, K. T., and Fleming, D. G., 1979, Effects of Ba^{2+} upon the dark adapted intensity-response curve of toad rods, *Vis. Res.* **19**:395–398.

Burns-Bellhorn, M. S., and Lewis, K. R., 1976, Localization of ions in retina by secondary mass spectrometry, *Exp. Eye Res.* **22**, 505–518.

Burns-Bellhorn, M. S., 1978a, A review of secondary ion mass spectrometry in biological research, MAS, 13th Annual Conference 28A.

Burns-Bellhorn, M. S., 1978b, in *Advanced Techniques in Biomedical Electron Microscopy* (J. Gennaro, ed.) Masson, Paris.

Cadle, R. D., 1975, *The Measurement of Airborne Particles*, Wiley, New York.

Delhaye, M., DaSilva, E., and Hayat, G. S, 1977, The molecular microprobe, *Am. Lab.* **May/June**:69–73.

Echlin, P., and Kaufmann, R. (eds.), 1978, Microprobe analysis in biology and medicine, *Microscopica Acta, Suppl. 2*, Hirzel-Verlag, Stuttgart.

Eicke, A., Sichtermann, W., and Benninghoven, A., 1980, Secondary ion mass spectrometry of nuclic acid components: Pyrimides, purines, nucleo-sides, and nucleotides, *Organic Mass Spectrom.* **15**:289–294.

Eloy, J. F., 1978, Chemical analysis of biological materials with the laser probe mass Spectrometer, *Microscopica Acta, Suppl.*, 307–317.

Estler, R. C., and Nogar, N. S., 1986, Mass spectroscopic identification of wavelength dependent UV laser photoablation fragments from polymethacrylate, *Appl. Phys. Lett.* **49**:1175–1177.

Etz, E. S., Rosasco, G. J., Blaha, J. J., and Hinrich, K. F. J., 1978, Particle analysis with the laser-raman microprobe, 13th Annual Conference of MAS, Ann Arbor, Michigan, June 19–23, pp. 66A–66L.

Feigl, P., Schueler, B., and Hillenkamp, F., 1983, LAMMA 1000, a new instrument for bulk microprobe mass analysis by pulsed laser irradiation, *Int. J. Mass. Spectrom. Ion Process.* **47**:15–18.

Fürstenau, N., 1981a, Investigation of laser induced damage, evaporation and ionization with homogeneous inorganic target foils, *Fresenius Z. Anal. Chem.* **308**:201–205.

Fürstenau, N., and Hillenkamp, F., 1981b, Laser-induced cluster-ions from thin foils of metals and semiconductors, *Int. J. Mass Spectrom. Ion Process.* **37**:135–151.

Gabriel, E., Kato, Y., and Rech, P. J., 1981, Preparation methods and LAMMA analysis of dental hard tissue with special respect to fluorine, *Fresenius Z. Anal. Chem.* **308**:234–238.

Gardella, J. A., Hercules, D. M., and Heinen, H. J., 1980, Mass spectroscopy of molecular solids: Laser microprobe mass analysis (LAMMA) of selected polymers, *Spectrosc. Lett.* **13**:347–360.

Graham, S. W., and Hercules, D. M., 1982, Laser desorption mass spectra of biomedical polymers. Biomer and avcothane, *Spectrosc. Lett.* **15**:1–19.

Haas, U., Wieser, P., and Wurster, R., 1981, A quantitative interpretation of LAMMA spectra based on a local thermodynamic equilibrium (LTE) model, *Fresenius Z. Anal. Chem.* **308**:270–273.

Hall, T. A., 1971, The microprobe assay of chemical elements, in: *Physical Techniques in Biological Research* (G. Oster, ed.), Academic, New York, Chap. 3, pp. 157–275.

Harding-Barlow, I., 1974, Quantitative laser microprobe analysis, in: *Laser Applications in Medicine and Biology* (M. L. Wolbarsht, ed.), Plenum Press, New York, Vol. 2, p. 133.

Harrington, J. S., Allison, A. C., and Badami, D. V., 1975, Mineral fibers, chemical, physicochemical, and biological properties, *Adv. Pharmacol. Chemother.* **12**:291–402.

Heinen, H. J., Meier, S., Vogt, H., and Wechsung, R., 1983, LAMMA 1000, a new laser microprobe mass analyzer for bulk samples, *Int. J. Mass. Spectrom. Ion Process.* **47**:19–22.

Heinen, H. J., Meier, S., Vogt, H., and Wechsung, R., 1980, Laser induced mass spectrometry of organic and inorganic compounds with a laser microprobe mass analyser, *Adv. Mass Spectrom.* **8**:942–953.

Hercules, D. M., Parker, C. D., Balasanmugam, K., and Viswanadham, S. K., 1983, 4.5 Laser mass spectrometry of organic compounds, in: *Ion Formation from Organic Solids* (A. Benninghoven, ed.), Springer-Verlag, Berlin, pp. 222–228.

Heresch, F., Schmidt, E. R., and Huber, J. F. K., 1980, Repetitive pulsing laser desorption mass spectrometry: A versatile technique for non-volatile organic compounds suitable for scanning type instruments, *Analyt. Chem.* **52**(12):1803–1807.

Hillenkamp, F., Bahr, U., Karas, M., and Spengler, B., 1987, Mechanisms of laser ion formation for mass spectrometric analysis scanning microscopy, Supplement **1**:33–39.

Hillenkamp, F., 1983, Laser induced ion formation from organic solids, in: *Ion Formation from Organic Solids* (A. Benninghoven, ed.), Springer Series in Chemical Physics, Springer-Verlag, Berlin, pp. 190–205.

Hillenkamp, F., and Kaufmann, R. (eds.), 1981, Sessions of the LAMMA Symposium, *Fresenius Z. Anal. Chem.* **308**(3):193–520.

Holm, R., Karas, M., and Vogt, H., 1987, Polymer investigations with the laser microprobe, *Anal. Chem.* **59**:371–373.

Kaduck, B., Metze, K., Schmidt, P. F., and Brandt, G., 1980, Secondary athrocytotic cardiomyopathy—Heart damage due to Wilson's disease, *Virchow's Archiv. A Path. Anat. Histol.* **387**:67–80.

Karas, M., Bachmann, D., and Hillenkamp, F., 1985, Influence of the wavelength in high irradiance ultraviolet laser desorption mass spectrometry of organic molecules, *Anal. Chem.* **57**:2935–2939.

Karas, M., Bachmann, D., Bahr, U., and Hillenkamp, F., 1987, Matrix-assisted ultraviolet laser desorption of nonvolatile compounds, *Int. J. Mass Spectrom. Ion Proc.* **78**:53–68.

Karas, M., and Hillenkamp, F., 1988, Laser desorption ionization of φproteins with molecular masses exceeding 10,000 daltons, *Anal. Chem.* **60**:2299–2301.

Kaufmann, R., 1980, Recent LAMMA studies of physiological cation distributions in retina tissues, *Scanning Electron Microsc.* **2**:641–646.

Kaufmann, R., Hillenkamp, F., and Wechsung, R., 1978a, Laser microprobe analysis, *ESN-Eur. Spectrosc. News* **20**:41–44.

Kaufmann, R., Hillenkamp, F., Nitsche, R., Schürmann, M., and Wechsung, R., 1978b, The laser microprobe mass analyzer (LAMMA): Biomedical applications, *Microsc. Acta, Suppl. 2* (P. Echlin and R. Kaufmann, eds.), Hirzel-Verlag, Stuttgart, pp. 297–306.

Kaufmann, R., Hillenkamp, F., Nitsche, R., Schürmann, M., Vogt, G., and Wechsung, R., 1978c, The LAMMA instrument: A new laser microprobe mass analyzer for biomedical purposes, *Proc. MAS*, 19–23.

Kaufmann, R., Hillenkamp, F., and Wechsung, R., 1979a, Laser microprobe mass analysis (LAMMA) for biomedical purposes, Abstracts of the 51. Meeting Dtsch. Phys. Ges., *Pflugers Arch., Eur. J. Physiol. Suppl.* **379**:R58.

Kaufmann, R., Heinen, H. J., Schürmann, M., and Wechsung, R., 1979b, Recent advances of laser microprobe analysis (LAMMA) as applied to biological and engineering specimens, in: *Microbeam Analysis* (D. Newbury, ed.), San Francisco Press, San Francisco, pp. 63–72.

Kaufmann, R., and Wieser, P., 1980, Laser microprobe mass analysis (LAMMA) in particle analysis, in: *Characterization of Particles* (K. F. J. Heinrich, ed.), National Bureau of Standards, Special publication 533, pp. 199–223.

Kaufmann, R., Wieser, P., and Wurster, R., 1980, Application of the laser microprobe mass analyzer LAMMA in aerosol research, *Scanning Electron Microsc.* **2**:607–622.

Kirk, R. G., and Lee, D., 1978, X-ray microanalysis of cation and hemoglobin contents in red blood cells, in: *Microprobe Analysis in Biology and Medicine* (P. Echlin and R. Kaufmann, eds.), *Microsc. Acta Suppl. 2*, Hirzel-Verlag, Stuttgart, pp. 102–112.

Kistemaker, P. G., Lens, M. M. J., Van der Peyl, G. J. Q., and Boerboom, H. A. J., 1980, Laser induced desorption mass spectrometry, *Adv. Mass Spectrom., Inst. Petroleum, London* **8**:928–934.

Klein, P., and Bauch, J., 1979, On the localization of ions in cell wall layers of treated wood based on a laser-microprobe-mass-analyzer (LAMMA), *Holzforschung* **33**:35–40.

Knox, B. E., 1971, Laser probe mass spectrometry, *Dynamic Mass Spectrom.* **2**:61.

Kovalev, I. D., Makstimov, G. A., Suchkov, A. I., and Lavin, N. V., 1978, Analytical capabilities of laser probe mass spectrometry, *Int. J. Mass Spectrom. Ion Process.* **27**:101–137.

Krueger, F. R., 1979, Fast ion induced and collective electronic perturbation at the surface, *Surf. Sci.* **86**:246–256.

Krueger, F. R., and Schueler, B., 1980, Organic mass spectra obtained by fission-fragment and pulsed laser-induced desorption, *Adv. Mass Spectrom.* **8**:919–927.

Kupka, K. D., Schropp, W. W., Schiller, C. H., and Hillenkamp, F., 1980, Laser micro mass analysis (LAMMA) of metallic and organic ions in medical samples, *Scanning Electron Microsc.* **2**:635–640.

Lauchli, A., 1972, in: *Micro-autoradiography and Electron Probe Analysis* (U. Luttge, ed.), Springer-Verlag, New York, Chap. 7, pp. 191–927.

Liebmann, P. A., 1978, Rod disk calcium movement and transduction: a poorly illuminated story, *Ann. N.Y. Acad. Sci.* **307**:642–644.

Lipton, S. A., Ostroy, S. E., and Dowling, J. E., 1979. Ca^{2+} and photoreceptor adaptation (a reply), *Nature* **281**:407–408.

Long, J. V. P., and Rockert, H. O. E., 1963, X-Ray fluorescence microanalysis and determination of pottasium in nerve cells, in: *X-Ray Optics and X-Ray Microanalysis* (H. Pattee, V. E. Cosslett, and A. Engstrom, eds.), Academic, New York, pp. 513–521.

Lorch, D. W., and Schafer, H., 1981, Localization of lead in cells of phymatodocis nordstedtiana (Chlorophyta) with the laser microprobe analyzer (LAMMA 500), *Fresenius Z. Anal. Chem.* **308**:246–248.

McCrone, W., and Delly, J. G., 1973, *The Particle Atlas*, Science Publ. Ins., Ann Arbor, Michigan.

Mercer, T. T., 1973, *Aerosol Technology in Hazard Evaluation.* (U.S. Atomic Energy Commission Monograph Ser.), Academic Press, New York.

Meyer zum Gottesberge-Orsulakova, A., and Kaufmann, R., 1985, Recent advances in LAMMA analysis of inner ear tissue, *Scanning Electron Microsc.* **1**:393–405.

Morton, R. R. A., and McCarthey, C., 1975, *Microscope* **23**:239.

Nicholson, W. A. P., 1974, in: *Microprobe Analysis as Applied to Cells and Tissues* (T. Hall, P. Echlin, and R. Kaufmann, eds.), Academic, New York, pp. 59–73.

Orsulakova, A., Kaufmann, R., Morgenstern, C., and D'Haese, M., 1981, Cation distribution of the cochlea wall (stria vascularis), *Fresenius Z. Anal. Chem.* **308**:221–223.

Palekar, L. D., Pooner, C. M., and Coffin, D. C., 1979, Influence of crystallization habit of minerals on in vitro cytotoxicity, *Ann. N.Y. Acad. Sciences* **330**:673.

Posthumus, M. A., Kistemaker, P. G., Meuzelaar, H. L. C., and Ten Noever de Brauw, M. C., 1978, Laser desorption mass spectrometry of polar non-volatile bio-organic molecules, *Analyt. Chem.* **50**:985–991.

Schiller, C. H., 1980, Untersuchungen zur laserinduzierten Massenspektrometrie: Ionenbildung und Fragmentierungsverhalten von Aminosäuren, Diploma thesis, Institute of Biophysics, University of Frankfurt am Main.

Schmidt, P. F., Fromme, H. G., and Pfefferkorn, G., 1980, LAMMA—Investigations of biological and medical specimen, *Scanning Electron Microsc.* **2**:623–634.

Schmidt, P. F., 1984, Localization of trace elements with the laser microprobe mass analyser (LAMMA), *Trace Elements Med.* **1**:13–20.

Schroeder, W., Frings, D., and Stieve, H., 1980, Measuring Ca-uptake and release of avertebrate photoreceptor cells by laser microprobe mass spectroscopy, *Scanning Electron Microsc.* **2**:647–654.

Schroeder, W. H., and Fain, G. L., 1984, Light-dependent calcium release from photoreceptors measured by laser micro-mass analysis, *Nature* **309**:268–270.

Seydel, U., and Heinen, H. J., 1979, First results on fingerprinting of single mycobacteria cells with LAMMA, Proceedings of the 6th International Symposium on Mass Spectrometry in Biochemistry and Medicine, Venice, Italy, 21–22 June.

Seydel, U., and Lindner, B., 1981, Qualitative and quantitative investigation on mycobacteria with LAMMA, *Fresenius Z. Anal. Chem.* **308**:253–257.

Seydel, U., and Lindner, B., 1983, Mass spectrometry of organic compounds (= 2000 amu) and tracing of organic molecules in plant tissue with LAMMA, in: *Ion Formation from Organic Solids* (A. Benninghoven, ed.), Springer-Verlag, Munich, pp. 240–245.

Simons, D. S., 1984, Isotopic analysis with the laser microprobe mass analyzer, *Int. J. Mass Spectrom. Ion Process.* **55**:15–30.

Spengler, B., Karas, M., Bahr, U., and Hillenkamp, F., 1987, Excimer laser desorption mass spectrometry of biomolecules at 248 and 193 nm, *J. Chem. Phys.* **91**:6502–6506.

Sprey, B., and Bochem, H. P., 1981, Uptake of uranium into the alga dunaliella detected by EDAX and LAMMA, *Fresenius Z. Anal. Chem.* **308**:239–245.

Spurny, K. R., Schormann, J., and Kaufmann, R., 1981, Identification and microanalysis of mineral fibers by LAMMA, *Fresenius Z. Anal. Chem.* **308**:274–279.

Spurny, K. R., 1983, Measurement and analysis of chemically changed mineral fibers after experiments in vitro and in vivo, *Environ. Health Perspect.* **51**:343–355.

Stoll, R., and Rollgen, F. W., 1979, Laser desorption mass spectrometry of thermally labile compounds using a continuous wave CO_2 laser, *Org. Mass Spectrom.* **14**:642–645.

Truchet, M., 1975, Application de la microanalyse par emission ionique secondaire aux coupes histologiques: Localisation des principaux isotopes de divers elements, *J. Microsc.* **24**:1–22.

Vastola, F. G., and Pirone, A. J., 1968, Ionization of organic acids by laser irradiation, *Adv. Mass Spectrom.* **4**:107.

Whitby, K. T., 1978, The physical characteristics of sulfur aerosols, *Atmosph. Environ.* **12**:135–159.

Wieser, R., Wurster, R., and Seiler, H., 1980, About the identification of airborne particles by laser induced mass spectroscopy, *Atmosph. Environ.* **14**:485–494.

Wilhelm, G., and Kupka, K. D., 1981, Identification of amino citric acid in biological peptides, *FEBS Lett.* **123**:141–144.

Yoshikami, S., and Hagins, W. A., 1973, in: *Biochemistry and Physiology of Visual Pigments* (H. Langer, ed.), Springer-Verlag, pp. 246–255.

Zebel, G., and Hochrainer, R., 1972, Zur Messung der Grössenverteilung des Feinstaubes mit einem verbesserten Spektralimpaktor, *Staub Reinhalt. Luft* **3**:91–95.

CHAPTER 3

Ultrashort Laser Pulses in Biomedical Research

Claude Reiss

Institut Jacques Monod
CNRS and
Université Paris VII
75251 Paris, France

1. INTRODUCTION

Modern methods, primarily intended for the study of short-lived events in physics and chemistry, have lately been transferred to biological systems, where they now enable the investigation of precise mechanisms of even very fast processes. This is an interesting illustration of constructive feedback in science, since centuries ago, physiology successfully assisted the birth of short-time measurement technologies. According to the story, Galileo, using his pulse as a time mark, established the isochronism of the pendulum, which found application in mechanically regulated clocks. This, and various other devices, based on auditory (de la Tour's siren, for instance) or visual comparison of fast events with selected standards, led to remarkable achievements (among them the measurement of the speed of light by Foucault); long before electric instrumentation photographic recording with electric sparks as light source allowed direct observation of events in nano- (10^{-9}), pico- (10^{-12}), and even femto- (10^{-15}) second time ranges. However, with lasers this can be done more easily and to a higher degree of reliability.

Will the race to ever shorter time measurements come to an end? In physics, events of much shorter duration are known to exist. In high-energy physics for instance, lifetimes of "strange" particles have been measured (indirectly) to be as short as 10^{-28} s. In chemistry, and especially in biological chemistry, however, almost all methods used to gain detailed information on transient species rely on their interaction with light. Since the interaction of a photon with a biomolecule takes place within a few *hundred Å* (typical size of a protein), the time of their interaction is limited to around a femtosecond, which is, on the other hand, the time during which a visible light wave performs only a few oscillations. Thus, as far as current spectroscopic technics are used to investigate short-lived species in biology, 10^{-15} s will be a practical limit.

Even so, one may wonder of what use picosecond investigations may be in biology. In nature, it is common knowledge that biological processes take minutes, days, or months to be completed. In molecular biology, enzymatic reactions have turnover times of milliseconds, perhaps even microseconds. However, such overall reactions rely ultimately on elementary steps, like bond linking, isomerization, proton and electron transfer, and molecular rearrangements—i.e., events that can occur in times ranging from 10^{-8} to 10^{-14} s. The chronology of these elementary steps in the overall reaction scheme is determined by diffusion of the intermediary products, steric factors, etc. Thus, in order to understand a biological process, to establish the reaction pathway leading from its initial to its final state with the structures and lifetimes of all intermediates, examination procedures with picosecond resolution may be required.

The main purpose of the present work is to give a practical introduction of picosecond techniques to investigators who want to enter the field. The first part will be devoted to a qualitative review of physical and technological aspects particular to the interaction of a strong electromagnetic field with matter. The second part will deal with instruments and experimental setups actually used in picosecond biological investigations.

1.1. Study of Fast Biological Events May Require Particular Laser Technologies

The native biological preparations required for meaningful investigations differ from most usual chemicals in at least two aspects: stability and amount.

Even when expertly handled in appropriate buffers, most biochemical reactions involving purified particles or molecules are basically irreversible.

This is due either to spontaneous modifications or degradation induced by the investigation technique itself. As an example, recovery after illumination of visual pigments or bacteriophotosynthetic reaction centers at room temperature requires minutes, and full recovery of physiological activity to the preillumination level is never reached again, probably owing to absence of the molecular framework and enzymatic machinery surrounding these particles *in situ*. This fact obviously determines the choice between the two philosophies that prevail for picosecond investigation in general, i.e., derivation of the required information either from just one single pulse, or from data averaging over several—usually very many—pulses. Although this does not imply that "cw" mode locked lasers are useless for biochemical investigation, still, single-pulse methods are ideally suited for the vast majority of such studies.

To avoid synchronization problems, induction and analysis of a biochemical reaction is best achieved by using a single pulse for both. Of course, this requires that the pulse contain enough photons for both purposes; perturbation may require 10^{16} photons/cm^3 for photochemical reactions at millimolar concentration, up to 10^{18} photons (hundreds of millijoules at 1.06 μm) for T-jump purposes. On the other hand, for fair photometric accuracy, spectroscopic analysis requires that enough photoelectrons reach the anode of the sensing device during the time interval of interest to yield a reasonable signal-to-noise ratio. The particular levels needed depend on the number of spectroscopic "channels" used (see below). For all these reasons, in the end, single-pulse techniques require giant pulses.

1.2. How Lasers Deliver Short Pulses

We will restrict this discussion to mode-locked lasers that emit pulses of 10^{-12} s duration or less, since relaxed or Q-switched lasers (10^6–10^9 s pulse duration) have reached a state of reliability such that their handling is straightforward even for inexperienced investigators, and their use does not require any special knowledge.

A mode-locked laser oscillator, in common with most of these lasers, is an ensemble composed of an active medium enclosed within two mirrors to form a cavity. The active medium harbors a population of particles in excited states. Spontaneous deexcitation can yield photons, which in turn may stimulate deexcitation of other excited particles and yield more photons back into the active medium. The frequency of the light emitted by the oscillator is subjected to several restrictions. Within the frequency band allowed by the active medium (its state and nature are relevant factors), the cavity geometry selects a set of modes (i.e., frequencies at which light energy can be detected). The modes are characterized by spatial distribution of luminous energy within the cavity, along its axis (longitudinal modes) or orthogonal to it (transverse modes). Proper spatial filtering allows usually single transverse mode operation of the amplifier as is needed to align the laser beam with the geometric axis of the oscillator. Longitudinal modes correspond to electric oscillations having nodes on the cavity surfaces (e.g., mirrors, exit windows): the allowed laser frequencies are then multiples of twice the cavity length, $2L$. The frequency interval separating allowed modes is then

$$\delta v = c/2L \qquad (1)$$

with c the speed of light. If the goal is to produce monochromatic light emission, the number of allowed modes should be reduced to a single one, for instance by introducing into the cavity selective elements—e.g., prisms, gratings, or Fabry–Perot etalons. However, the interest here is in producing short pulses; in this case, the Fourier transforms, which link the time and frequency domains, indicate that any phenomenon of duration Δt is to be described by a superposition of a set of waves with frequencies lying within a band Δv such that

$$\Delta v \cdot \Delta t \simeq 1 \qquad (2)$$

This means that short light pulses and high monochromaticity are incompatible requirements. To produce short pulses, the amplification bandwidth Δv of the active medium has to be broad. For instance, the familiar neodymium laser (either glass or YAG), which is the most common source for picosecond, giant laser pulses, has an active bandwidth of Δv around 10^{12} Hz, centered at 1.06 μm. If the laser cavity length L is 1.5 m, $\delta v \simeq 10^8$, the active bandwidth may harbor as much as $\Delta v/\delta v$, or 10^4 proper modes. For short light pulses, any selection among these modes has to be avoided. Thus, no optical surface inside the cavity should be orthogonal to the beam axis except for the cavity mirrors, of course. In particular, the active medium components (rod) are usually shaped so as to intercept the laser beam at the Brewster angle, which has the additional advantages of polarizing the laser emission and of avoiding reflection losses.

Nonselection among the proper modes is necessary for short pulse production, but this does not mean that sizable energies can readily be extracted from the laser. The output energy is the sum of the energies produced by each mode and by their interferences. Since no phase relationship between the modes exists *a priori*, the output energy is low and fluctuates randomly. This can be avoided by locking the relative phases of the modes. Several methods of mode locking exist, both active (triggered) and passive. Passive mode locking of the Nd^{3+} laser can be achieved by inserting in the cavity a saturable dye, which absorbs the laser wavelength. Spontaneous deexcitation produces photon emissions (fluorescence) that are amplified by stimulation inside the active medium. If the resulting light pulses travel along the cavity axis, they will encounter the dye, where they will simply be absorbed, unless a particular pulse harbors enough photons for exciting all the dye molecules present (dye saturation). The dye becomes transparent to the remainder of this pulse, which can then reach the cavity end reflector and gain amplification by passing through the amplifying medium again. Since deexcitation of the dye is very fast, absorption is restored quickly and the dye is very likely to be saturated again only when this pulse comes back from the other mirror of the cavity (after one more amplification) and so forth. The result is a train of pulses that can be extracted from the cavity, for instance through a semitransparent mirror. A pulse selector, using the Pockels effect, for instance, can isolate a single pulse of the train, which can be further amplified in passive amplifiers.

Femtosecond Laser Pulses. Pulses of ps to a few tens of ps duration are readily produced by mode locked lasers. To understand how pulses of shorter duration can be obtained, let us have a closer look at the interaction of an individual pulse and the mode-locked cavity components. As the pulse enters the absorbing dye cell, photons in its leading edge are efficiently absorbed by the dye, which becomes increasingly saturated; photons travelling in the main part or the trailing edge of the pulse mostly escape absorption. As a result, the pulse length is shortened (the leading edge has been "eaten" away by the dye) and its new leading edge is steepened. This pulse now enters the active medium. The steep leading edge meets the full population of particles in excited state and will be strongly amplified, whereas the remainder of the pulse feeds on partly depleted population of excited particles and will be less amplified. The pulse sharpening and narrowing effect just sketched is opposed by the broadening induced by dispersion in optical components (e.g., rod, windows, or Fabry–Pérot selectors). After a few round trips in the cavity, the pulse shape equilibrates by adopting a compromise between these antagonistic effects. The shortest pulse obtained this way are barely below 1 ps.

A way to decrease the pulse duration would be to use two colinear pulses, propagating in opposite direction and brought to collide in the dye. The standing wave produced upon collision saturates the dye (transitory generation of a spatial absorption grating), hence provides transmission for pulses crossing the dye cell during the time of minimal dye absorption. The device works roughly like a shutter with opening time $1/c$, 1 being the optical path in the dye, c the speed of light (the spectral width of the locked modes is $c/1$). In practice, a thin jet ($\sim 10\ \mu$m) of the dye solution crosses the cavity axis, providing both small 1 ($1/c \sim 30$ fs) and steady renewal of the dye in the beam. With a linear laser cavity, the jet must be positioned with micrometer precision at the place where both pulses, reflected by the cavity minors, collide. This adjustment is automatic with a ring cavity, where pulses propagating in opposite directions travel the same distance between two consecutive jet crossings, hence the cavity oscillates spontaneously in a configuration where both pulses collide in the dye jet. Commercial ring-lasers deliver pulses below 100 fs at mJ level (~ 10 GW power) in the visible.

Before discussing various aspects of the interaction of a strong electromagnetic field with matter, it is useful to point to several physical parameters connected with a giant light pulse. Emphasis will mainly be set on practical aspects, with eventually references to quantitative data or theory.[1]

1.3. Physics of Giant Light Pulses

Characteristic figures of a light wave can be derived from Maxwell's equations. The most accessible parameter of a laser pulse being its power, Maxwell's equations can express the associated electric field E (V/cm), magnetic field H (G), and radiation pressure p (atm) as a function of the power density W (W/cm)^2 of the radiation burst:

$$\log V \sim \tfrac{1}{2} \log W + 1.5 \tag{3}$$

$$\log H \sim \tfrac{1}{2} \log W - 1 \tag{4}$$

$$\log p \sim \tfrac{1}{2} \log W - 9 \tag{5}$$

If a square pulse of 0.2 J, lasting for 20 ps, is focused on a 0.1-cm^2 surface, $W \simeq 10^{+11}$ W/cm^2; $V \simeq 10^7$ V/cm; $H \simeq 3 \times 10^4$ G; and $p \simeq 10^2$ atm. If the same pulse is focused on a spot 10 μm in diameter (easily achieved by a conventional lens), $W \simeq 10^{+16}$; $V \simeq 3 \times 10^9$ V/cm, $H \simeq 10^7$ G; and $p \simeq 10^7$ atm. V and H oscillate at the laser frequency, i.e., around 10^{15} Hz for visible radiations. If the laser wavelength is 1 μm, a 0.2-J pulse will harbor about 10^{18} photons. As an example, the value of the electric field of such a pulse is of the order of magnitude of the electric field extended by the nucleus on a bound electron. It is thus not surprising that the interaction of a giant laser pulse with matter gives rise to various effects that are not observed at usual electromagnetic power levels. Such effects have to be

considered quite seriously by investigators, first to avoid erroneous inter-
pretations of experimental results, but also because the transient interaction
of a giant light burst with matter may be very useful to induce or analyze
very short events, as will be shown in the following sections.

1.4. Nonlinear Technologies

When an electromagnetic wave interacts with some nonabsorbing
material, its electric field E forces the charged particles inside the material to
move so that a polarization P is induced in the latter. For weak fields, P is
proportional to E, and the incident and transmitted waves have the same
frequency. Strong fields, however, may exert forces on electrons and nuclei
that more or less strongly modify their relative motions, so that the linear
relation between P and E no longer holds. Terms in E^2, E^3,... contribute to
P, which introduces multiple frequencies $2v$, $3v$,... of the E field frequency v:

$$P = K^{(1)}E + K^{(2)}E^2 + K^{(3)}E^3 + \cdots \qquad (6)$$

The susceptibility $K^{(1)}$ accounts for the classical optical properties of
matter (refractive index, etc.), whereas $K^{(2)}$ is involved in second-harmonic
generation (production of light at twice the laser frequency or half its
wavelength) and sum and difference frequency ($v_{laser} \rightarrow v_1 + v_2$ or $v_1 - v_2$).
$K^{(3)}$ accounts for third-harmonic ($3v$) generation, four-photon parametric
interaction ($v_1 + v_2 \rightarrow v_3 + v_4$ in particular $2v_{laser} \rightarrow v_1 + v_2$), and stimulated
Raman scattering, SRS, $v_{laser} \rightarrow v_{SRS} + v_M$ (v_M being a characteristic
frequency of the material); self-focusing of the pulse, which is mediated by
the nonlinear index of refraction of the medium, is also governed by $K^{(3)}$.
The following review will be restricted to the most commonly encountered
—and used—nonlinear effects (Austin, 1977).

1.4.1. Second Harmonic Generation

Frequency doubling is very useful in generating intense pulses with
temporal, spectral, and spatial characteristics comparable to those of the
central laser pulse. The yield of second harmonic generation is of course controlled by the particular value of $K^{(2)}$ for the material considered. $K^{(2)}$ is not
zero only if the material has no inversion symmetry, either on a molecular
level (e.g., benzene has a $K^{(2)}$ of 0), or in the crystalline state. Actually, the
amplitude of $K^{(2)}$ is directly controlled by the asymmetry of the charge

distribution in the crystal: the larger the dissymmetry on the distribution map, the stronger the nonlinear behavior. Mineral crystals (ADP, KDP), and, recently, organic materials (nitroaniline derivatives, for instance) are efficient nonlinear materials.

Besides the influence of $K^{(2)}$, the intensity of the frequency-doubled light pulse depends on several other factors: it is proportional to the square of the pumping light pulse intensity, the square of the interaction length l of the light beam with the frequency doubling material and the square of a diffraction contribution, $\sin(\Delta k \cdot l/2)/(\Delta k \cdot l/2)$. Δk is the phase mismatch, $\Delta k = k_{2v_L} - 2 \cdot k_{v_L}$, where k_{2v_L} and k_{v_L} are the wave vectors of the frequency-doubled and laser waves, respectively. The sin term strongly affects the yield and stems from the different refractive indices n_1 and n_2 of the material at frequencies v_L and $2v_L$. The waves propagate at speed $v_1(c/n_1)$, and $v_2(c/n_2)$, c speed of light so that they reach a given point in the crystal at times differing by an interval proportional to $n_2 - n_1$. The waves interfere randomly (i.e., with poor conversion efficiency) unless n_1 and n_2 are equalized. This can often be realized in birefringent crystals, where both ordinary and extraordinary indices of refraction n_0 (isotropic) and n_e (anisotropic) coexist, both depending on the wavelength (Fig. 1). Equalization of n_1 and n_2 is based on the nonisotropy of the extraordinary index, which varies between n_e (along the optical axis of the crystal) and n_0, when the angle θ between the direction of light propagation with the optical axis varies. For a proper value θ_0, n_0 at v_L may be equal to $n_{\mathrm{ex}}(\theta_0)$ at $2v_L$, so that Δk may be θ^0 (phase matching).

The characteristics of the fundamental beam, such as divergence and emission modes, are also critical parameters for the yield of frequency doubling. With 30-ps pulses of 1 J at 1.06 μm, conversion to 530 nm with yields in excess of 50% have been reported.

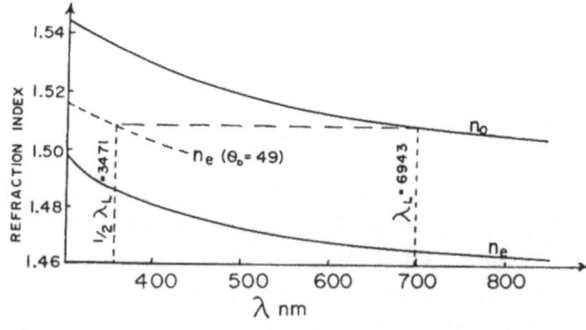

FIGURE 1. Ordinary (n_0) and extraordinary (n_{ex}) dispersion curves of KDP. The dotted curve is the extraordinary dispersion curve along a direction $\theta_0 = 49°$ from the crystal axis.

The bandwidth and duration of the frequency-doubled pulse are controlled by group velocity matching. The critical parameter is l^*, the ratio between the fundamental pulse duration, and the difference between the reverse group velocities at v_L and $2v_L$ in the material considered. For KDP, a 1-ps pulse has an l^* of 3.7 cm at 1.06 μm, but l^* is only 0.3 cm at 5.3 μm. l^* is to be compared with l: when l is less than l^*, the frequency-doubled pulse is shortened and its bandwidth increases. For instance, if the fundamental pulse has a Gaussian temporal intensity profile, the frequency-doubled pulse is shortened by $\sqrt{2}$. If l is greater than l^*, the duration of the second harmonic pulse is stretched by l/l^* with respect to the fundamental, and its bandwidth reduced accordingly (Comly and Garmire, 1968).

1.4.2. Higher-Order Harmonics Generation

Laboratory setups have produced up to the seventh harmonic conversion yielding soft x-ray radiation, but commercial devices produce the third harmonic by mixing (in a nonlinear, phase-matched crystal) the fundamental and second harmonic, or the fourth harmonic by doubling again the second harmonic. Yields of a few percent are readily obtained. The fourth harmonic of the Nd^{3+} laser (265 nm) is especially valuable for nucleic acid or protein studies.

1.4.3. Parametric Emission

Parametric emission is becoming a convenient means for generating intense light pulses with spectral and temporal characteristics similar to those of the laser pulse but at nonharmonic frequencies. In a first step of the process, the spontaneous decay of a laser photon into two new photons (signal and idler) occurs in some nonlinear crystal: $v_L \rightarrow v_s + v_i$. The precise value of v_i and v_s are selected by the particular phase-matching conditions in the crystal (as seen in Section 1.4.1 for frequency doubling). In a second step, if the laser pulse is intense enough, v_i and v_s can mix in the crystal with v_L, producing, respectively, more v_i and v_s ($v_s - v_L \rightarrow v_i$ and $v_L - v_i \rightarrow v_s$, if $v_s > v_L > v_i$), i.e., a gain at signal and idler frequencies is observed. If, in addition, appropriate feedback is provided through the cavity, the system behaves as a parametric oscillator.

The actual phase-matching condition in the crystal may be changed by some external means, for instance, by changing the crystal temperature, which shifts its ordinary and extraordinary dispersion curves. This allows continuous change of the frequencies v_i and v_s over rather sizable ranges.

For instance, using a 1.06-μm pump pulse, temperature tuning of a commercial parametric oscillator (KDP crystal) allows the production of pulses with wavelengths from 0.87 to 1.2 μm and reasonable percent conversion efficiencies. In another example (ADP) the pump is the fourth harmonic of 1060 nm, 265 nm, which allows temperature tuning between 420 nm and 720 nm, with 10% conversion efficiency.

1.4.4. Stimulated Raman Scattering (SRS)

When photons of frequency v_L interact with matter, some scattered photons have new frequencies, v_S and v_{AS} as a result of the interaction with matter-characteristic excitations of frequency v_M: $v_L + v_M \rightarrow v_{AS}$, the anti-Stokes frequency, blue-shifted with respect to v_L; $v_L - v_M \rightarrow v_S$, the Stokes frequency, red-shifted with respect to v_L. The matter characteristic excitation waves may result from collective behavior of particles (thermal or entropy waves) or may reflect molecular characteristics (nuclear vibration, rotation, electronic excitation). To visualize the coupling of v_L and v_M, the susceptibility $K^{(1)}$ can be considered as a linear function of the normal coordinates Q of the molecule,

$$K^{(1)} = K_0^{(1)} + \frac{\partial K^{(1)}}{\partial Q} \cdot Q \tag{7}$$

Q is a periodic function of time, oscillating at frequency v_M. When an electric field of frequency v_L interacts with this material, a polarization P is produced:

$$p = K^{(1)} \cdot E = K_0^{(1)} \cdot E + \frac{\partial K^{(1)}}{\partial Q} \cdot Q \cdot E \tag{8}$$

i.e., a sum of two terms, the first oscillating at frequency v_L (Raleigh scattering), and the second at frequencies $v_L + v_M$ and $v_L - v_M$ (Raman scattering).

It immediately follows that Raman scattering monitors both the frequency and the (squared) amplitude of the Raman-active mode ($\partial K^{(1)}/\partial Q$ is not equal to 0) of the irradiated material. Notice that anti-Stokes scattering ($v_L + v_M$) is usually much weaker than Stokes scattering, because only excited molecules (generally due to thermal excitation induced by the sample temperature) contribute to the former.

When a strong laser pulse is sent through a Raman-active material, a nonlinear mixing process can occur, by which a pump quantum, hv_L, mixes with either an excitation quantum, hv_M, to yield a Raman quantum, hv_{SRS}, or with a spontaneous Raman quantum, to drive the molecular vibration. Mixing thus amplifies the spontaneous Raman wave, i.e., $I_{SRS}(\lambda)$ is $I_{SRS}(0) e^g$, where the gain g increases linearly with the interaction path length λ of the laser light in the Raman material. The gain is also proportional to the laser pump light intensity, and to $((\partial K^{(1)}/\partial Q)^2)$. Within a few centimeters, the SRS wave carries an energy comparable to that of the pump wave, which is of course depleted accordingly. v_{SRS} is determined by v_M, so that it may be selected by the proper choice of the Raman-active medium (see Fig. 12 and Table I). In case the selected medium has several Raman-active modes, a rule of thumb is that the strongest, best polarized mode dominates the stimulation process. Thus, SRS scattering is a cheap, convenient, and very efficient way of generating strong light bursts at a wealth of wavelengths. Durations are comparable to those of the pump pulse or usually slightly less. Most often, the SRS pulse is to intense that it can act itself as a pump that excites again SRS, and so on. Proper filtering allows the selection of the desired wavelength (Valat et al., 1978).

In conjunction with frequency multipliers, SRS allows the production of intense pulses all over the visible and near UV with a Nd^{3+} laser. In many cases, SRS generates a large excess of the excited population, so that stimulated Raman anti-Stokes scattering (blue shifted) is also observed with intensities sufficient for many purposes. However, at variance with SRS scattering, which is collinear with the pump direction, SRAS is forward scattered along a cone, as a result of momentum conservation.

The intense material vibration v_M accompanying SRS can also be used to study the energy transfer between the molecules, for instance, by measuring dephasing times and energy relaxation times of individual modes of the molecules after the excitation pulse ceases (Von der Linde, 1977). So far, only simple molecules have been studied by this method, but its application to energy transfer analysis in biological particles and complexes should be of great interest.

1.4.5. Self-Focusing and Self-Phase Modulation

An intense light beam may have its spatial, temporal, and spectral properties altered as it travels through a medium. This arises from the fact that the refractive index of the medium is a weak function of the electric field E carried by the beam. Neglecting transient effects, only even powers of E contribute by their time averages to the refraction index (Svelto, 1974):

$$n = n_0 + n_2 \langle E^2 \rangle + \cdots \tag{9}$$

Before reviewing briefly the most important mechanisms that couple n

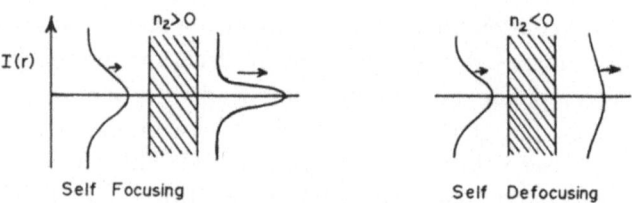

FIGURE 2. Modification of the spatial envelope of the incoming pulse due to the nonlinear refraction index of the material.

and E, the main effects of self-focusing and self-phase modulation will be described. Assume that in the beam cross section the intensity $E^2(r)$ (determined by the transverse modes of the laser) is a Gaussian with the maximum at the beam axis (r at 0). The nonlinear index of refraction gives rise to an index gradient $n_2 \cdot E^2(n)$ the effect of which is determined by the sign of n_2. For $n_2 > 0$ (the most frequent case), the gradient diminishes away from the axis, producing thus a lenslike effect, which focuses the beam. Conversely, $n_2 < 0$ produces a self-defocusing. Thus, the light beam geometry (alignment with a cw laser, for instance) may be modified in the pulse mode.

The refractive index nonlinearity will also give rise to a phase shift:

$$\delta\psi = \delta(n/c \cdot v \cdot \zeta) \tag{10}$$

A temporal variation of $\delta\psi$ induces a frequency modulation.

$$\delta v = -\frac{\delta}{\delta t}\delta\psi = n_2/c \cdot v \cdot \zeta \cdot \frac{\delta}{\delta t}(E^2) \tag{11}$$

Assuming, for instance, a Gaussian temporal profile of E^2 and $n_2 > 0$, the leading edge of the pulse will experience a negative frequency shift, the trailing edge, by contrast, a positive one. The result is a continuous frequency shift across the pulse ("chirp"). The broadening is usually of a few wave numbers only, but must be considered when the precise laser wavelength is of importance—for instance, in Raman scattering.

What are the origins of the coupling of n with E? For anisotropic molecules like CS_2, n_2 may be due to the so-called orientational Kerr effect.

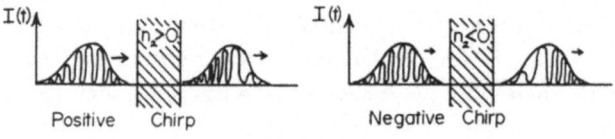

FIGURE 3. Wavelength sweep ("chirp") inside the pulse, as it crosses a medium with the non-linear refraction index.

For a symmetric top molecule, for instance, the susceptibility along the top axis, α'_{zz}, differs from the orthogonal one, α'_{xx}. In the absence of an external field, the molecules are oriented at random and no birefringence is observed. When a field E is applied, a dipole moment is induced in each molecule, sharing direction and frequency with E. Interaction of the induced dipole with E results in a torque with a DC component $(\alpha'_{zz} - \alpha'_{xx}) E^2$ tending to align the molecule so that the axis with greatest polarizability comes into the plane of E. As a result, the medium becomes birefringent, the index of refraction difference being proportional to $(\alpha'_{zz} - \alpha'_{xx})^2$. This effect is used in the optical Kerr cell, an optical shutter with picosecond speed (see below).

Electronic hyperpolarizability is usually less efficient in coupling n to E. This effect arises from a distortion of the electronic charge distribution by E, inducing birefringence, which depends on the detailed symmetry characteristics of the material. For instance, in isotropic molecules, the change of the index of refraction induced by hyperpolarizability along the electric field is three times that induced perpendicular to it. Natural conjugated polyenes (carotenoids), which play important roles in vision or photosynthesis, have among the highest known n_2 values, to which the electronic hyperpolarizability contributes strongly.

Thermal effects also contribute to n_2, with two mechanisms acting successively. First, partial (even weak) absorption of the light leads to heating along the light path (see laser T-jump, Section 2.3) giving rise to a local change of entropy and consequently change of refractive index, $(\partial n/\partial T)_v$. This effect is usually small, but may build up very quickly. A second step is diffusion of the heat away from the light path. The resulting thermal expansion contributes an index change $(\partial n/\partial p)_T \cdot (\partial p/\partial T)$, which may be rather large. Electrostriction (transient density change induced by the high E field of the pulse) gives a barely perceptible effect on a picosecond time scale. It may, however, induce a delayed action on the next laser pulse, which may experience self-focusing or (more probably) self-defocusing if the index gradient induced by the previous pulse has not been dissipated meanwhile.

Finally, when pulse levels close to breakdown are reached, plasma formation by avalanche and multiphoton ionization may give very strong alterations of n. This can produce a very large frequency modulation δv; indeed, broadenings exceeding by far 10^3 cm^{-1} are generated, which then have been used as continuum light sources of picosecond duration.

1.5. Detection of Picosecond Light Events

In the present state of mode-locked laser technology, steady control of the output is recommended, to secure both effective mode-locking operation and power level constancy. In many experiments, the pulse itself is used as a time-mark, so that double-pulse generation or failure to extract a single pulse from the train may easily lead to gross erroneous results. Furthermore, a particular application may require pulse modification by some nonlinear effect. As shown in Section 1.4, these effects depend on higher powers of the electric field and are directly related to the laser pulse power. Thus, even small fluctuations of the pump energy (around 5% is routine in the mode-locked regime) may greatly influence the yields of the nonlinear effects. For instance, 20% energy fluctuation of the second harmonic is common.

Two parameters characterize a picosecond light pulse: its temporal envelope $I(t)$ and its frequency characteristics. The methods of monitoring one or both of them are of two kinds, depending on whether they involve optoelectronic devices, or optical elements only, with eventually hybrid combinations of both.

1.5.1. Optoelectronic Methods

The photoelectric effect is thought to be faster than 10^{-14} s, which, in principle, enables investigation of subpicosecond phenomena.

Direct methods make use of a photosensor associated with a fast measuring device. In the present state of technology, the fastest photomultipliers available have rise times of a few hundred picoseconds. For photodiodes, rise times are more than 70 ps. On the other hand, oscilloscopes using the traveling wave technique (i.e., the signal deflects progressively the electron beam along the CRT) have up to 5-GHz bandwidth (70 ps rise times). Intensification of the CRT screen image permits moderate acceleration of the CRT beam, hence fair sensitivity (0.1 V/cm) and high writing speed (10 cm/ns). Thus, the combination of even the fastest sensor and oscilloscope would have its rise time limited to $[(70)^2 + (70)^2]^{1/2}$, similar to 100 ps only. Any faster event will be integrated by the direct method in its present state.

Streaking is a powerful method, by which the temporal profile of an event, $I(t)$, is displayed in space as $I(x)$. As shown in Fig. 4, the train of photons $I(t)$ is converted on the photocathode FK of the streak tube into a train of photoelectrons, which are accelerated along the zz' axis of the tube, toward its fluorescent screen FS. During the time of flight of the photo-electrons, a voltage ramp, directed orthogonal to zz', builds up progressively from 0 to several kilovolts (streak). The photoelectrons are thus deflected with steadily increasing amplitudes. If correct synchronization is provided, the photoelectrons generated by the front of the light pulse are not deflected, and those by the rear mostly are. Along the streak direction xx', the screen FS can be calibrated in time from the known slope of the voltage ramp. In

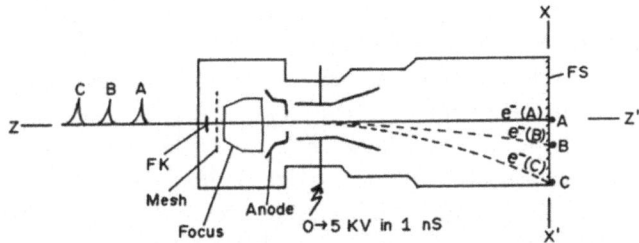

FIGURE 4. Principle of a streak camera. Each incoming pulse, A, B, and C, produces an electron burst on photocathod (FK), the burst is accelerated along ZZ' by the anode potential (positive relative to that of FK), deflected by the E-field (xx' direction) created by the capacitor plates and focused on the fluorescent screen (FS). The potential of the plates is risen linearly, from 0 to 5 KV in 1 ns (voltage ramp). The ramp is triggered by the pulse itself (see Fig. 5). In the case sketched in the figure, trigger could be by pulse A; the trigger delay is adjusted to start the ramp after the electron burst produced by pulse A has crossed the deflection plates; the electrons are not deviated and focus on point A of FS. Pulse B is delayed with respect to A, its electron burst will experience the E-field and will be deflected along direction xx', since it crosses the plate as the ramp builts up, and will hit FS in B. The same occurs for the electron burst produced by pulse C, but since the voltage ramp is near its upper end, the E-field is stronger and deviation larger (point C on FS).

most streak cameras, electron focusing allows streaking of images, with spatial resolution up to 10 line pairs per millimeter at highest streak speed.

The use of fast, high potential switching circuity (Marx bank generators) allows generation of smooth, linear potential ramps from 0 to 5 kV in 1 ns. Triggering is usually achieved optically by part of the laser pulse itself, with delays around 10–20 ns before streaking (jitter around

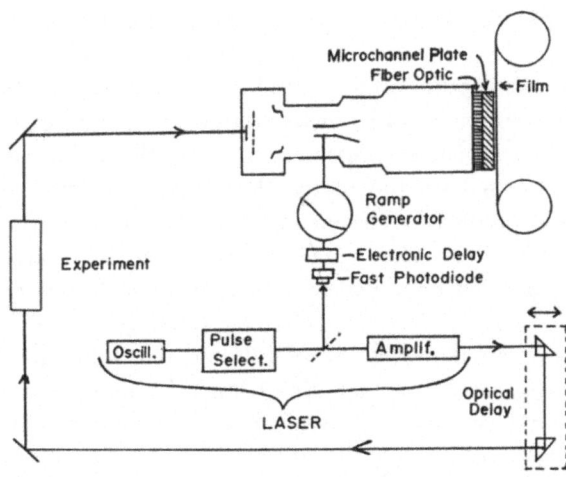

FIGURE 5. Electrooptic triggering of the voltage ramp generator and set-up of the optical delay.

150 ps), so that the optical event to be streaked must be properly delayed (in 1 ns, light travels about 30 cm in vacuum).

The time resolution achieved can be better than 1 ps, being ultimately limited by the spread of energy (velocity) of the photoelectrons as they are extracted from the photocathode FK, and the time dispersion they experience during transit in the tube. High photocurrent density in the tube is detrimental for both spatial and temporal resolution. Thus, the operation of the FK at low light level is recommended, which requires intensification of the FS image, possibly by a microchannel plate.

The dynamic range of streak cameras is usually low. Applications where photometric accuracy is important (spectrophotometry) can hardly be achieved with these devices in the present state of technology, even though improvements have been claimed recently by several manufacturers. Commercial streak cameras (Hadland Photonics, Thomson CSF, Hamamatsu, General Engineering, and Applied Research) are reliable and easy to use. However, these devices are rather expensive, as are fast oscilloscopes.

1.5.2. Optical Methods

Optical methods can yield either the temporal envelope of the pulse, or its frequency characteristics, or both. They take advantage of correlation techniques, which do not directly depend upon detector response time. The second-order correlation function ($\langle I(t) \cdot I(t+\tau) \rangle$ for autocorrelation) provides the duration of the pulse; the third-order correlation function ($\langle I^2(t) \cdot I(t+\tau) \rangle$ for autocorrelation) yields information about the pulse shape.

(a) A temporal pulse profile (amplitude structure) can be derived from two-photon fluorescence (TPF) or second harmonic generation (SH), if the pulse characteristics (power, spectrum) allow the induction of nonlinear effects. Devices based on the optical Kerr effect permit the study of $I(t)$ versus t for light pulses with more common characteristics. In all three cases, two pulses are needed, which may be either the two parts of one single original pulse, divided by a beam-splitter, or separate pulses from different sources.

The TPF method makes use of a liquid dye solution which does not absorb at the frequencies of the pulse (v_L and v'_L) but does so at the summed frequency ($v_L + v'_L$ usually $2v_L$), with subsequent fluorescence emission. The two pulses are brought to meet head-on in the solution. Provided the pulses are intense enough, sizable two-photon absorption induced fluorescence emission occurs in the region of overlap. Recordings from the track along which fluorescence is emitted and densitometry of the fluorescence intensity profile permit an estimation of the pulse duration (see below).

In the SH method, the two pulses are polarized differently (for instance orthogonally) and brought to overlap in a SH generating crystal. The polarization of the pulses and the crystal axis orientation are arranged so that harmonic generation occurs only when both pulses add in the crystal. The curve of the SH signal intensity versus delay is the convolution of both pulses. If the frequencies of the pulses differ, the group velocity mismatch must be taken into account.

Kerr shutters make use of the optical Kerr effect (briefly described in Section 1.4.5). In the traditional Kerr effect, birefringence is inducted by an applied electric field, owing to partial alignment of (usually anisotropic) molecules along the field direction. The electric field associated with very intense, polarized light pulse can induce a short-lived birefringence in an appropriate medium. The amplitude and response time r of the effect depend on the chemical nature, state, and orientational relaxation time of the molecules: $r \propto 2$ ps for liquid CS_2, often used because of its high non-linear index contribution n_2 (the birefringence is proportional to n_2).

In a simple optical setup, two pulses are polarized at $45°$ from each other; one pulse (pump) is devoted to gating the cell; the other (probe) crosses the Kerr cell along the path of the gate pulse and is normally blocked by a crossed polarizer located behind the cell (Duguay and Hansen, 1969). When the probe pulse crosses the activated Kerr cell, its polarization is rotated by the induced birefringence. Partial transmission through the analyzer can then be detected with maximum intensity when probe and gate pulse nearly overlap (actually when the probe pulse follows the gate pulse with a delay amounting to the Kerr effect response time). The main requirement for this method is that the gate pulse intensity I_G is high enough to induce sizable birefringence, which, at time t, is proportional to

$$\int I_G(t') \exp[-(t-t')/r] \, dt'/r \tag{12}$$

The three methods just sketched have in common that they do not directly yield the temporal pulse intensity profile $I(t)$, but give the correlation function of it. For two identical pulses, SH and TPF give the second-order correlation function of $I(t)$, $\langle I(t) \cdot I(t+r) \rangle / \langle I^2(t) \rangle$. The brackets indicate averages over the time intervals deemed sufficient. For the Kerr shutter, if r is much shorter than the gate pulse and birefringence modulation, the (integrated) probe intensity transmitted by the analyzer is $\langle I_{\text{probe}}(t+r) \cdot I^2(t) \rangle$, i.e., a third-order correlation function. Formally, the knowledge of the second- and the third-order correlation functions allows description of $I(t)$ in

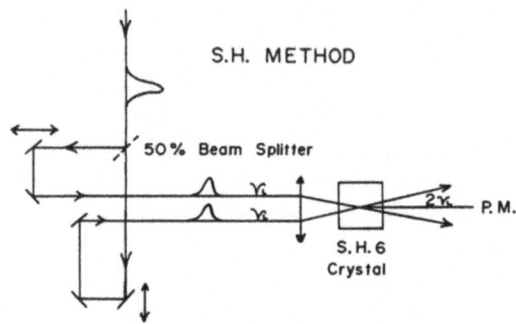

FIGURE 6. The dye absorbs at ω_{2L} (but not at ω_L), i.e., in the area when both pulses collide head-on. The absorption induces fluorescence at a red-shifted wavelength, which can be filtered out and recorded. Changing the relative delay of the pulses allows registration of their convolution.

all cases. This can also be achieved by just the second-order correlation function, which is always symmetric, provided one knows *a priori* $I(t)$ is symmetric.

(b) The frequency characteristics of short optical pulses are of two classes, depending on the width of the frequency spectrum across the pulse. Only small, monotonic frequency sweeps ("chirps"), as induced, for instance, in a laser pulse by self-phase modulation (see Section 1.4.5), will now be considered. Broad frequency spectra pulses deserve different analysis methods, which will be examined in Section 3.1 together with picosecond spectroscopy techniques. Control of the chirp is important in cases where the precise knowledge of the pulse frequency is required (Raman spectroscopy).

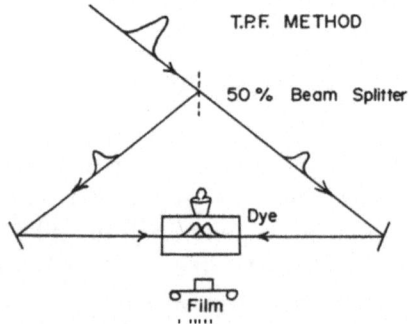

FIGURE 7. The incoming pulses (frequencies ω_L) have different polarization planes. In the overlapping area, second harmonic generation (ω_{2L}) occurs only at the spot where the resulting *E*-field exceeds the SH generation threshold. Changing the relative delays of the pulses allows one to register their convolutions.

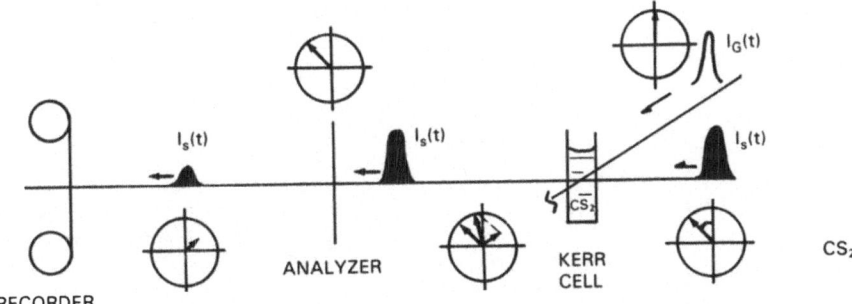

FIGURE 8. Kerr Shutter. Pulse I_G induces transient birefringence in the Kerr cell. Pulse I_S is normally blocked by the analyzer, but if I_S crosses the cell during the time it is birefrigent, due to the action of I_G, its polarization plane is partly rotated, hence part of I_S crosses the analyzer and can be recorded. Changing the relative delay of both pulses allows to register their 3rd order correlation function.

A pulse with small, positive chirp (often the case with Nd^{3+} pulses, owing to self action in the laser rods) can be tested by a pair of simple, matched gratings as shown in Fig. 9. The gratings delay the low-frequency components with respect to the higher-frequency ones. Since in positive chirped pulses, the former are located in the pulse front and the latter in the rear, the pulse is compressed to an extent determined by the distance between the gratings. Comparison of TPF measures of the duration of both compressed and uncompressed pulses gives a rough estimate of the amount of frequency sweep in the laser pulse.

Conversely, this method can be used for deliberate reduction of the pulse duration. The compression can be further enhanced either by filtering out spectral components, or by exaggerating the frequency chirp prior to compression, for instance, by having the pulse travel through a medium with a high nonlinear index contribution (CS_2). Tenfold compression of a picosecond pulse has been achieved by this method. Pulses as short as 0.3 ps ("cw," mode locked Ar^+ lasers) are produced in commercial instruments based on this principle.

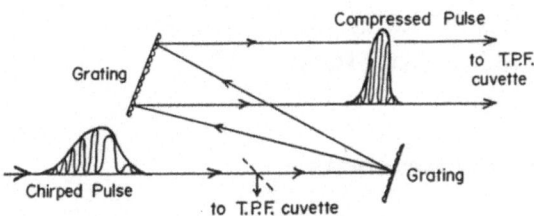

FIGURE 9. Compression of a pulse with positive chirp by a pair of matched gratings; correlation of the compressed pulse with the uncompressed pulse yields an estimate of the extend of chirp.

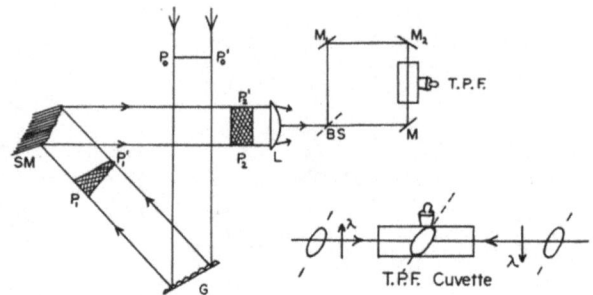

FIGURE 10. Set-up of the TREACY experiment (see text).

A more accurate estimation of the frequency sweep can be made with setup combining a grating G, a stepped mirror SM, a lens L, a beam splitter BS, three mirrors M, and a TPF cuvette, as shown in Fig. 9 (Tracy, 1971). Ignoring first the effect of SM, the grating G disperses spatially the various frequency components of the chirped incoming pulse $P_0 P_0'$, which then crosses the exit focal plane of L at different times. The sweep rate can be derived from the differential arrival time, which is measured by the TPF method. After beam splitting by BS, inversion of one of the pulses by double reflexion on M_1 and M_2 allows crossing of one of the pulses with an inverted image of itself in the TPF cuvette, centered at the exit focus of L. Since the chirp in both pulses is then oriented in opposite directions, the TPF display will be tilted by an angle directly related to the frequency sweep rate. In order to reduce the response time of the spectrometer $(G + L)$ (i.e., the time during which the dispersed pulse $P_1 P_1'$ crosses the exit focus of L), the stepped mirror SM (a tipped stack of microprobe slides) transforms the canted pulse $P_1 P_1'$ into a set of pulses located within $P_2 P_2'$.

2. USE OF LASER PULSES TO INDUCE BIOCHEMICAL REACTIONS

2.1. Photobiological Reactions

Photobiological reactions result from the direct interaction of photons with one or several reaction participants. Although simple in principle, the procedure may involve sophisticated experimental arrangements to produce photons at the frequency required for inducing the reaction. Using harmonic generation, parametric oscillation emission, stimulated Raman scattering, self-phase modulation, and combinations of these methods, the wavelength of a laser pulse can be changed to almost any value of interest to photobiologists (265–1800 nm with an Nd^{3+} laser, for instance). Also, dye lasers (driven by a Nd:YAG pulse, for instance) have been commercially available recently (Quantel) that, with properly chosen dyes, give continuous adjustment to any wavelength between 350 and 700 nm. The use of these high-power pulses requires extra care in addition to the usual precautionary measures in photochemistry such as spatially homogeneous irradiation of the sample using cylindrical optics, the irradiation of all particles that are subsequently present in the analyzed reaction volume, and sufficient number of photons to photoreact all molecules present.

The high photon density of a giant laser pulse results in a high probability of having two or more photons present simultaneously, or at short time intervals, within the interaction distance of a particle, thus inducing multiphotonic effects, as already mentioned, or cascade effects by which a photon interacts with an already excited molecule before its return to the ground state. For instance, the quantum yield and fluorescence lifetime of reaction centers from photosynthetic bacteria with picosecond pulses were found to be anomalously low as compared to figures derived from experiments with nanosecond pulses, an effect ascribed to the occurrence of nonlinear optical effects. A simple means to check for such an effect is to repeat the experiment with varying pulse power levels.

2.2. Chemical Relaxation

Chemical relaxation is a powerful method to study reaction kinetics. In this method, some parameter governing a reaction equilibrium (temperature, pressure, pH, etc.) is suddenly jumped. A detailed analysis of the subsequent return of the participants to a new equilibrium yields information on reaction pathways, intermediates, lifetimes, rate constants, etc. Conventional temperature, pressure, and electric field jump methods have been extensively used to study chemical and biochemical relaxations down to microseconds. Now giant laser pulses can induce such jumps with subnanosecond temporal resolution, thus allowing the study of ordinary (i.e., nonphotochemical) reactions on time scales that previously almost exclusively reserved to radio or photochemical studies.

Strikingly, however, very few "ordinary", as opposed to photobiochemical, relaxation studies have so far been performed with laser pulses, although such reactions are by far the most frequent in living species.

(a) We have seen that giant laser pulses carry electric and magnetic fields with enormous amplitude; thus, E and H jumps can be induced, but then these fields oscillate at some 10^{15} Hz, whereas, for instance, classical E jump devices deal usually with constant fields. Optical rectification of light pulse fields, i.e., a two-photon process producing a static (ω at 0) polarization by nonlinear frequency subtraction ($\omega_L - \omega_L$), could be used to produce picosecond jumps to high static fields.

(b) Laser-induced *pressure jumps* have been obtained by directing the light pulse on thin metal foil layered on a crystalline quartz bar (Key, 1980). A high-pressure plasma develops on the foil surface, creating a shock wave with picosecond rise time, transduced by the bar into the reaction sample. Picosecond p jumps to terapascals (10^7 bars) have been used in this fashion.

(c) In contrast to the laser-induced electromagnetic field and pressure jumps, which have not yet been applied to biochemical relaxation studies, laser *temperature jump* experiments with simple biochemical systems have been described (Campillo *et al.*, 1978; Dewey and Turner, 1978). In principle, the T-jump method allows a study of the kinetics of any reaction involving an enthalpy change. As the configurations of most biological macromolecules change (usually cooperatively) with temperature, this method is extremely valuable for studying configuration dynamics of those species that are thought to be of key importance in the structure–function relationships of biological systems.

2.3. T Jumps

In principle, the production of a T jump through absorption of a light pulse is simple. In general, almost all absorbed light quanta enter into a nonradiative decay scheme, which, for instance, ends up in transferring molecules to the higher vibration levels. The net result is heat deposited in the medium.

Lambert's law states that the intensity $I(x)$ of a light beam at depth x in a medium of absorbance D at the light wavelength decreases exponentially with x:

$$I(x) = I(0) \exp(2.3Dx) \tag{13}$$

where $I(0)$ is the light intensity at the entrance of the beam into the medium. The light-induced temperature variation at depth x, $\Delta T(x)$ follows a law of the same form:

$$\Delta T(x) = \Delta T(0) \exp(-2.3Dx) \tag{14}$$

$\Delta T(0)$ being temperature rise at the sample surface. If C is the heat capacity of the medium:

$$\Delta T(0) = C^{-1}(\partial I/\partial x)_{(x=0)}$$

$$= 2 \cdot 3D \cdot W/C \qquad (15)$$

with W as the energy density of the laser beam. Figure 11 shows the variation of $\Delta T(x)/\Delta T(0)$ as a function of x for various values of D. The approximately uniform temperature rise required for proper thermal pertur- bation of the reaction mixtures is produced only for moderate values of D. On the other hand, the amplitude of the jump is proportional to D. Thus, a compromise in the value of D is necessary to restrict the unwanted non- uniformities in temperature while at the same time achieving a usable amplitude to the jump. A D of 0.5–1 is generally satisfactory. Small values of x (of the order of 1 mm) are recommended anyway.

The jump amplitude is also controlled by the energy density W. However, for picosecond pulses, as we have seen, the upper limitations are

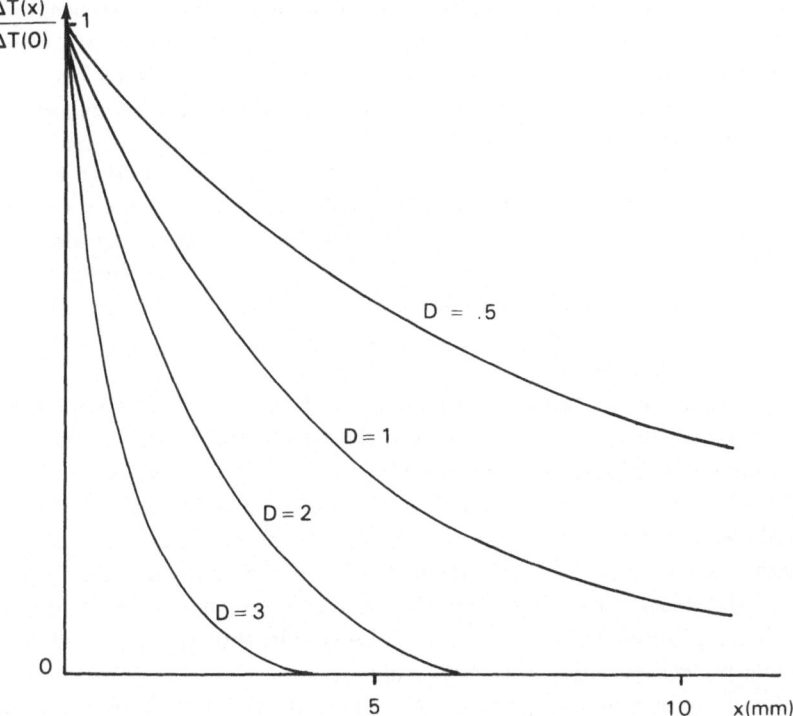

IGURE 11. Temperature gradient induced at depth x by light absorption in a medium of ptical density D.

imposed by the onset of nonlinear effects. To summarize, $\Delta T(x)$ is limited by the characteristics of the medium, such as heat capacity, absorbance at the particular light wavelength, and any nonlinear parameters. The pulse power available also imposes its own limit on the sample volume which can be subjected to the T jump.

What is the best route for the energy transfer? Owing to the high pulse power, direct absorption of the quanta by a reacting species would probably give rise to photochemical effects (either directly, or through some multi-photonic process), which are usually not desirable in the T-jump experiments. Rather, it is desirable that the energy be deposited in some molecule that is present in the reaction mixture but does not directly participate in the reaction. Collisional transfer of heat from this substance to the reactant species would then give an elevated reaction temperature.

The addition of "dyes" to absorb the laser pulse and then release their heat to the reaction mixture seemed at first a good way of transferring energy. This is especially so since the dye and its concentration can be conveniently selected to have any desired absorbance at the laser wavelength. However, dye molecules usually have high extinction coefficients, so that the required moderate values of D impose low dye concentrations. Hence, only a relatively few molecules are available for the energy transfer. Furthermore, the power handling capacity of a given dye molecule is limited because its quantum pickup vibrational-release turnover time usually exceeds the picosecond laser pulse duration. Thus a given molecule can, at best, transfer energy only once during the passage of the pulse. Yet another difficulty comes from the fact that the dyes are usually irradiated in their electronic absorption bands. Thus, even, if the dye withstands photodegradation by multiphotonic effects, for instance, it is most often not inert vis-à-vis biological molecules when in its excited state. This situation is likely to be encountered in all cases when a T jump is induced by energy transfer of quanta in the UV or visible.

These difficulties can be overcome by using less energetic photons in the infrared where they can be absorbed in the vibrational bands of the solvent. For aqueous solutions, the overtone vibrational bands of water are located in the IR between 1.2 and 2 μm (Fig. 12), the absorbance ranging from 0.5 to 100 per cm. From available data it can be shown that the route from quanta absorption and vibrational relaxation of water to collisional transfer toward the solute does not exceed 10^{-10} s on the average for mM solutions. A T jump induced by a 30-ps light pulse should thus be perceived by the reacting molecules within less than 1 ns.

Giant pulse-delivering commercial lasers do not emit between 1.2 and 1 μm, but the pulse length can be easily shifted from the IR or the visible toward this range by nonlinear technology. In the case of a 1.06-μm laser

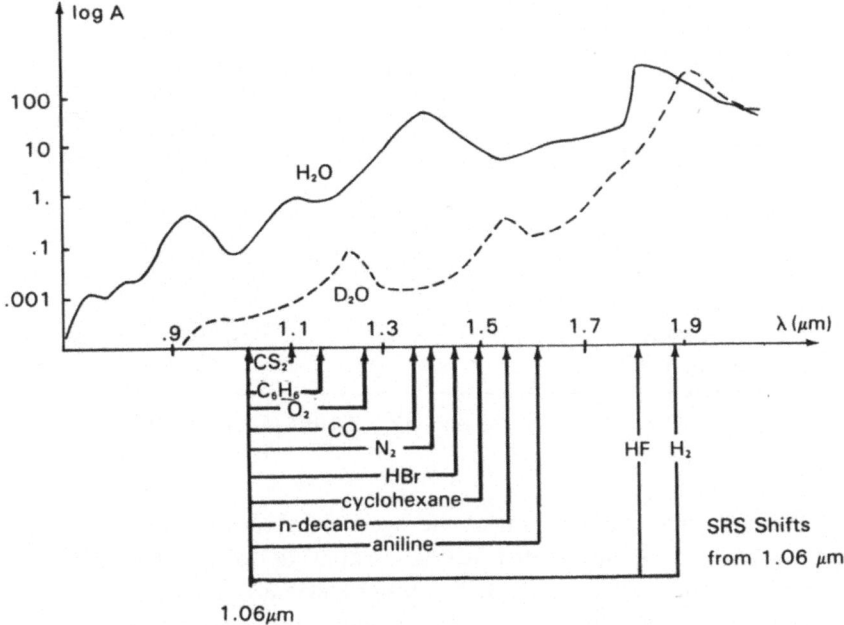

FIGURE 12. Dispersion of water and deuterated water in the near IR (Colles and Walrafen, 1976). Stimulated Raman Stokes (SRS) shifts which can be achieved from 1.06 μm to higher wavelength with various chemicals as indicated under the abscissa axis.

TABLE I. Some Simple Molecules and Their Stimulated Raman Shift Frequencies, $\omega_M{}^a$

Material	ω_M, cm^{-1}	Material	ω_M, cm^{-1}
H_2	4155	$CaCO_3$	1086
HF	3962	C_6H_6	992
n-Decane	2900	F_2	893
HBr	2558	CS_2	655
N_2	2326	Cl_2	554
CO	2143	$SiCl_4$	425
O_2	1522	$SnCl_4$	368
Nitrobenzene	1345	CaF_2	322

[a] Molecules were selected for fair I_{SRS}/I laser ratio, as estimated from scattering cross-section data found in Schrötter and Klöckner (1979).

TABLE II. Data for T jumps in Aqueous Solutions Using an 0.5-J, 27-ps, 1.06-μm Laser Pulse[a]

Medium	C_6H_6(liq)	N_2(liq)
Shift	991 cm^{-1}	2326 cm^{-1}
S.R. λ	1.18 μ	1.41 μ
Absorb. (H_2O)	0.8 cm^{-1}	50% H_2O–D_2O 4.5 cm^{-1}
$\overline{\Delta T}$ Calc.	0.9° C $L = 1$ cm	8.0° C $L = 1$ cm
$\overline{\Delta T}$ Meas.	0.8° C $L = 1$ cm	6.9° C $L = 1$ cm
$\overline{\Delta T}(0)$ Calc.	1.8° C	10.3° C
Opt. Yield	>80%	>80%
Energ. Yield	89%	86%

[a] $I_0 = 0.1$ cal. focused on 0.1 cm^2; $\overline{\Delta T} = \Delta T(0)/L \int_0^L \exp(-2.3\,Dx)\,dx$. From Reiss (1974).

wavelength, almost any region of the overtone absorption profile of water can be reached by stimulated Raman scattering (SRS) in various pure liquids or compressed gases. If SRS transfer efficiency considerations impose a given scattering medium, i.e., by wavelength shift, it may happen that the absorption of water at this precise wavelength is outside the desirable 0.5 to 1.0 OD range. As the overtone absorbance profile of deuterated water is red-shifted with respect to that of ordinary water (Fig. 12), a mixture of both species in amounts can reduce D to the desired value. For instance, SRS in liquid nitrogen can transfer a 1.06 μm picosecond pulse to 1.41 μm with 80% efficiency (see Table I). At 1.4 μm the optical density of H_2O is about 9 per cm, that of D_2 only 0.01. A mixture of 1 to 10 H_2O to D_2O would have the required absorbance. Table II gives some data for T jumps in aqueous solutions achieved with 0.5-J, 27-ps, 1.06-μm laser pulse. The heat transfer equation allows us to estimate a jump "lifetime" in the millisecond range, which has just been observed experimentally.

3. USE OF LASERS FOR MONITORING FAST BIOCHEMICAL REACTIONS

As stated in Section 1.1, a major advantage of using a single giant laser pulse to both induce and analyze biochemical reaction is to avoid repetitive studies on more or less rapidly denaturing samples. This approach assumes that all the information needed to understand the reaction scheme can be gathered from a single experiment.

Detailed insight into the reaction is usually derived from the spectral data, which reflect the electronic or vibrational states of the reaction participants, i.e., absorption or emission spectroscopy in the UV and visible, and Raman spectroscopy. For aqueous solutions IR absorption spectroscopy is obscured by the wealth of IR active modes of water. However, as a rule most biological species have rather complex spectral patterns, so that it is most likely that a whole set of spectral bands will have to be monitored to gather the needed information. This can be achieved in so-called multichannel spectroscopic setups, which enable simultaneous recording of large portions of a given spectrum. This is at variance with conventional spectroscopy, where spectral elements are analyzed in a one-by-one sequence. The multichannel spectral analyses give a larger gain in time and, possibly, accuracy.

3.1. Multichannel Spectral Analyses

In principle, devices associating a dispersive and a focusing element could allow the observation of a multichannel spectrum. However, two basic difficulties have so far opposed a development of multichannel spectrometers. First, the spectral images delivered by standard design spectrometers using prisms or flat gratings as dispersive elements remain coplaner in the exit focal plane over small areas only. The photosensing device must thus fit a given curvature. This can be easily accomplished by photographic plates, but hardly by standard electronic tubes. The second difficulty comes from the photosensing device, which in multichannel spectroscopy has to measure the light intensity for each individual element. Photomultipliers are useless for this purpose because they scramble all photoelectrons coming from the various spectral elements falling onto the photocathode, thus yielding only the overall intensity of the explored spectrum. Photographic plates have been extensively used by astronomers for recording multichannel spectra. They yield a precise value of the spectral band frequencies but are rather poor and unreliable for photometric purposes.

As compared to photomultiplying devices, photographic techniques have low sensitivity and low dynamic range. They also need precise calibration. In addition to these drawbacks, there are the problems of cumbersome handling, processing, and data exploitation.

4. NEW TECHNOLOGIES IN MULTICHANNEL SPECTRAL ANALYSIS

Three technological improvements aid in overcoming the difficulties associated with multichannel spectral analysis: holographic ruled gratings, which can be deposited on blanks with almost any desirable curvature; image intensifiers, which provide high photon or photoelectron multiplication with fair conservation of the spatial information carried by an image; and image converter tubes (TV cameras) linked to minicomputers enabling high-speed readout and digital processing of an image.

(a) Polychromators. Using blanks with carefully adjusted curvatures, holographic concave gratings can be manufactured that can be used in the construction of polychromators with an extended planar, focal surface at the exit. In addition, since both the focusing and dispersion are provided by a single device, only a minimum number of optical components are necessary (usually a single flat mirror in addition to the grating), thus reducing stray light. Such polychromators may cover the whole visible, the near IR, or the near UV domain.

Stock polychromators are available, although instruments with custom-tailored gratings to meet particular requirements of aperture, dispersion,

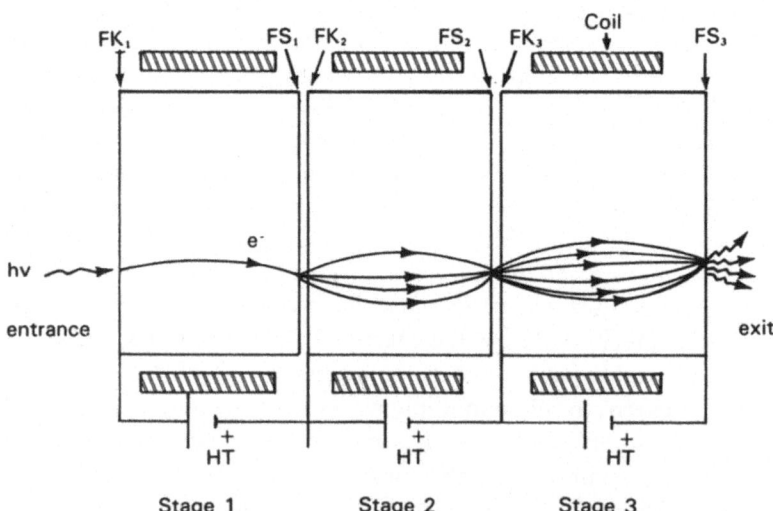

FIGURE 13. Magnetically focused, three-stage image intensifier tube. FK, photocathode; FS, fluorescent screen; Coil, magnetic coil for electron focus; HT, high-voltage electron acceleration potential.

spectral interval, etc. can also be obtained (Instruments S.A.). For instance, in multichannel Raman spectroscopy, where rejection of the excitation frequency ω_L is essential, a premonochromator, using a pair of matched concave gratings in substractive configuration, permits suppression of the excitation line with 10^{-12} efficiency a few angstroms away from ω_L (see below). In particular, this allows the investigation of "soft" vibrational

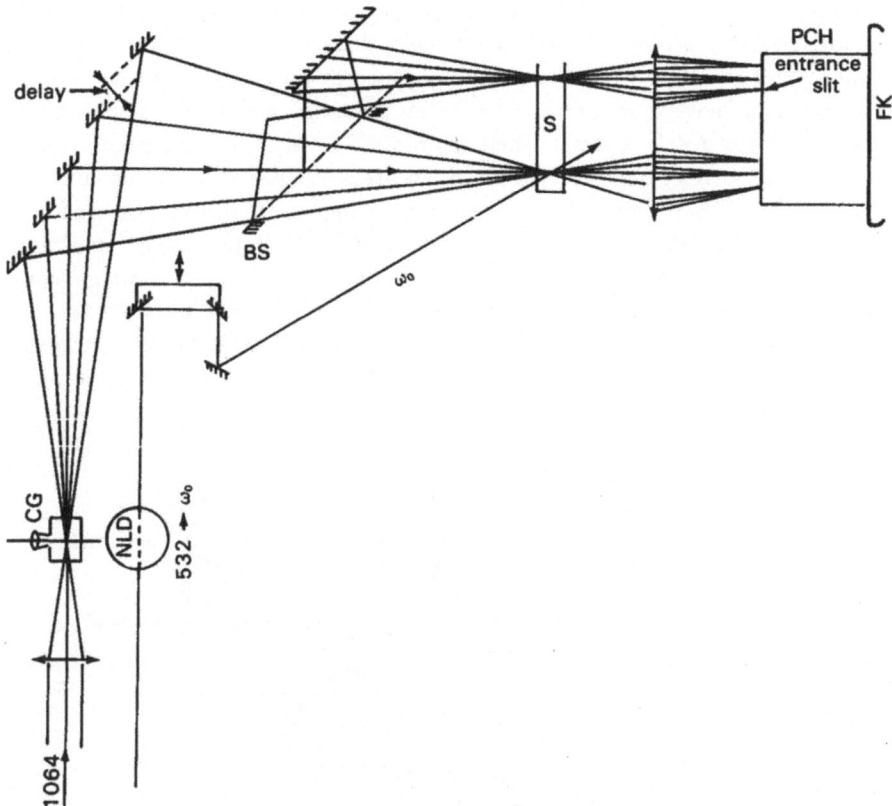

FIGURE 14. Set-up for registration, at picosecond to nanosecond intervals, of complete, double beam absorption spectra. The 1064- and 532-nm pulses, delivered by a mode-locked laser and SHG attachment, are separated by prism P1. The 532-nm frequency is further shifted by a selected nonlinear device NLD, to the desired wavelength, ω_0, which induces a photochemical reaction in sample S. The 1064-nm pulse is used to generate a strong, structureless continuum upon focusing in a selected medium (H_2O, D_2O...), CG. The resulting continuum pulse is subdivided into a set of subpulses, by a stappled set of mirrors, so that each subpulse is delayed, with respect to the preceding one, by a selected time interval. The resulting train of pulses is further subdivided by a 50% beam splitter (BS) into a sample and reference train. Only the sample train crosses the sample along the path of the excitation pulse ω_0. After dispersion of both trains by a polychromator PCH, matched sample and reference subpulses are focused side by side on an appropriate photodetecting device FK.

modes ($1–100\ \text{cm}^{-1}$), which characterized overall vibrations and rotations related to size and shape of molecules.

Depending on the light intensity of spectral elements, multichannel spectroscopy may require intensification of the spectral image produced in the polychromator exit focal plane, prior to data acquisition. In image intensification, ideally, each individual pixel of the image has its intensity magnified by a constant gain factor. This is achieved by a cascade of photocathodes and fluorescent screens (as shown in Fig. 13). The image to be intensified is focused on the first photocathode FK 1; the primary photoelectrons produced there are accelerated and directed onto the first fluorescent screen FS 1. Each photoelectron has an energy high enough to generate a few tens or hundreds of photons on FS 1. The second photocathode FK 2 is sandwiched behind FS 1, so that these photons will produce a burst of photoelectrons, which are accelerated and so on. Photon gains reach 10^8 (i.e., one photon after generating one photoelectron on FK 1 goes on to generate 10^8 photons on the exit fluorescent screen), enabling single-photon detection. Besides the difficulty of achieving homogeneous gain throughout the image, the main difficulty with image intensification is in retaining the correct spatial relation for information transferred through the device. The resolving power is defined simply by the maximum number

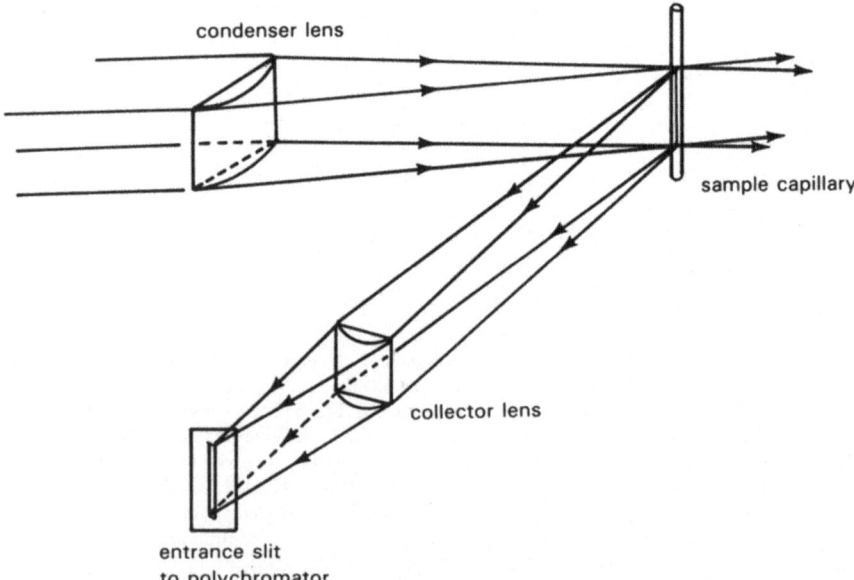

FIGURE 15. Condenser and collector optics useful in Raman spectroscopy.

of pixels of the incoming image (usually the number of black and white line pairs of a grid) which can still be identified in the amplified image.

Two techniques for image intensification have been successively developed. The first, and to date most accurate, makes use of a cascade of photocathodes and fluorescent screens housed in a single vacuum tube. The image to be intensified is focused on the first photocathode. The photoelectrons produced there are strongly accelerated and focused on the first fluorescent screen, where the impact of each photoelectron generates a few tens to hundreds of photons. The second photocathode is layered on the back of the first fluorescent screen, so that these photons immediately generate again photoelectrons, which are accelerated, and so on. These devices have good intensification homogeneity throughout the image surface; spatial resolution of 35 line pairs per millimeter (1 pm) are standard; photonic gains of 10^6 are produced with less than one non-photonic event (mainly due to residual ion impact) per second and cm^2, on the exit fluorescent screen. However, these tubes are cumbersome to handle, and for magnetically focused tubes need cooling of the coils. Extreme care must be exercised to avoid exposure of the photocathodes to undue light

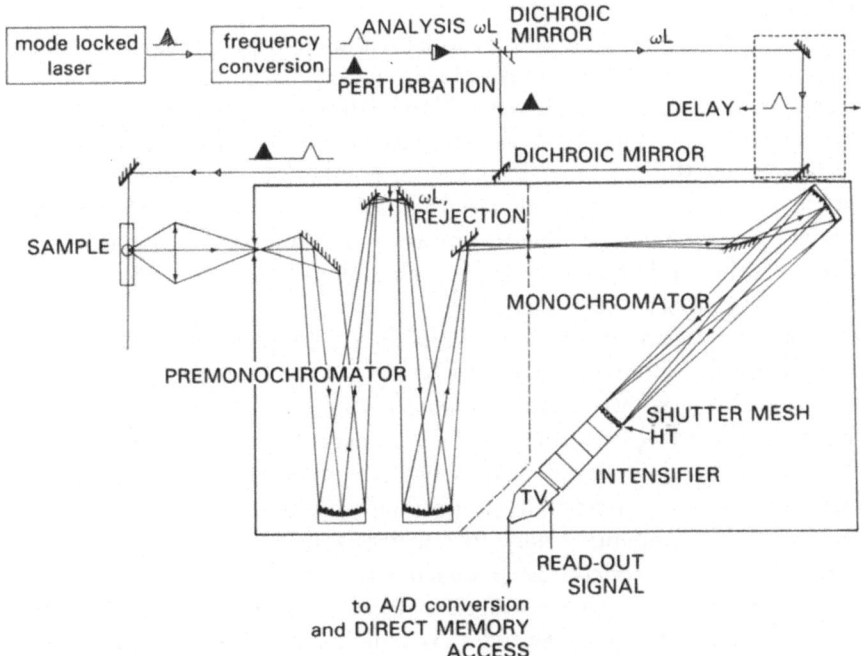

FIGURE 16. Set-up of a multichannel, picosecond Raman spectrophotometer.

level. The risks of damage are minimized by progressively increasing the HT (up to 40 kV) during a series of preliminary experiments, until the value of the HT proper for the actual light level of the image is reached (EMI).

Recently, new devices—so-called multichannel plates—have been developed (especially for night vision). Here, photoelectron multiplication is achieved by secondary electron emission on the walls of tiny channels (a few micrometers in diameter), into which the accelerated photoelectrons are shot. The photocathode and fluorescent screen are layered directly on plates harboring a large number of channels in a honeycomb arrangement. Spatial resolution is about that of a channel diameter and photonic gain is usually limited to 10^4. The microchannel plates are 3–2 cm in diameter and a few millimeters thick, and, thus, are much easier to handle than tubes. On the other hand, intensification homogeneity is less good than for tubes, and ion trapping in channels is still a problem. However, the technology is improving rapidly, so that the microchannel plate characteristics should soon compete with those of conventional tubes.

We come finally to the last stage of image treatment before its analysis: electronic image read-out (address and intensity of each pixel), provided by image converter tubes. The most widely used image converter is the TV camera. The photoelectrons generated on a photocathode are accelerated and focused on a semiconductor (SC) target charged positively. Each photoelectron generates several hundreds or thousands of negative charge carriers as it enters the SC, thus reducing locally the positive charge of the target. An electron beam of well-calibrated intensity scans the target in a series of lines. The emerging beam intensity (video signal) at a given spot of the target (i.e., the difference between the number of electrons of the read-out beam and the number of electrons that have recombined with the positive charge carriers in the SC) measures the number of photoelectrons having reached this spot, hence the light intensity on the corresponding locus of the photocathode.

After amplification, the video signal can be applied to a TV monitor for visual display of the image, or can be digitized and fed to a computer for processing, storage, prints, etc. A spatial resolution of 30 ppm is standard. This technology has been adapted to the particular requirements of spectroscopy: high photometric sensitivity (integrated image intensification stage) and accuracy (secondary impact target technology, providing S/N ratios of 10^4); temporary, noise-free memorization of the image on the target, allowing extremely fast events to be registered and, subsequently, read-out at speeds compatible with memory access times (secondary element conduction technology). These tubes (Westinghouse, RCA, Siemens, Thomson CSF) are available with partly integrated A/D conversion and digital addressing of the electron beam, enabling autonomous monitoring of the read-out by microprocessor (PAR, EMR, Spektrum Technik). However, no tube combines all of these features.

Charge-coupled devices (CCD), introduced very recently, have several advantages over conventional TV tubes; besides small size and fully integrated electronics directly computer compatible, they have extremely low noise figures, allowing very accurate photometry within images. In these devices, charge arrays are produced within the (silicon) target, either directly by photon (quantum efficiency as high as 80% over the visible and near IR), or by properly focused photoelectron (from image amplifier stage, for instance). The target is beset with arrays of electrodes polarized so as to confine the charge carriers locally in potential wells. The charges can be quickly moved from well to well by switching electrode polarities, so that an initial readout of the charges spread over the target can take place. The charge transfer is almost capacity free and proceeds without loss; the charges can thus be directly injected into the junction of an amplifying transistor; the readout noise, determined by the transfer capacity, is negligible. For instance, in a CCD produced by Texas Instruments, the pixel readout noise is of a few electrons only, so that the photoelectron noise is by far predominant, an almost ideal situation allowing extremely high photometric accuracy.

The circuitry needed for monitoring the CCD and transferring the data to the computer are integrated on the silicon wafer supporting the CCD, resulting in a very low amplification noise, high reliability, and easy handling and interfacing. The spatial resolution, mainly determined by the packing density of the electrode arrays, is presently comparable to that of a conventional TV tube; CCD with 800×800 pixels on a one-inch-square surface are produced (Texas Instruments, Fairchild); they are likely to soon replace conventional TV tubes for spectroscopic purposes.

REFERENCES

Austin, D. H., 1977, in: *Topics in Applied Physics*, Vol. 18 (S. L. Shapiro, ed.), Springer-Verlag, Berlin, p. 123.

Campillo, A. J., Clark, J. H., Shapiro, S. L., and Winn, K. R., 1978, in *Picosecond Phenomena* (C. V. Shank and E. R. Ippen, eds.), Springer-Verlag, Berlin, p. 319.

Colles, M. J., and Walrafen, G. E., 1976, *Appl. Spect.* **30**:463.

Comly, J., and Garmire, E., 1968, *Appl. Phys. Lett.* **12**:7.

Dewey, T. G., and Turner, D. H., 1978, *Adv. Mol. Relax. and Interact. Proc.* **13**:331.

Duguay, M. A., and Hansen, J. W., 1969, *Appl. Phys. Lett.* **15**:192.

Reiss, C., 1974, in *Lasers in Physical Chemistry and Biophysics* (J. Joussot-Dubien, ed.), Elsevier, Amsterdam, p. 239.

Schrötter, H. W., and Klöckner, H. W., in *Raman Spectroscopy of Bases and Liquids* (A. Weber, ed.), Springer-Verlag, Berlin, p. 129.

Svelto, O., 1974, in *Progress in Optics*, Vol. XII (E. Wolf, ed.), North-Holland, Amsterdam.

Tracy, E. B., 1971, *J. Appl. Phys.* **42**:3848.

Valat, P., Tourbez, H., Reiss, C., Gex, J. P., and Schelev, M., 1978, *Optics Commun.* **25**:1072.

Von der Linde, D., 1977, in *Topics in Applied Physics*, Vol. 18 (S. L. Shapiro, ed.), Springer-Verlag, Berlin, p. 203.

CHAPTER 4

The Excimer Laser

A New Ultraviolet Source for Medical, Biological, and Chemical Applications

Richard M. Osgood, Jr.

Department of Electrical Engineering and Applied Physics
Columbia University
New York, New York 10027

1. INTRODUCTION

Within the last decade there has been a marked increase in the pace of development of ultraviolet (UV) lasers and their technology. In part, this is a result of the demand for ultraviolet wavelengths in the developing fields of laser photochemistry and photobiology, and to some extent the interest in materials processing and microfabrication. In part, it is a result of a natural tendency for the laser physicist to turn in a direction in which there has been heretofore little success, that is, in developing efficient ultraviolet lasers. The most striking result of this activity has been the class of UV excimer lasers.* As a result of this work, these lasers have been engineered to the

---

* Three excellent reviews have been published by Ewing and Brau (1976), Rokni *et al.* (1978), and Brau (1979).

point where they can provide a variety of intense coherent sources with wavelengths from 400 to less than 200 nm. Compared to other laser systems at the same state of development, they are surprisingly simple and inexpensive. Already they can provide average powers up to 50 W and pulse powers up to 100 MW. The ability to produce such power levels at variable wavelengths through the UV will clearly have a major impact on the use of lasers in medical or biological research.

Currently, the most widely used monochromatic sources in the UV are ion lasers (e.g., Ar^+, Kr^+) and resonance lamps. In addition, the Hg arc lamp, which generates a broad continuum from 200 nm to the visible, has been used in conjunction with a monochromator to supply a wavelength-tunable source in the UV (Nakayama *et al.*, 1975). The properties of these sources are summarized in Table I, which also includes data on a representative, excimer-laser system. Excimer lasers have comparable efficiency and spectral purity to the resonance lamps, but the superior spatial coherence and peak power of the laser allows attainment of vastly higher power densities. The spatial coherence means that only a simple, optical train is needed to allow the laser to be located at some distance from the experiment. Because of the short laser pulse, significant doses of UV energy can be delivered in time scales that are rapid compared to the response times of living organisms. In addition, the high laser intensities allow investigation of the validity of reciprocity (that is, the linearity of the effect of a given energy versus its delivery time) over a wide range of pulse

TABLE I. Comparison between an Excimer Laser and Common Coherent and Incoherent Ultraviolet Sources

Source	Wavelength (nm)	Power (W)	Typical cost (thousands of $)
He: Cd laser	325	0.010	20
Argon ion laser	351–364 (several lines)	2.3 (total)	60
Frequency-doubled Ar ion laser	257	0.100	80
N_2 laser	337	0.030 (∼1-mJ pulses)	25
Hg low-pressure resonance lamp (30-W input)	254	0.040	1
High-pressure Xe arc lamp (25-kW input)	<220	0.1 (per nm interval)	10
KrF excimer laser	449	20 (∼200-mJ pulses)	60

energies. For some materials, it is to be expected that nonlinear optical effects will be important at high intensities, > 10 kW; and thus reciprocity will fail. In a more practical connection, such nonlinearities have already been used as a technique for discretely shifting the wavelength of each type of excimer laser.

Ion lasers, in comparison to excimer sources, possess the same, or superior, spatial coherence. In addition, they can also be operated on a continuous basis, which is not possible for excimer lasers. However, in general excimer lasers offer much higher average and peak output power over a wider wavelength region. This is particularly true for the wavelength region below 300 nm, which is important for experiments in photochemistry. Ion lasers have been commercially available for 22 years and thus are at a much more mature stage of development. Nonetheless, the high current density in an ion-laser plasma requires sophisticated materials engineering of the laser-plasma tube. As a result, the cost per watt for an excimer laser is much greater than for an ion laser.

In short, the excimer laser provides an ultraviolet source that possesses greater intensity and wavelength agility than other conventional coherent and incoherent sources, and at a competitive cost. These features mean that applications are imminent in disciplines that rely heavily on UV sources such as dermatology, molecular physics, photochemistry, and materials processing. In fact, even in the first two years of excimer-laser development, research in all these areas had already blossomed.

The purpose of this chapter is to give the reader insight into the underlying physics and the technology of excimer lasers, and to provide a few examples of their applications. To that end, the following sections will be provided: Physics of excimers, and fundamental limitations on their use as laser medium; technology and capabilities of excimer lasers; applications; future improvements.

2. PHYSICS OF EXCIMERS, AND LIMITATIONS ON THEIR USE AS LASER MEDIUM

There are several basic requirements that any atomic or molecular system must meet in order for it to be suitable as a high-power laser. First, the deexcitation of the lower level of the laser transition must be faster than the excitation of the upper level. This permits the relative level population to remain inverted during stimulated emission. Second, the processes that lead to excitation of the upper level must be sufficiently efficient that both the requirements on the laser power supply and cooling of the laser envelope

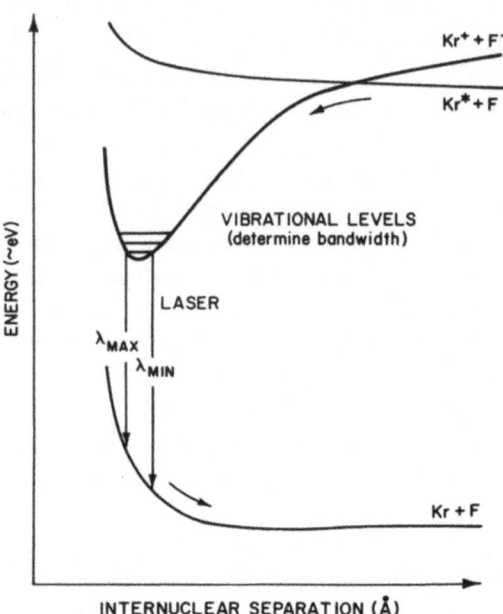

FIGURE 1. Interatomic potential energy versus internuclear distance for a typical excimer molecule. The horizontal lines in the potential well of the upper laser level denote the first three vibrational states of that laser level. The lower level has no vibrational states since it is not bound.

are reasonable for an acceptable output power. Third, the stimulated-emission cross section of the laser transition must be high enough to allow the laser small-signal gain, at a typical inversion density, to dominate any incidental intracavity losses. An important aspect of the use of excimers as laser medium is that all three of these requirements are satisfied by virtue of the unique, molecular structure of the excimer. To illustrate this point, a set of potential-energy curves for a representative excimer are shown in Fig. 1. Each curve represents the variation in the potential energy between the two constituent atoms of the molecule as a function of their internuclear separation. Note that the negative derivative of the potential-energy curve gives the internuclear force. For each molecular energy state there is a separate curve that arises from the interaction between atoms in the particular electronic states, which are shown on the right-hand side of the figure.

The most striking feature of the excimer structure is that only the upper (excited) states are bound. Thus, when the molecule makes a radiative transition to the ground state, the molecule dissociates of on the order of 10^{-13} s. In fact, the term "excimer" derives from the fact that such a molecule is only stable (i.e., bound) while it is excited. Because of its dissociative ground state, an excimer has an inherent, molecular basis for population inversion.

In Fig. 1, the upper laser level is shown to be the lowest excited state. Such a feature is a desirable aspect of an efficient excimer laser, since generally the lowest excited-level receives most of the excited-state population during the excitation process. A sample kinetic chain for this process is illustrated in Fig. 2. This is a result of the rapid, collisional coupling between the various excited atomic and molecular states that occurs at high laser-gas pressures. Coupling is aided by the numerous crossings (upper arrow in Fig. 1) in the potential-energy curves of the excited states. Because of the coupling, the relative populations of the molecular states are generally given by a thermal distribution that has an effective temperature close to or slightly above the gas temperature. Since the majority of molecular levels are spaced at energies much greater than the typical thermal energy, the lowest excited state is the most populated. The energy difference between the first excited state and the ground state is generally such that it gives rise to visible or ultraviolet emission. In fact, however, all efficient excimer lasers discovered thus far emit in the UV. Because the rate of spontaneous emission between two energy levels is inversely proportional to the cube of the wavelength of the transition, the radiative lifetime for the upper excimer level is typically short ($\sim 10^{-8}$ s) if it is connected to the ground state via a strongly allowed transition.

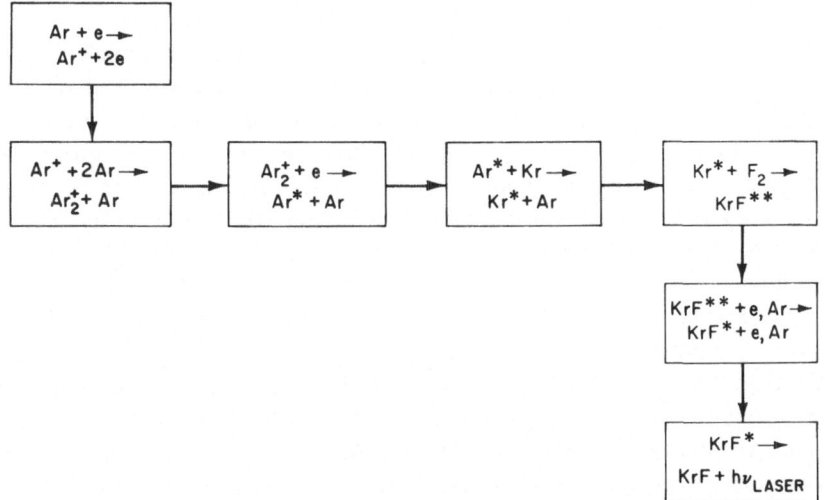

FIGURE 2. A typical kinetic chain, which leads to excimer formation in a discharge or electron-bombarded gas. The following symbols are used: e, electron; M^+, ion of species M; M^*, M^{**}, first and higher excited state of species M; $h\nu$, laser–laser photon.

Excimer lasers possess an extremely broad emitting linewidth compared to gas lasers based on molecules which are bound in both excited and ground states. The origin of the linewidth can be shown by considering laser action from a single, upper-level, vibrational-electronic state (horizontal lines in Fig. 1). Because of the dissociative lower state, a molecule can emit a distinct UV wavelength for each value of the internuclear coordinate that it traverses during its vibrational motion. The laser bandwidth will be given approximately by the difference between the wavelengths that correspond to the right and left (λ_{min} and λ_{max} in Fig. 1) turning points of the vibrational motion. For the potential curves shown in the figure, the laser bandwidth will be proportional to the height (in energy) of the vibrational state in the upper well and to the slope of the repulsive dissociative curve in the ground state (Mies, 1973).

The optical gain of a laser transition is inversely proportional to the bandwidth of that transition. Since the excimer linewidth is broad, only strongly allowed transitions (hence those possessing short radiative lifetimes) can have sufficient gain to be above threshold. Further, in order to achieve a large gain in the laser medium it is also necessary to form a substantial population density of excimers. This can only be accomplished by overwhelming the rapid upper-level radiative decay rate with a faster formation rate. Since excimers are formed via three-body collisional channels, high pressures ($\gtrsim$ 2–3 atm) are necessary before the upper level is formed faster than it decays.

Thus far in this section all excimers have been treated as a single class of laser molecules. However, it is more appropriate to divide excimers into three subclasses: homonuclear rare gas, rare-gas halide, and quasiexcimer. The first type is the homonuclear rare-gas excimers, Xe_2, Ar_2, and Kr_2, for example. Historically these rare-gas molecules are important because they were the first excimer molecules that exhibited laser action. This group can be an important source of coherent far-UV emission, since all members emit at wavelengths less than 180 nm. However, because of the sensitivity of these systems to even slight amounts of impurities ($\sim$ 10 ppm), they have not, as yet, found widespread applicability.

The second and certainly the most important excimer group is that of the rare-gas halides. It is made up of molecules that contain one rare-gas and one halide constituent atom, ArF or KrF, for example. These molecules emit throughout the near and medium UV from 358 nm (XeF) to 193 nm (ArF). Members of this class have been found to be the most efficient and powerful excimer-laser species to date. The typical commercial excimer laser is a rare-gas halide laser.

A third and also important class is the quasiexcimers. This includes molecules such as HgCl, Br_2, and I_2 which have strongly bound ground states. However, owing to the sizable shift in the minimum in the upper and lower state potential wells, emission from the upper state still terminates on energy levels above the dissociation limit of the ground state. Thus the radiative properties of quasiexcimers are very similar to those of an ordinary excimer molecule.

3. TECHNOLOGY AND CAPABILITIES OF EXCIMER LASERS

While excimers are in many ways the ideal laser molecule, the technology required to excite them and to enable well-controlled optical extraction was, until recently, not realized. The laser medium, generally a rare-gas halogen mixture, does not readily form a stable gas discharge if only classical, positive-column-discharge excitation techniques are used. In addition, these gas mixtures are reactive and sensitive to small amounts of impurities. The unusually high optical-gain to optical-specific-energy ratio for excimers makes optical extraction within a clean Gaussian mode particularly difficult.

In this section we will discuss four areas in which excimer-laser technology differs considerably from that encountered in the other laser types that are used in biology and medicine. These are excitation techniques, gas mixtures, optical cavities, and laser-wavelength tuning.

3.1. Excitation Techniques

Although a wide variety of schemes have been used to excite excimer lasers, the only two techniques that are more than academic novelties are electron beam (E-beam) and externally ionized-discharge excitation. While both techniques will be discussed here, discharge excitation will be heavily emphasized because it is the most commonly used of the two techniques in small-scale laboratory and commercial laser devices.

E-beam excitation has had its greatest utility in generating spectacular output energies in large-scale devices. The highest-energy excimer laser built to date has been an E-beam-excited krypton fluoride laser (see, for example, Hunter, 1976). In addition, this excitation technique has historical importance because it was used in demonstrating the feasibility of virtually all important excimer lasers.

Figure 3 shows a schematic of a typical E-beam-driven laser. Electrons are emitted from a series of carbon or metal blades in a gun diode. During

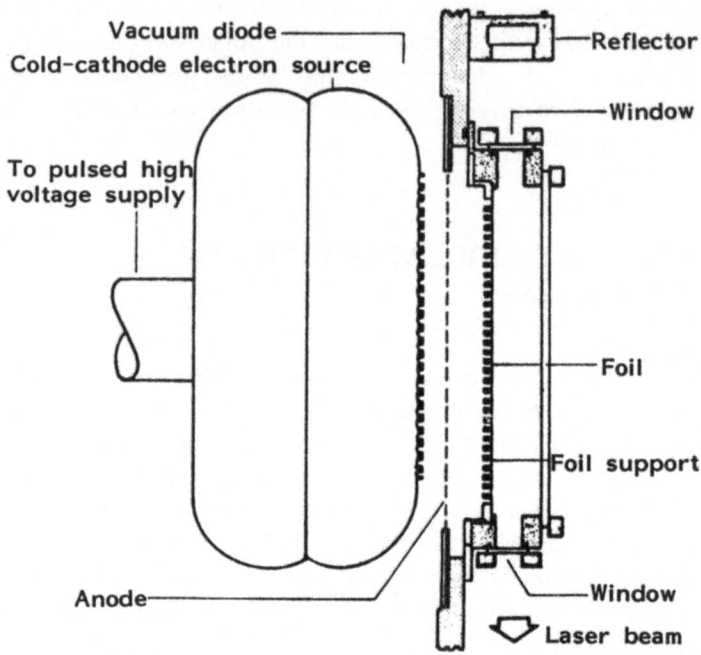

FIGURE 3. Schematic diagram of a typical electron-beam-excited excimer laser (J. Goldhar, Lawrence Livermore Laboratory).

the application of a fast high-voltage pulse, electrons leave the gun after being accelerated to energies of 0.5 to 1.5 MeV. Typically, they enter the laser medium transverse to the optical axis, after first passing through a thin metal or plastic foil which presents little energy loss.

Electron energy is transferred to the gas chiefly through electron collisional ionization of the dominant gas species. Other collisional processes then convert this energy into that stored in the desired excimer species (Fig. 2). The entire transformation process from incident electron energy into optical energy can be highly specific. Thus, conversion efficiencies of up to 50% have been observed in the Xe_2 excimer system (Turner *et al.*, 1975). Note, however, that in this case the optical energy appeared as spontaneous fluorescence only. Xe_2 laser efficiencies are generally much lower because of the presence of intracavity absorption.

There are several practical limitations to more widespread use of electron-beam excitation. One difficulty is that the combination of high voltage and short pulse times that are required for practical inversion densities necessitate the use of large and cumbersome hardware—even for comparatively low-energy lasers. Second, during the passage of the electron pulse through the foil, some energy deposition and subsequent foil heating

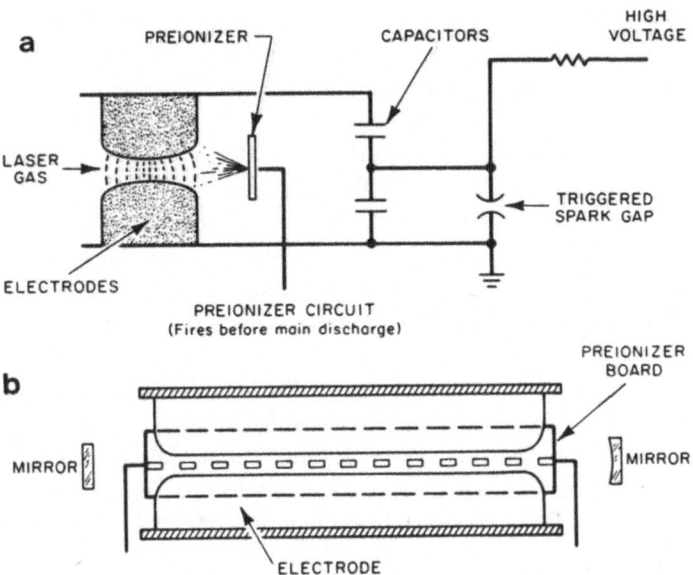

FIGURE 4. Circuit diagram and sketch of small, discharge-excited excimer laser (a, end view; b, side view).

occur. The cooling time of the foil is slow as heat flow occurs primarily along the thin foil cross section to the nearest support rib. As a result, foil cooling forms a fundamental limitation on the repetition rate for the technique of E-beam excitation (Schlitt, 1976).* Both of the above points are sharply illustrated by the fact that the electron beam, which excited the high-power, KrF laser mentioned earlier, had dimensions of $3 \times 4 \times 10\,\mathrm{m}^3$ and a repetition rate of $3 \times 10^{-4}\,\mathrm{Hz}$ (Hunter, 1977).

Discharge excitation circumvents the problems encountered in E-beam lasers by depositing energy in the gas with a low-voltage electric discharge. The chief difficulty with discharge excitation in excimer gas mixtures is that these mixtures are particularly electrically unstable and tend to form arcs rather than a glow discharge. A twofold approach has been used to solve this problem successfully. First, the gas is preionized; that is, a low density of electrons is seeded in the gas mixture between the two electrodes. This reduces the tendency to form a local arc center. Second, the discharge circuitry is designed to be sufficiently fast that the stored energy can be deposited prior to the arc-formation time (Brau, 1979). Coincidentally, for a typical excimer laser this time scale is of the same order as that necessary to

* There are also significant practical limitations on the repetition rate of E-beam-driven lasers which stem from the standard low pulse rate design of the electron beam high-voltage circuitry.

overwhelm the upper-state radiative or collisional quenching, and thus produce a substantial upper-level population. It is the requirement for an extremely rapid discharge that distinguishes excimer-laser discharge techniques from related high-pressure CO_2 laser excitation schemes.*

Figure 4 shows a sketch of a device that is representative of the class of discharge-excited excimer lasers. In this laser, the electrons for preionization are produced by using a linear array of arc sources (within the laser gas) that emit sufficient UV to photoionize various laser-gas impurities with low ionization thresholds. Other preionization schemes have been used, including those utilizing electron beams, radioactive sources, and corona discharges.

A few hundred nanoseconds after the gas is preionized, the main discharge circuit is fired. As in the E-beam laser, excitation is transverse to the optical axis in order to reduce the inductance in the discharge. The electrode contouring, shown in the sketch, is an additional design feature that has been introduced in order to reduce the tendency to arc. The circuit shown is sometimes called a Blumlein circuit, because it is the lumped-circuit equivalent of a distributed-impedance device, with the same name, which is used to produce a fast pulse at twice the power-supply voltage. Since the laser discharge is extremely fast, electrical interference can be a problem in any electronic equipment operated adjacent to the laser.

A photograph of an early, commercial laser is shown in Fig. 5. The charging voltage of the laser is 28 kV, which is approximately doubled by the Blumlein circuit. The usable duration of the laser pulse is approximately 25 ns. A metal casing, which is partially removed in the photograph, is used to reduce the radiated electrical discharge noise. The laser, which has a discharge length of 60 cm, is driven by 16 J of stored, electrical energy. With krypton fluoride in the laser approximately 60–80 mJ of optical energy can be extracted; the efficiency is thus 1.5%, typical for an excimer laser. Discharge lasers as long as 1 m have been reported (Goldhar, 1977); however, scaling to greater lengths is not practical because of the resulting increase in discharge inductance and because of the extremely high single-pass optical gain, which may actually lower the laser brightness (see Section 3.3).

Throughout this subsection we have mentioned only short pulse excitation. One can extend the laser pulse width to some extent by adjusting the gas mixture and increasing the degree and length of the externally generated ionization. However, the short lifetime of the upper laser level and the increased requirements of the energy in the ionization arcs limit the

---

* High-pressure transverse discharge lasers are sometimes called TEA (transverse excitation atmospheric) lasers. While in CO_2 lasers the total gas pressure is generally one atmosphere, it is somewhat higher in excimer lasers.

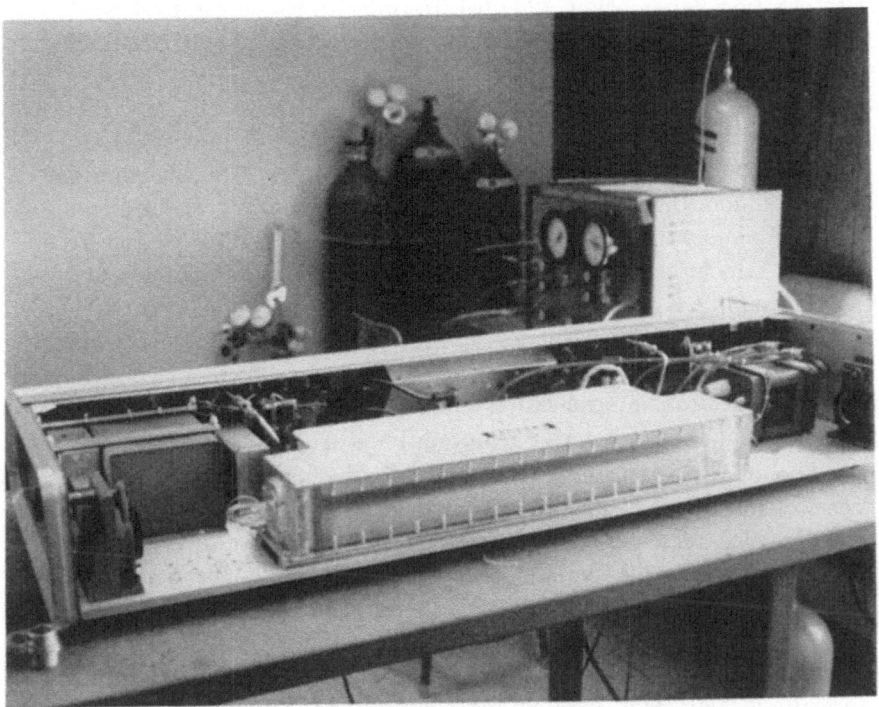

FIGURE 5. Photograph of small, "table-top," discharge-excited excimer laser (Tachisto, Inc.).

extent to which this is possible. The longest pulse width yet reported for a discharge-pumped KrF laser is 100 ns (Goldhar *et al.*, 1977).

High average power can be obtained from excimer lasers by multiple pulsing. In a discharge-excited laser, the maximum repetition rate is determined, fundamentally, by gas heating and subsequent arcing. Practically, the limit is usually set by gas degradation or the power-handling capability of the pulsing circuitry. By using gas-recirculation schemes (cf. Section 3.2), commercial firms have realized repetition rates of $\lesssim 200$ pps, with 50 W average power for a KrF laser mixture. Recently a laser of 50 W average power with a 1 kHz repetition rate has been described (Fahlen, 1978). Heavy-duty commercial XeF lasers have also been developed for use in the semiconductor industry; however, the purchase price for such devices increases by a factor of approximately 4 over laboratory-qualified models.

3.2. Gas Mixtures

The laser discharge shown in Fig. 4 can be used as a general excitation technique for all but a small subclass of excimer systems (those excimers with emission bands below 190 nm). In order to convert from one excimer to another, using the discharge system, one must change only optics and gas mixtures.

The gas mixture for the most generally useful class of lasers, the rare-gas halides, consists typically of 89.8 % He, 10 % active rare gas, and 0.2 % halogen donor. The helium serves as a diluent to prevent gas heating and to help condition the discharge. In addition, it provides a third collision partner for many of the teratomic collision processes, such as ionic recombination, which feed the upper laser level. The pressure of the active rare gas, for example Kr in a KrF laser, is adjusted so that it will receive most of the discharge excitation. The percentage of the halogen donor controls the formation rate of the excimer. In addition, as is shown schematically in Fig. 6, the electron multiplication is reduced by its presence. This in turn tends to stabilize the discharge temporarily (Rokni *et al.*, 1978; Brau, 1979).

The total gas pressure in discharge lasers is typically 2 atms. Higher gas pressures can be used, but since the voltage to pressure ratio is fixed by the discharge characteristics, an increase in gas pressure requires a corresponding increase in laser voltage. For example, the 6 atm excimer laser described by Sarjeant *et al.* (1977) required a 100-kV charging voltage.

The chemical reactivity of the halogen compounds typically results in the disappearance of these compounds and the formation of impurities, especially if the system is sealed and contains only a single fill of gas within the laser. This problem is exacerbated by the absorption bands of the impurities, which often overlap. Systems that can use relatively inert halogen donors, such as Cl_2 in XeCl lasers (Sze *et al.*, 1978), have had sealed-off lifetimes greater than 16,000 shots. Longer laser-operation times in

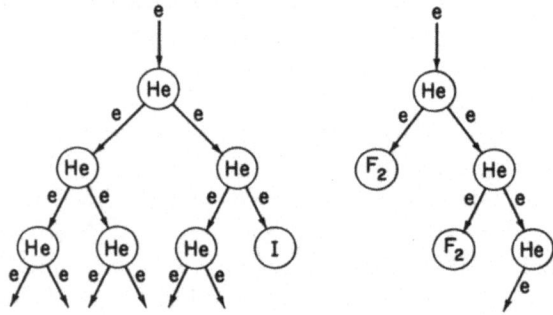

FIGURE 6. Simplified illustration of how the electron (e)-attaching properties of fluorine (F_2) prevent runaway electron multiplication in a discharge. In a pure rare-gas discharge (left figure) an electron may collide with a rare gas (helium in the figure) and produce another electron. This process can proceed to the pathological limit of an arc. If a halogen (F_2 in the figure) is added, low-energy electrons can be captured (attached) by this additive species, and their mobility reduced to the point where electron multiplication in the rare gas is balanced.

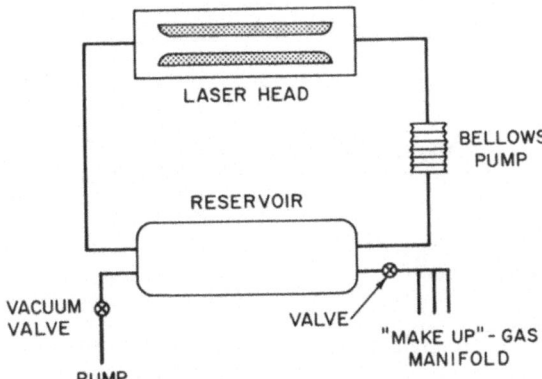

FIGURE 7. Schematic diagram of the gas recirculating system which is common in discharge excimer lasers.

all excimer systems can be achieved with gas-recirculation systems or single-pass flow systems. The latter approach is practical in systems that use inexpensive constituent gases, such as the ArF system. There are several approaches to gas recirculation, but most (Fig. 7) rely on trapping out the spent halogen and replacing the same gas with a small input stream (Johnson *et al.*, 1978).

The use of some halogen donors can present a potential health hazard. However, both premixed laser gases, and diluted helium–halogen mixtures, which are considerably safer to handle, are available commercially. Because of the low percentage of halogen in the laser mix, the health hazard due to slight leaks is minimal.

3.3. Optical Cavities

The high optical gain that is characteristic of excimer lasers makes it particularly difficult to obtain the laser output in the form of a focusable, uniphase beam. For example, even in small discharge lasers the gain is 100 % for a single pass through the laser. Since the cross section (transverse to the optical axis) of the optical medium has dimensions that are large compared to the beam waist for the TEM_{00} mode in the optical resonator, conventional stable resonators are inadequate to provide efficient and reliable mode selection in excimer lasers. In fact, the beam from an excimer laser with a stable resonator has a cross section that is determined primarily by the gain profile and the resonator magnification. Thus, the usual beam cross section is rectangular. A slight skewing toward the preionization source is seen if one side of the discharge has slightly more preionization than the

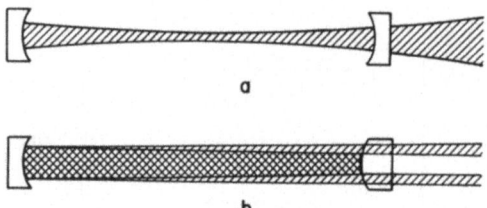

FIGURE 8. Diagram showing stable (a) and unstable (b) optical cavitites. The cross-hatched area represents the laser beam.

other. For many excimer-laser applications, this rectangular beam pattern is adequate. It does provide a spatially uniform cross section, which is useful for the investigation of laser effects over a large planar area (see Section 3.4). However, for other applications, particularly involving nonlinear effects, high intensities are needed. In this case, improved laser brightness can be achieved by use of an unstable resonator.

An unstable resonator, an example of which is shown in Fig. 8, is an optical cavity which magnifies the beam as it reflects between the mirrors. By adjusting the mirror magnification and cavity output coupling, one mode can be selected to reach sufficient intensity to extract the energy from the laser medium. Generally, the maximum laser power from a given device is reduced by the use of an unstable resonator. However, despite this reduction in total power, the gain in laser brightness that is achieved with an unstable resonator makes it a useful technique for excimer lasers (Goldhar *et al.*, 1977).

3.4. Wavelength Tuning

One of the most appealing aspects of excimer lasers is the wide variety of UV wavelengths which they can produce. In more conventional gas lasers such as CO_2 or Ar^+, wavelength selection is over a limited range or is non-existant. Wavelength tunability in excimer lasers can be obtained by three separate techniques: tuning across the broad excimer gain profile; use of a series of excimer gas mixtures in a single discharge device; and wavelength shifting through stimulated Raman scattering. These three techniques have enabled the development of one generic excimer-laser excitation system to provide a wide variety of intense coherent sources throughout the ultraviolet.

The dissociative nature of the excimer lower level provides a broad laser-gain profile. Depending on the distribution of the upper-state vibrational population, and shapes and positions of the upper and lower

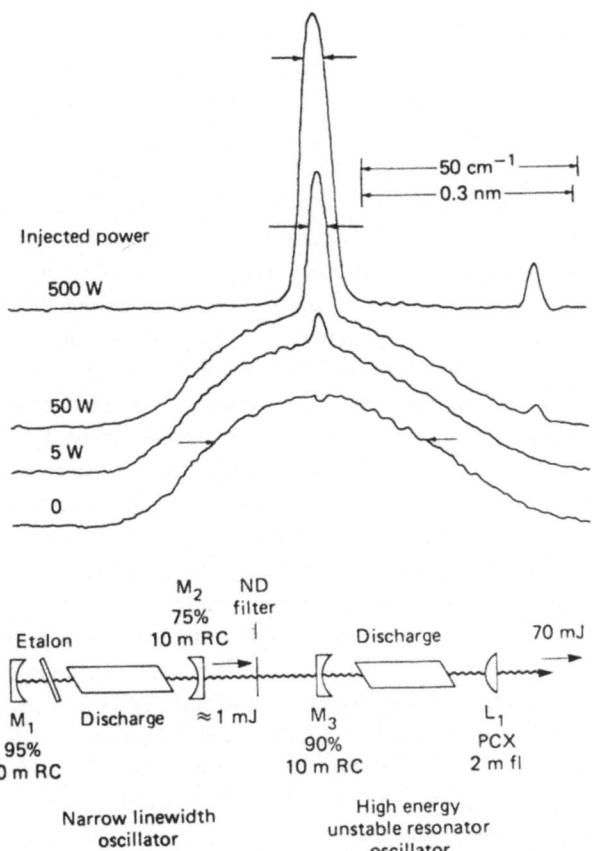

FIGURE 9. Narrowing in output spectrum of KrF laser for various values of injected signal incident on the laser. The lower diagram shows the apparatus used in the experiment (J. Goldhar, Lawrence Livermore Laboratory).

potential curves, the laser gain profile can be 1–2 nm wide. In order to tune across this linewidth, it is necessary to allow feedback only within a selected narrow-wavelength region. In a small-scale excimer laser, this can be readily accomplished with the use of an intracavity dispersive element. Both an etalon and a tandem combination of two prisms have been successfully used for this purpose (Goldhar, 1977; Loree *et al.*, 1978). Incidental intracavity optical losses limited the tuning range in both of these experiments to 0.5–1 nm.

In large lasers, the high single-pass gain means that it is nearly impossible to obtain adequate wavelength discrimination* with intracavity

---

* By "wavelength discrimination" we mean the ratio of round-trip transmittances at two different wavelengths.

elements. For example, if in a 1-m device, a small-signal gain of 10% cm^{-1} exists for line center, a factor of 150 in discrimination is required before laser action can be forced to occur at a wavelength one half-width from the peak in the gain profile. In such a large laser, the technique of injection locking has been used to realize a narrow-bandwidth with tuning across the whole laser gain profile (Goldhar *et al.*, 1977). In this technique, a smaller excimer laser (of the same species) is used to provide a narrowed and tunable signal, which is then injected into the larger free-running laser. This signal will typically be of much higher intensity than any unamplified spontaneous emission within the larger laser. Thus, as long as the tuned signal is injected prior to the firing of the large device, its output will build up from the narrowed source and will be locked to the wavelength of the initial signal. Figure 9 shows the narrowing obtained by using this technique.

Cavity tuning and injection locking can provide only fine-scale tuning of an excimer laser. In order to reach a wider range of wavelengths, one can, as a first step, select a different excimer system. Figure 10 shows the wavelengths of excimer systems that can be realized with a discharge laser similar to that shown in Fig. 4. Typical pulse energies for these systems, which have an active discharge volume of 60 cm^3, are 20–40 mJ. The wavelength available from each of these systems can, in turn, be extended by using stimulated Raman scattering to shift the natural laser frequency. This is most readily accomplished by stimulated scattering using the Raman-active vibrational modes of various high-pressure molecular gases. During the scattering process the vibrational mode is excited, and the frequency of the laser photon is shifted by an amount corresponding to the frequency of the vibrational mode (see Fig. 11). Since the gain for the stimulated scattering process is proportional to the incident laser intensity, the high peak power and the spatial coherence in excimer lasers make this practical option.

Figure 10 shows the matrix of wavelengths that can be obtained by

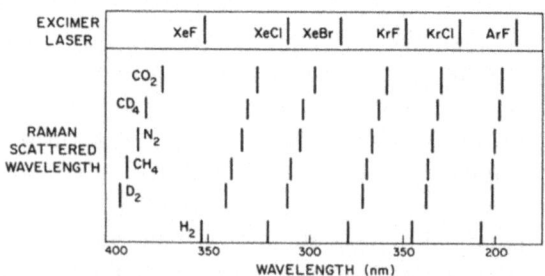

FIGURE 10. Wavelengths obtainable from various rare-gas–halide excimer lasers (top line) and those obtainable by Raman shifting the output of these lasers in various molecular gases.

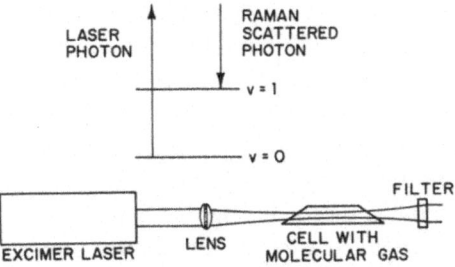

FIGURE 11. Energy level diagram (top) depicting Raman scattering process, and experimental apparatus (bottom) used in realizing Raman scattering.

Raman shifting the output from various excimer lasers in a number of common molecular gases. High conversion efficiencies are possible; for example, 40% conversion of the laser energy into energy at the Stokes-shifted wavelength has been accomplished for the output of a KrF laser when scattered in molecular hydrogen (Loree *et al.*, 1978). In addition, under some conditions either four-wave mixing or higher-order Raman scattering

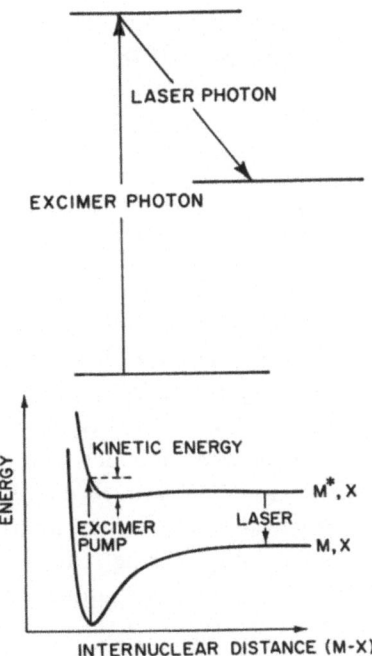

FIGURE 12. Energy level diagrams showing shifting of the wavelength of an excimer laser by optical pumping (top) and photodissociation (bottom).

will generate additional Stokes or anti-Stokes wavelengths. Conversion efficiencies as high as 10%–20% have been obtained for such higher-order processes (Loree *et al.*, 1978). Efficient Raman shifting has also been observed in atomic metal vapors (Djeu *et al.*, 1977), thus further increasing the wavelengths available from shifted excimer sources. The experimental apparatus that is necessary for obtaining Raman shifting is particularly simple, and an example is shown in Fig. 10. A laser beam is weakly focused into a cell containing 5–20 atm of the desired molecular gas. Typical conditions are a focused intensity of several GW/cm^2 over a 30-cm sample path. By using intermediate state resonances, stimulated Raman scattering in metal vapors can occur at much lower densities and optical intensities.

Optical pumping and photodissociation are additional approaches to generating shifted excimer-laser wavelengths (Fig. 12). In the former approach, a laser is used to populate an excited state in an atom or molecule. Laser action then occurs to a lower-lying excited state. The photodissociative process is similar except that the excited state that is pumped is dissociative. If this dissociative state feeds into at least one electronically excited fragment, then laser action can occur in that fragment. The photodissociative process is better matched to the broad linewidth of an unnarrowed excimer laser, and thus does not require narrowing of the pump source. Both techniques have been used to realize efficient laser-pumped UV sources (Burnham, 1978; Ehrlich *et al.*, 1978).

4. APPLICATIONS

Although excimer lasers of practical scale and expense have been in existence for only a limited period, applications in many diverse fields are already emerging. For example, because UV photons are sufficient to rupture many common molecular bonds, excimer lasers, which have a unique combination of tunability and intensity, have found immediate use as a research tool in photochemistry. Further, the recent development of excimer lasers with high average powers (Brau, 1978) means that these lasers are potentially practical sources for industrial photochemistry. Research into the feasibility of utilizing excimer lasers for the gas purification of silane, for semiconductor applications, and the isotope separation of O_2 has already been reported (Clark *et al.*, 1978; Sander *et al.*, 1977, respectively).

Nonlinear optical effects have been used to investigate various phenomena in physics such as collisional processes in mixtures of noble and molecular gases. In one instance, two-photon excitation of H_2 with the output from an ArF laser was used to observe the collision rate for conversion of one symmetry species of H_2 into the other (Klinger and Rhodes, 1978).

Finally, because of the ease of ultraviolet-induced, photochemical reactions, UV lasers have potential for materials processing. The short wavelength of the ArF laser (193 nm) may make it a particularly good source for microcomponent fabrication.

The use of the excimer laser in physics and chemistry has suggested allied applications in biology and medicine. A tunable UV laser can provide a detailed mapping of the wavelength response of tissue (v.i. Section 4.2). In more sophisticated experiments, it may be possible to obtain information on the dynamical response of various organic molecules, such as that recently reported for irradiation of a DNA-psoralen solution at 347 nm (Johnson *et al.*, 1977).

For the remainder of this section we will discuss, in detail, the results of two experiments that illustrate the utility of excimer lasers in the fields of photophysics and medicine.

4.1. Photoassociation

Since excimer lasers have an attractive wavelength range for applications in photochemistry, they have already been used in several areas of photodissociative photochemistry, including studies of the photolysis of alkali salts (Ehrlich *et al.*, 1979) and UV multiphoton dissociation of complex organic molecules (Jackson *et al.*, 1978). In addition these lasers have been used to photodissociate various metallic salts to produce laser action in one of the resulting photofragments (Schimitschek *et al.*, 1977; Ehrlich *et al.*, 1978). Recently, however, it has become clear that excimer lasers can be used fruitfully for the converse photochemical process —namely, photoassociation.

An example of the utility of photoassociative techniques has been the investigation by D. Ehrlich and the author into collisional and spectroscopic phenomena in the excimer species, Hg_2, which was associated with the 193-nm output of an ArF laser (Ehrlich *et al.*, 1978). In general, when studying such phenomena for any molecular species, such as Hg_2, it is desirable to examine the molecule in a single well-characterized state, if ambiguities in data interpretation are to be minimized. However, for an excimer molecule this requirement is at odds with the usual production technique, which utilizes the relatively unselective collisional-formation channels (cf. Section 2.1). Optical excitation out of the very weakly bound ground state can be used to produce selectively a particular excited state.

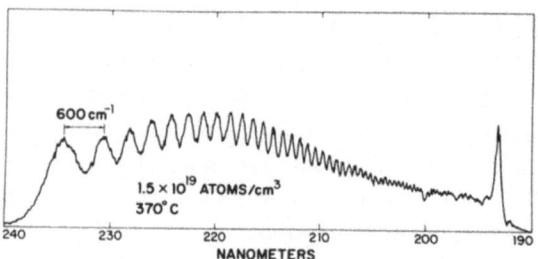

FIGURE 13. UV fluorescence emission from the photoassociated-excimer Hg_2. The oscillations in the spectrum are a manifestation of the quantum wave properties of the excited Hg_2 state.

But for excimers, few ground state molecules are bound and the approach is therefore limited. Now, however, it has been found that under certain conditions, photoassociation with ultraviolet light can produce a surprisingly well characterized Hg_2 excited-state distribution.

For the particular case of atomic mercury vapor illuminated with the light from an ArF laser, an extremely narrow band of highly excited vibrational states are produced in the $Hg_2(O_u^+)$ electronic state. Because the distribution states are so nearly monoenergetic, information on the phase of the average upper-state vibrational wave function can be seen in the UV fluorescence from this state to the ground state. This phase information is manifested by the oscillations that appear in the spectrum of the UV fluorescence (see Fig. 13) due to maxima and minima in the O_u^+ state vibrational wave function. If the upper state were collisionally produced, many upper-state vibrational levels would be excited, and virtually all of the oscillatory structure would disappear. On the other hand, the clear oscillations that are seen in the photoassociated Hg_2 fluorescence give otherwise unobtainable details on the ground- and upper-state potential curves.

4.2. Dermatology Response to Excimer Radiation

Turning now to medical applications, Dr. John Parrish and Rox Anderson of Massachusetts General Hospital in collaboration with Thomas Deutsch and the author, while both were at MIT Lincoln Laboratory, have recently performed experiments designed to determine the threshold for dermatological response in guinea pig and human skin at wavelengths less than 200 nm, and to ascertain the validity of reciprocity for excimer-laser pulse intensities, at both 193 and 249 nm (Parrish et al., 1978). The experiment used ArF and KrF laser to produce the 193- and 249-nm

TABLE II. Dermatological Experiments with Excimer Lasers

248 KrF (10^6 W/cm^2)	254 nm Hg Arc (4×10^{-4} W/cm^2)
Human M.E.D. 19–25 mJ/cm^2 (forearm)[a]	34–48 mJ/cm^2
Guinea pigs	
No. 2 186–248 mJ/cm^2	317–440 mJ/cm^2
No. 3 69–73 mJ/cm^2	87–113 mJ/cm^2
No. 4 366–416 mJ/cm^2	Not done
193 nm ArF (2×10^6 W/cm^2)	
Guinea pig (No. 1)	M.E.D. 2.1 J/cm^2
Human	M.E.D. 0.5 J/cm^2

[a] After exposure, reactions to more than M.E.D. doses are comparable for both sources.

wavelengths. The laser-output pulse energies were typically 5–10 mJ at 193 nm and 30–50 mJ at 249 nm, and were approximately 20 ns in duration. The laser output was weakly focused to achieve the desired maximum-per-pulse irradiance for a given experimental run. Attenuators were then used to vary the laser intensity over one order of magnitude. The laser repetition rate was approximately 1.5 Hz.

Experimental runs were performed on depilated albino guinea pigs and Caucasian humans. In essence, the experiment consisted of varying the total delivered laser energy until the threshold for a minimum erythemal dose (M.E.D.) could be determined. By irradiating the same guinea pig and human subjects with the 254-nm output from a Hg arc lamp, the validity of linear reciprocity relation between the two sources could be determined. In our case the pulse rate of the laser was 1 Hz and its pulse width was 2×10^{-8} s; and thus, a comparison of the M.E.D. dosage, which is in units of energy, of the continuous and pulsed sources implies a comparison over a factor 5×10^7 in the intensities of the two sources. Table II gives a representative sample of the experimental results. No erythemal response was observed at 193 nm despite the fact that the maximum dosage was 100 times that at 249 nm. This result can probably be attributed to the opacity of the stratum corneum for wavelengths near 200 nm. The fact that there is reasonable agreement in the measured M.E.D. for both the Hg arc lamp and the KrF laser sources shows that reciprocity holds for even the factor of 10^8 difference in the intensity of the two sources. The cause of the comparatively large variation in the individual guinea-pig response is, at present, unknown. Note, however, that this variation holds for both light sources, and is thus attributable to some variability in the animals.

5. FUTURE IMPROVEMENTS

Because of the many unique capabilities of the excimer laser, marked improvements in the engineering of small-scale devices seem certain over the next few years. In particular, the current effort to boost laser repetition rates by designing improved gas-recirculation systems and electrical charging circuits should yield a commercial laser system with 20 W average power and 1–2 kHz repetition rates.

In addition, improved resonators and beam quality should become standard on all commercial lasers in the near future. Great progress has already been made in developing discharge configurations with more uniform gain. In addition, data on which cavity design works best for the gain distribution that is peculiar to excimer lasers are now being gathered (cf. Section 3.3). The demand for high UV intensities in experiments requiring nonlinear effects will help spur the incorporation of unstable resonators in commercial lasers.

Further improvements in the very compact discharge excimer lasers also can be anticipated (Goldhar *et al.*, 1978). For example, there have been reports of small-diameter-wall capillary lasers for both XeF and KrF lasers. While these devices are all specialized experimental lasers, their design is sufficiently straightforward that some commercial production of such lasers has already occurred.

Finally, in the longer term the most important development in excimer technology will be the availability of intense sources with wavelengths < 190 nm. This will be accomplished either by using nonlinear-optical effects, such as frequency tripling the 350-nm output of an XeF laser to get a 120-nm source; or by using improved gas purification and handling techniques in order to realize discharge-pumped laser action on species such as F_2 at 158 nm.

ACKNOWLEDGMENTS. I would like to thank the following colleagues for particularly useful conversations during the preparation of this manuscript: J. Goldhar, R. Anderson, D. Ehrlich, T. Deutsch, C. Brau, R. Sze, and P. Kelley. Of these, I am particularly indebted to Tom Deutsch, who has recently demonstrated many medical and biological applications of excimer lasers at his laboratories in The Welman Clinic at Massachusetts General Hospital. This work was sponsored by the Department of the Air Force and the Department of Energy.

REFERENCES

Brau, C., 1979, *Rare Gas Halide Excimers*, in *Excimer Lasers* (Charles K. Rhodes, ed., with contributions by Charles A. Brau), Springer-Verlag, New York.

Burnham, R., 1978, Optically pumped bismuth lasers at 472 and 475 nm, in talk delivered at 10th International Quantum Electronics Conference, Atlanta, Georgia.

Clark, J., and Anderson, R. G., 1978, Silane purification via laser-induced chemistry, *Appl. Phys. Lett.* **32**:46.

Djeu, N., and Burnham, R., 1977, Efficient Raman conversion of XeF laser output in Ba vapor, *Appl. Phys. Lett.* **30**:160.

Ehrlich, D., and Osgood, R. M., 1978a, Efficient thallium photodissociation laser, to be published, *Appl. Phys. Lett.* **33**(11):931–933.

Ehrlich, D., and Osgood, R. M., 1978b, Condon internal diffraction in the $O_u^+ \rightarrow O_g^+$ fluorescence of photodissociated Hg_2, *Phys. Rev. Lett.* **41**:547.

Ehrlich, D., Osgood, R. M., and Deutsch, T., 1979, personal communication.

Ewing, J. J., and Brau, C., 1976, High efficiency UV lasers, in *Tunable Lasers and Applications* (A. Mooradian, T. Jaeger, and P. Stokseth, eds.), Springer-Verlag, Berlin, p. 21.

Fahlen, T., 1978, High pulse rate 10 W KrF laser, *J. Appl. Phys.* **49**:455.

Goldhar, J., 1977, personal communication.

Goldhar, J., and Murray, J., 1977, An injection-locked narrow-band KrF discharge laser using an unstable resonator cavity, *Opt. Lett.* **1**(6):149.

Goldhar, J. and Rapport, R., 1978, personal communication.

Hunter, R., 1977, High energy KrF laser, Talk delivered at 6th Winter Colloquium on High Power Visible Lasers, Park City, Utah.

Jackson, W., Halpern, J., and Lin, C., 1978, Multiphoton ultraviolet photochemistry, *Chem. Phys. Lett.* **55**:254.

Johnson, P., Keller, N., and Turner, R., 1978, A closed-cycle gas recirculating system for rare-gas halide excimer laser, *Appl. Phys. Lett.* **32**:291.

Johnston, B., Johnson, M., Moore, C., and Hearst, J., 1977, Psoralen-DNA photoreaction: Controlled production of mono- and diadduct with nanosecond ultraviolat laser pulses, *Science* **197**:906.

Klinger, D., and Rhodes, C., 1978, Observation of two photon excitation of the H_2E, $F^1\Sigma_g^+$ state, *Phys. Rev. Lett.* **40**:309.

Loree, T., Sze, R., Ackerhalt, J., and Barker, D., 1978, Broadband UV stimulated Raman scattering with excimer lasers, presentation at the 10th International Quantum Electronics Conf. Atlanta, Georgia.

Mies, F., 1973, Stimulated emission and population inversion in diatomic bound-continuum transitions, *Mol. Phys.* **26**:1233.

Nakayama, Y., Morikawa, F., Fukuda, M., Hamano, M., Toda, K., and Pathak, M., 1975, Monochromatic radiation and its application—Laboratory studies on the mechanism of erythema and pigmentation induced by psoralen, in sunlight and man (T. Fitzpatrick, ed.), University of Tokyo Press, Tokyo, p. 100.

Parrish, J., Anderson, R., Deutsch, T., and Osgood, R., 1978, personal communication.

Rokni, M., Mangano, J., Jacob, J., and Hsia, J., 1978, Rare gas fluoride lasers, *IEEE J. Quantum Electron.* **14**:464.

Sander, R., Loree, T., Rockwood, S., and Freund, S., 1977, ArF laser enrichment of oxygen isotopes, *Appl. Phys. Lett.* **30**:150.

Sarjeant, W., Alcock, A., and Leopold, K., 1978, *IEEE J. Quantum Electron.* **14**:177.

Schimitschek, E., Celto, J., and Trias, J., 1977, Mercuric bromide photodissociation laser, *Appl. Phys. Lett.* **31**:608.

Schlitt, L., 1976, Laser Program Annual Report, 1976, UCRL-50021-76, University of California, Lawrence Livermore Laboratory.

Sze, R., and Scott, P., 1978, Intense lasering in discharge excited noble-gas monochlorides, *Appl. Phys. Lett.* **33**:419.

Turner, C., Hoff, P., and Taska, J., 1975, Electron beam energy deposition and VUV efficiency measurements in rare gases, presented at the International Topical Conference on Electron Beam Research and Technology, Albuquerque, New Mexico.

CHAPTER 5

The Photopathology and Nature of the Blue Light and Near-UV Retinal Lesions Produced by Lasers and Other Optical Sources

William T. Ham, Jr., and Harold A. Mueller

Department of Biostatistics
Biomedical Engineering Division
Virginia Commonwealth University
Richmond, Virginia 23298-0001

1. BACKGROUND AND CURRENT STATUS

The literature prior to 1916 has been reviewed exhaustively by Walker (1916) in his systematic account of the effects of radiant energy upon the eye. Birch-Hirschfeld (1912) had postulated that the visible portion of the solar spectrum was responsible for solar retinitis and/or eclipse blindness and that the effects of sunlight on the retina were abiotic in nature. However, his opinions were contested by Verhoeff and Bell (1916), who produced what they considered incontrovertible evidence that solar retinitis and/or eclipse blindness resulted from thermal effects on the retina.

In retrospect it is difficult to understand why these authors ignored the possibility of actinic injury from short-wavelength light. Verhoeff and Bell were well aware that photochemical effects played a major role in many ocular pathologies:

all radiation of whatever wavelength is convertible into heat when absorbed by material
bodies and may produce chemical changes as well. As a matter of fact these latter show
a general tendency to increase with the frequency of the oscillations, so that chemical
changes are rare in the infrared and increasingly frequent as one approaches the
extreme ultraviolet.

They also knew that solar radiation reached peak irradiance at the short-
wavelength end of the spectrum: "it should never be forgotten that the solar
energy lies well toward the blue end of the spectrum and media which
successfully cut out the red and infrared are of very little service in protection against solar radiation." Finally, they understood that the small image
size of the sun on the human retina facilitated the dissipation of heat and
that miosis in bright sunlight drastically reduced retinal irradiance:

> the secret of the relative resisting power of the naked eye is that usually in observations
> of the sun, the pupil is in extreme miosis, so that the amount of energy received is
> probably not more than 6 % of that computed for the normal pupil, while the extremely
> small area of the solar image favors rapid dissipation of the energy not found when a
> considerable area is attacked, as in the case of the mirror experiments.

Nevertheless, their dictum that visible and infrared radiation produce
exclusively thermal damage to the retina has prevailed among most
clinicians for the past fifty years despite mounting evidence to the contrary.

In 1920 van der Hoeve (1920) published a paper entitled "Eye Lesions
Produced by Light Rich in Ultraviolat Rays. Senile Cataract, Senile
Degeneration of Macula." He wrote "It is no wonder that the macula should
be affected most because when light falls into the eye the macula is usually
one of the most illumined parts. The macula as the highest organized part, is
also the most vulnerable, and moreover, it is the least protected part."
Among his conclusions were: "Senile cataract and senile degeneration of the
macula exclude each other in a certain measure; if one of those affections is
present the other is absent or less developed." "One of the principal causes
for the appearance of senile macular degeneration is the influence of
light rich in ultraviolet rays that have passed the lens." It is noteworthy that
van der Hoeve did not reference the 1916 publication of Verhoeff and Bell.
Research over the past two decades has confirmed the conclusions of Birch-
Hirschfeld and van der Hoeve, as will be shown in this review.

During War II the damaging effects of sunlight on the retina became a
major concern for military and naval personnel operating in the Pacific and
Mediterranean theatres. The syndrome of foveomacular retinitis observed in
the U.S. Navy was first described by Cordes (1944). It consisted of a
macular edema with loss of foveal reflex, often developing later into a foveal
cyst. He reported 176 cases occurring among both Naval and Marine
personnel stationed in the Hawaiian Islands and the South Pacific combat
zone. Dispute over the etiology of this disease has been reviewed by Marlor

(1973), who concluded that solar retinitis could not be ruled out as a probable cause.

Smith (1944) reported actinic pigment degeneration in the macula of 150 servicemen stationed on a tropical island in the Pacific during World War II. This maculopathy occurred only among outdoor personnel and not among yeomen and pharmacist mates who served indoors. Smith observed gradual decrease in visual acuity and pigmentary changes in the macula as characteristic for military personnel who served outdoors in the tropics for four months or longer. He attributed the effect to the infrared rays in the solar spectrum and described the photopathology as similar in appearance to senile macular degeneration.

Another aspect of chronic light damage to the retina during World War II occurred in the prison camps in Southeast Asia where thousands of allied servicemen were interned under severe malnutrition or starvation conditions and daily exposure to bright sunlight (Dekking, 1947; Churchill, 1945). These prisoners of war experienced a loss of central vision and in many cases macular lesions similar to those observed after sun gazing. Recent research suggests that this effect was due, in part, to the lack of antioxidants like vitamins E and C that protect the retina from oxidation products of light and oxygen. The effect, which came to be known as "camp eyes," was much more prevalent among blue-eyed prisoners than among those with dark pigmentation and was referred to as a blue-eyed white man's disease. The difference in effect between lightly and darkly pigmented individuals was probably because the more densely pigmented choroid in dark-eyed individuals protected their retinae from excessive exposure to light, whereas the retinae of blue-eyed individuals were exposed to additional light reflected back from the sclera and choroid. Also melanin is thought to protect the retina from light and oxygen damage in a number of ways that will be brought out in this chapter.

These early findings showing that sunlight had long-term chronic effects on the retina that could not be explained in terms of thermal injury have been followed by a series of investigations leading to a more modern and complex concept of retinal light damage involving mechanical and photochemical effects as well as thermal effects (Ham, 1980a).

Studies of light damage to the retina received renewed stimulus in 1966 when Noell (1966) published his paper, "Retinal Damage by Light in Rats." Albino rats exposed in their cages to low levels of constant fluorescent light for extended periods of time develop a retinal degeneration, which in its early stages is limited to the rod photoreceptors. Noell's findings have stimulated a large number of investigations on the effects of environmental lighting on albino, pigmented, and RCS rats as well as on other mammals. Lanum (1978) has reviewed this literature, which has brought forth valuable insights on light damage. Unfortunately, the environmental exposure of nocturnal rodents to fluorescent light has doubtful significance for the exposure of the primate retina to optical sources which irradiate a limited or discrete portion of the fundus. The Noell syndrome concerns early or initial degeneration of the rod photoreceptors and its action spectrum follows the absorption curve for rhodopsin (Kaitz, 1979), whereas retinal damage in primates from extended exposure to blue light involves initially the retinal pigment epithelium (RPE), a.id its action spectrum continues to rise exponentially as the wavelength approaches the near ultraviolet, simulating the absorption curve of melanin (Ham et al., 1976a, 1981, 1986a). Failure to distinguish the differences between the Noell syndrome in rodents and the blue light lesion in primates has resulted in considerable confusion in the literature, especially in the assumption made by most investigators that the initial site of damage must of necessity be the photoreceptor cell (Kremers and van Norren, 1988). A major purpose here is to remove this confusion by demonstrating that the blue light minimal lesion is confined initially to the RPE, and only later do the photoreceptor cells undergo damage. Obviously, the close relationship between RPE and photoreceptor cells makes it difficult to define the original site of injury (Ham et al., 1986a).

Friedman and Kuwabara (1968) exposed the retinae of rhesus monkeys to the light from an indirect ophthalmoscope for exposure times from 5 min to 1 h. They estimated that a 15-min exposure produced a minimal lesion. The irradiance on the retina was 370 mW cm^{-2} and the temperature above ambient during the exposure was 3° C as measured with a thermocouple. Recognizing that thermal injury alone could not account for such a lesion they concluded that "the precise mechanism responsible for this retinal damage was not ascertained but was felt to be a light effect which was potentiated by heat." Tso (1972a) in a similar experiment exposed the retinae of rhesus monkeys to the indirect ophthalmoscope for one hour. Damage to the retina was moderately severe. Histological and ultra-structural studies performed at different stages after exposure demonstrated partial recovery of the photoreceptor cells and the RPE. In both these experiments with the indirect ophthalmoscope the white light source was a tungsten bulb at 2900 K.

In another study Tso et al. (1972b) exposed 10 adult rhesus monkey eyes to the 488-nm line of the argon laser. The retinal spot size was 600 μm in diameter, exposure time was 32 s, and the corneal power ranged from 13 to 300 mW, corresponding to radiant exposures between 120 and 2800 J cm^{-2}. Damage to the outer segments (OS) of the photoreceptor cells, the RPE, and the choriocapillaris was noted. The RPE showed depigmentation and pyknosis. Follow-up from 3 to 5 months produced regeneration of the OS of both rods and cones but depigmentation of the RPE persisted.

These radiant exposures were well above threshold for blue light lesions, involving damage to the choriocapillaris as well as the RPE and the photoreceptors. In a more recent study of repair and late regeneration of the primate foveola after exposure to the 488-nm line of an argon ion laser, Tso and Fine (1979) reduced their radiant exposures to levels close to the threshold for blue light. Observations from ophthalmoscopy, fundus photography, and fluorescein angiography were correlated with light and electron microscopic studies. Changes during the first 6 months after exposure consisted of initial vacuolar changes of the RPE followed by persistent hypopigmentation with slow development of membranous bodies or lipoidal degeneration. Initial alteration and subsequent incomplete reformation of cone OS were also noted. Animals sacrificed 3 to 4 years postexposure showed separation of RPE from Bruch's membrane, abnormal basement membranes, and cystoid changes in the overlying retina. These observations suggested that while the epithelial cells may recover after a mild blue light lesion they develop functional incompetence at a later date which may result in serous detachment of the RPE and foveal (macular) edema.

Vassiliadis (1972) exposed the rhesus retina for 100 and 1000 s to the He–Ne (632.8-nm) laser and to the argon ion laser (514.5-nm). He noted a precipitous drop in the level of irradiance required to produce a minimal lesion in 1000 s when the 514.5-nm wavelength was compared to the 632.8-nm wavelength. The irradiance was 30 times lower than would be expected from a thermal damage model. Vassiliadis recognized immediately that a mechanism different from thermal injury must be involved, and in an appendix he proposed a "chronic damage mechanism" which assumed that the rate of depletion of an essential constituent of the retinal system was proportional to the power density of the light on the retina. This represented one of the earlier attempts to develop a model for photochemical damage to the retina, but it did not include wavelength as a parameter.

Harwerth and Sperling (1974) reported permanent blue blindness in the rhesus monkey after prolonged intermittent exposure to blue light (463 nm). They used a 1600-W xenon lamp with a 10-nm interference filter and an $18°$ Maxwellian exposure on the retina. Intermittent exposure times ranged from a total of 1680 min at a retinal irradiance of 1 mW cm^{-2} to 280 min at an irradiance of 1.9 mW cm^{-2}. These correspond to total radiant exposures of 101 and 31 J cm^{-2}, respectively. The authors were of the opinion that blue light destroyed the cones containing the photopigment whose absorption peaked at 445 nm in the rhesus monkey. In their words: "This is a new kind of lesion. It is different from the thermal lesion which is accounted for by energy absorption in the nonselective pigment epithelium of the retina."

In a later publication, Sperling (1979) sums up his conclusions as follows:

> We conclude that retinal damage from light in the visible region of the spectrum can be classified into three distinctly different kinds of lesions: (1) the thermal lesion characterized by nonselective absorption of energy in the melanin granules of the RPE and resulting in grossly observable damage; (2) the photochemical lesion described by Ham, which results from continuous exposure to energies of short wavelength light less intense than those which produce gross thermal lesions and primarily damaging the pigment epithelium; (3) the "color-blinding" lesion reported here which is wavelength selective, damaging blue-sensitive cones with blue light and green-sensitive cones with green ligtht and depends upon repeated, intermittent exposure.

Lawwill, first in the rabbit (1973) and later in the rhesus monkey (1977), provided evidence that extended blue light (514.5 nm originally and later 488 and 458 nm) exposure caused damage at power levels much below those needed for thermal effects. He used extremely large angle Maxwellian retinal exposures (50 and 102 degrees solid angle corresponding to 1.5 cm^2 and 2.5 cm^2 of retinal area), a 4-h exposure time, and electrophysiological (ERG) techniques plus LM and EM studies to assay the damage. His histological findings were nonspecific, in that damage was spread throughout the retinal layers. In his words, "Susceptibility varies throughout the retina, with higher thresholds in the center of the macula and around the optic disk. These findings suggest that visual pigment is not the sole mediator of retinal damage in chronic light exposure." Lawwill (1982) proposes three distinct pathophysiologic mechanisms of light damage to the retina. The first is visual pigment, rhodopsin-specific and occurs in rats and other nocturnal animals but is almost nonexsistent in primates. The second is cone pigment-specific and occurs in combination with the third mechanism, the blue light or short-wavelength effect. It is characterized by functional loss of specific cone populations and is overshadowed in the primate by short-wavelength light damage, i.e., mechanism three. This mechanism three is responsible for light damage in the primate in single exposures and has an action spectrum peaking in the short-wavelength (blue) end of the visible spectrum. It damages all layers of the retina from the RPE through the nerve fiber layer. In Lawwill's words, "The hypothesis is that the effect is caused by a direct action of light on the mitochondria in the different retinal layers, which inactivates the respiring enzymes, other effects being secondary to or additive with this major mechanism."

The lack of quantitative data on solar retinitis and eclipse blindness prompted Ham *et al.* (1973) to investigate this phenomenon in the rhesus monkey. By means of special filters and a 2500-W xenon lamp they simulated the solar spectrum with reasonable accuracy. The angular divergence of the radiation beam entering the eye of the anesthetized animal was adjusted to produce an image diameter corresponding to that of the solar disk on the human retina (159 μm). Retinal irradiancies required to inflict a minimal lesion visible with the fundus camera at 24 h postexposure were determined in 5 monkeys (10 eyes) for exposure times ranging from 1 to 180 s. The retinal irradiance ranged from 131 W cm^{-2} for a 1-s exposure to 18.9 W cm^{-2} for a 180-s exposure. Corresponding maximum temperatures at the RPE–photoreceptor interface as calculated from the solar mathematical model of White et al. (1971) ranged from 19°C for a 1-s exposure (131 W cm^{-2}) to 2.8°C for a 180-s exposure (18.9 W cm^{-2}). The authors concluded that "the distinction between thermal injury, thermally enhanced photochemical injury and photochemical injury becomes blurred as retinal irradiance is decreased below the so-called burn threshold while increasing the exposure duration." In this same paper the effectiveness of the full solar spectrum between wavelengths 400–1200 nm to inflict a minimal lesion was compared to that of a near infrared spectrum, 700–1200 nm. The ratio of near infrared spectral irradiance to the full solar spectral irradiance required to produce a minimal lesion ranged from a factor of 5 for a 180-s exposure to a factor of 2 for a 10-s exposure. This demonstrated conclusively that the short wavelengths in the solar spectrum were more effective than the near-infrared wavelengths in producing retinal damage. In a later and more elaborate investigation of solar retinopathy as a function of wavelength, Ham et al. (1980b) showed that solar retinitis was photochemical in nature, resulting from exposure to the short wavelengths in the solar spectrum and that the infrared component played a negligible role.

The series of investigations outlined above had made it abundantly clear by the early 1970s that prolonged exposure to white light produced a type of retinal damage that could not be explained in terms of thermal injury. The development of mathematical models of heat conduction in the retina made it possible to calculate time and space profiles of temperature with considerable accuracy so that it could be shown that retinal damage occurred at temperatures well below those needed to denature even the most sensitive macromolecules. Vos (1962), for example, had shown on theoretical grounds that the temperature rise in the retina associated with eclipse blindness was only 2°C above ambient. More sophisticated models of heat conduction in the retina were developed by a number of investigators, notably Mainster et al. (1970a, b) and White et al. (1971). Temperature predictions from these models received experimental confirmation from measurements with extremely small thermocouple probes developed by Welch and colleagues, Cain and Welch (1974), and Priebe, Cain, and Welch (1975). Temperatures measured in the retina agreed within 25% with theoretical predictions.

2. LIGHT TOXICITY AS FUNCTION OF WAVELENGTH

The first definitive study of photic damage in the retina as a function of wavelength was performed by Ham *et al.* (1976a) using the rhesus monkey as the experimental animal. They exposed the rhesus retina to 8 monochromatic laser lines extending from 1064 nm in the near infrared to 441 nm in the visible blue. Exposure times were 1, 16, 100, and 1000 s. The criterion for minimal damage was the appearance of a visible lesion as seen in the fundus camera at 48 h postexposure. Each laser beam was optically adjusted to produce a Gaussian distribution on the retina that was 500 μm in diameter at the $1/e^2$ points. The action spectrum for minimal retinal damage rose exponentially toward the short wavelengths in the visible spectrum as shown in Fig. 1, where the reciprocal of retinal irradiance required to produce a minimal lesion in a given exposure time is plotted against wavelength in nanometers. The corneal power required to inflict a minimal lesion with 1064-nm radiation was three orders of magnitude greater than for blue light at 441 nm when the exposure time was extended to 1000 s. During irradiation the calculated maximum temperature in the retina was 23°C for the infrared beam and less than 0.1°C for the blue light beam. It was obvious that thermal injury resulted from near infrared exposure while the blue light lesion was caused by some type or types of photochemical damage. These two types of retinal lesions not only differ in the basic mechanisms producing them but lead to entirely different biological effects, which are distinguishable both *in vivo* with the fundus camera and histologically with the light microscope (Ruffolo *et al.*, 1981).

In another publication Ham *et al.* (1980a) summarized their conclusions on retinal damage as follows:

> There are at least three types of radiation insult in the spectral range 400–1400 nm (Ham, 1976b). These are: *mechanical disruption* of retinal structure resulting from sonic transients or, at very high irradiance levels, shock waves engendered by extremely short pulses of radiation which are absorbed in the RPE and choroid (Ham *et al.*, 1974); *thermal insult* (independent of wavelength to a first approximation) resulting from absorption of energy in the RPE and choroid sufficient to produce temperatures greater than 10°C above ambient in the RPE, neural retina and choroid; *actinic insult* from the photochemical effects of extended exposure to the short wavelengths in the visible spectrum (400–550 nm) at irradiance levels too low to produce temperatures of more than a few degrees Celsius above ambient.

Sperling (1979) has discovered a fourth type of photochemical damage caused by repetitive exposures to blue light which stress the blue photoreceptors resulting in blue blindness. For completeness, a fifth type of photochemical damage, the Noell albino rat syndrome (Lawwill's

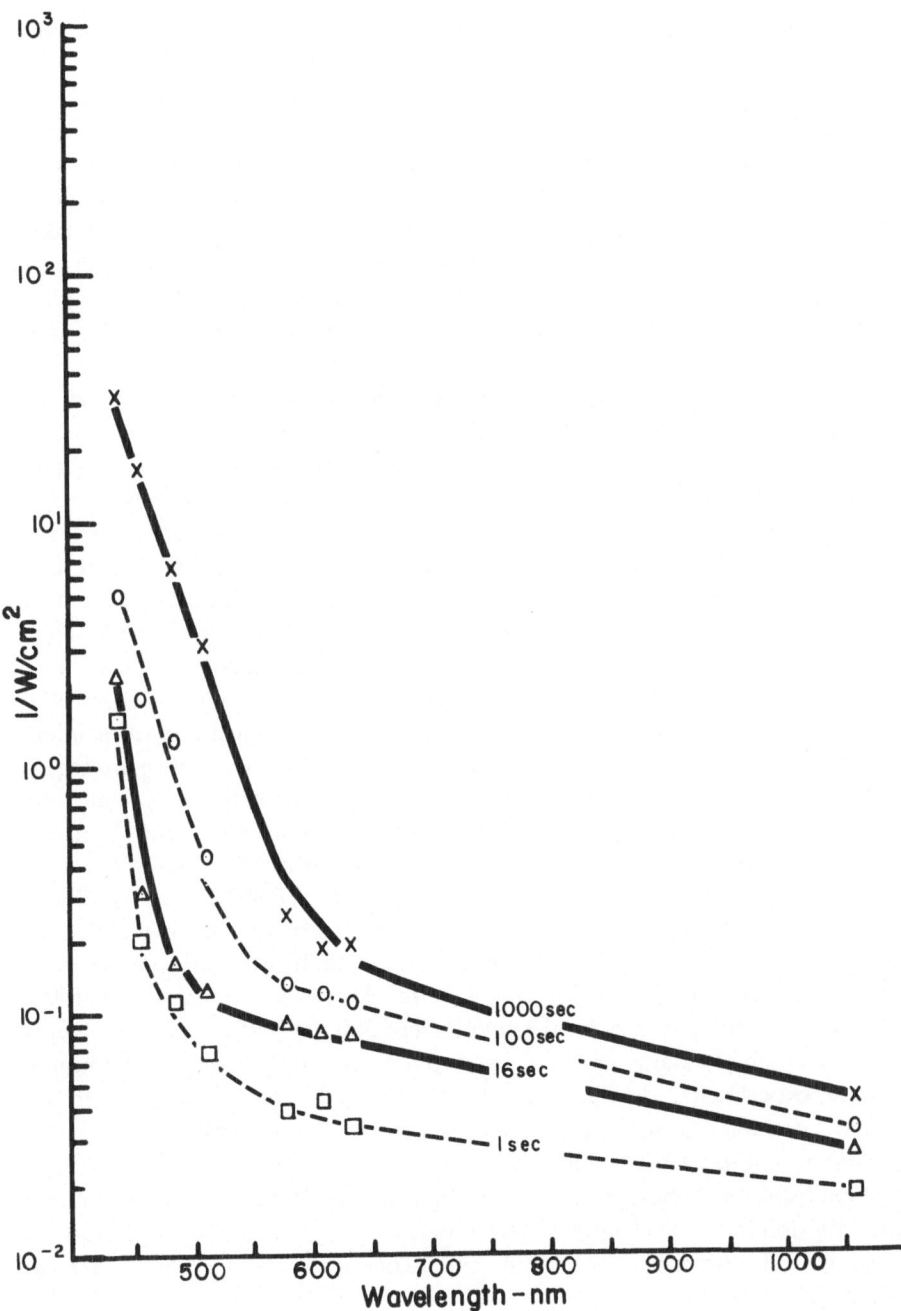

FIGURE 1. Action spectrum for sensitivity of rhesus retina to photic damage from eight laser lines between 441 and 1064 nm. The reciprocal of retinal irradiance in W cm^{-2} needed to produce a minimal lesion in 1, 16, 100, and 1000 s is plotted against wavelength in nm. Diameter of retinal spot size is 500 μm to the 1/e^2 points of Gaussian distribution. □, 1 s; △, 16 s; ○, 100 s; ×, 1000 s.

mechanism one) should be mentioned. Sykes *et al.* (1981), have shown this type of damage in primates, as will be discussed later.

Power level, wavelength, and exposure time are the important parameters determining the type of damage. There is no sharp demarcation between these types of retinal injury. Nonlinear phenomena associated with picosecond pulses of mode-locked laser radiation merge into thermal effects as exposure times approach the microsecond range for Q-switched laser pulses. Only power level and exposure duration determine whether the damage is mechanical or thermal in nature. Wavelength is relatively unimportant except insofar as transmittance through the ocular media and absorption by the RPE and the choroid are concerned. Rate of delivery and amount of energy absorbed are the dominant factors. As irradiance on the retina is further reduced and exposure duration extended, a point is reached where thermal effects become minimal or even completely negligible and wavelength becomes the dominant factor for photochemical effects.

In another study Ham and colleagues (unpublished) compared the efficiency of short versus long wavelengths of light to produce minimal retinal lesions in the rhesus monkey as detected with the fundus camera at 48 h after exposure to cw coherent and incoherent optical sources. They compared the retinal response of seven nearly monochromatic laser lines ranging from 441 to 632.8 nm with the incoherent light from a 2500-W xenon optical source using sharp cutoff filters and 80-nm bandwidth interference filters. The three types of radiation exposure are shown in Fig. 2, where the radiant exposure in J cm^{-2} is plotted logarithmically along the ordinate versus wavelength in nanometers along the abscissa. The sharp cut filter data designated by squares represent bandwidths of 435–735, 455–735, 485–735, 515–735, 545–735, 575–735, 625–735, and 675–735 nm; the 80-nm bandwidth data represented by the ×'s were peaked at 450, 500, 550, 600, 650, and 700 nm; the laser wavelengths denoted by the solid circles were 441, 458, 488, 514, 580, 610, and 633 nm. All exposure times were 100 s. The incoherent spot sizes on the retina were 500 μm and the laser spot sizes were 500 μm in diameter to the $1/e^2$ points of the Gaussian distribution. These experiments demonstrate clearly the influence of wavelength on photic damage to the retina and indicate that the biological effects of coherent and incoherent light are entirely similar. The nature of the damage ranges from pure photochemical at the shortest wavelengths to pure thermal at the longest wavelengths. The intermediate wavelengths produce a mixture of thermally enhanced photochemical damage and thermal damage.

Retinal thresholds in the rhesus monkey at wavelengths beyond 600 nm have been obtained by Ham *et al.* (1984b). Using a 2500-W xenon lamp with interference filters, the threshold radiant exposures in J cm^{-2} were

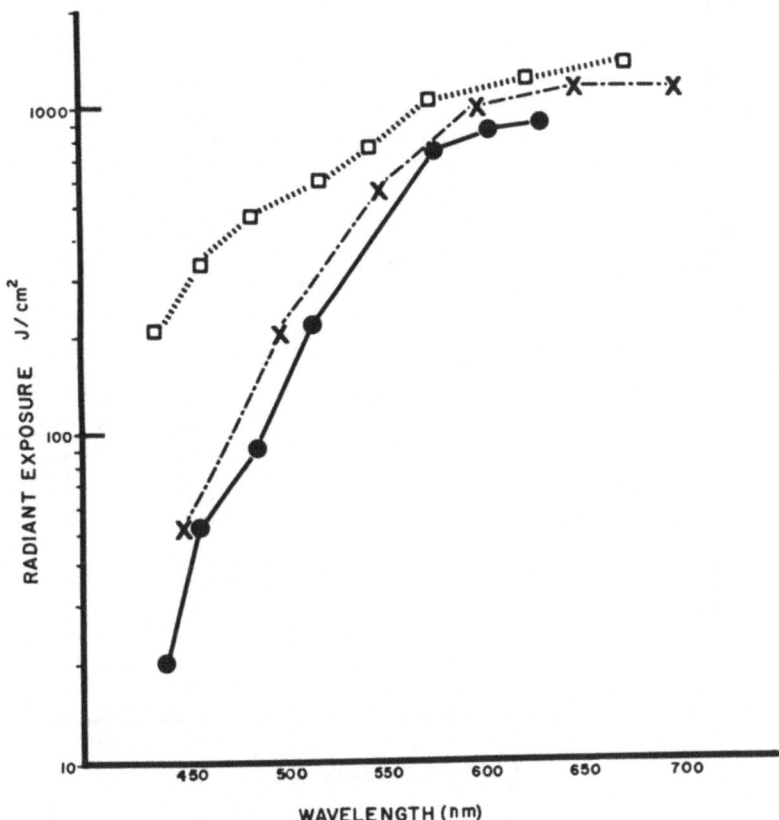

FIGURE 2. The radiant exposure in J cm^{-2} needed to produce a minimal retinal lesion for a 100-s exposure is plotted against wavelength in nanometers. □, sharp-cut filters 435–735, 455–735, 485–735, 515–735, 545–735, 575–735, 625–735, and 675–735 nm; ×, 80-nm bandwith filters 450 ± 40, 500 ± 40, 550 ± 40, 600 ± 40, 650 ± 40, and 700 ± 40 nm; ●, laser lines 441, 458, 488, 514, 580, 610, and 633 nm. Retinal spot diameter for laser lines was 500 μm to the $1/e^2$ points of Gaussian distribution. For the sharp-cut and 80-nm filters the retinal spot diameter was 500 μm as produced by a collimated xenon optical source.

determined for wavelengths 820 ± 5 nm, 860 ± 5 nm, and 910 ± 25 nm. Exposure times ranged from 1 to 1000 s and image diameters on the retina were 500 μm. No significant difference in threshold was noted for these wavelengths. In Fig. 3 these near infrared thresholds are compared with similar data previously obtained for laser wavelengths 1064 nm (Nd:YAG), 647 nm (Ar–Kr), and 632.8 nm (He–Ne). Radiant exposures in J cm^{-2} are plotted logarithmically against exposure time in seconds. All three lines are straight and approximately parallel, indicating a similar type or mechanism of injury for wavelengths greater than 600 nm. The type of injury is thermal

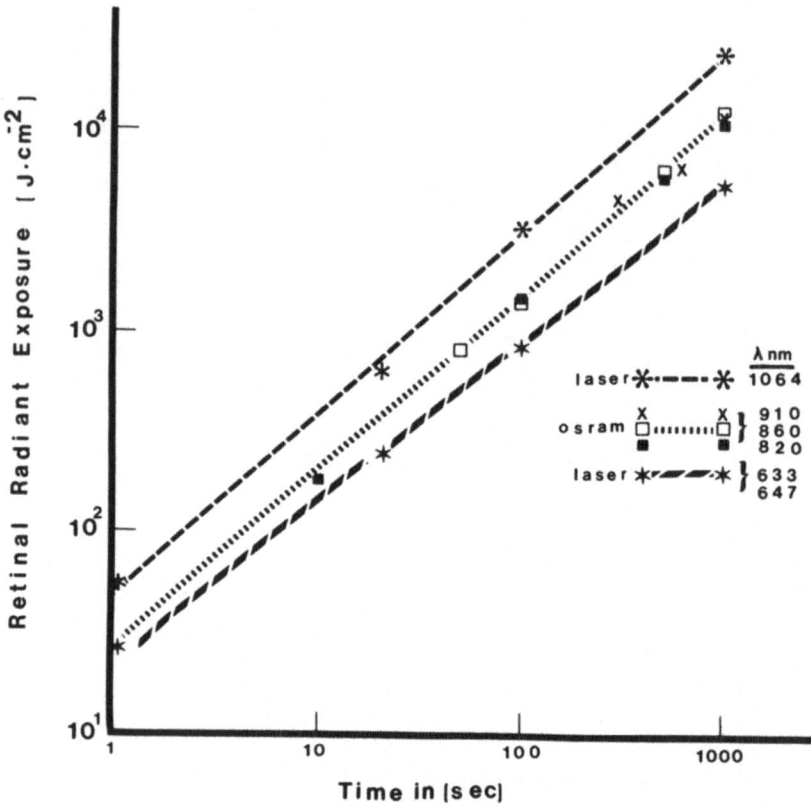

FIGURE 3. Retinal thresholds (J cm^{-2}) in the rhesus monkey for infrared wavelengths 820 ± 5 nm, 860 ± 5 nm, and 910 ± 25 nm as produced by a 2500-W xenon lamp and inter- ference filters are compared with similar data obtained with Nd:YAG, Ar–Kr, and He–Ne lasers at wavelengths 1064, 647, and 632.8 nm, respectively. Retinal spot diameter for the xenon lamp data is 500 μm, whereas the laser data are 500 μm in diameter to the $1/e^2$ points of the Gaussian distribution. Threshold radiant exposure in J·cm^{-2} versus exposure time in s is plotted in log–log units.

in nature as verified by funduscopy and histological analysis (Ruffolo *et al.*, 1981). Histological examination at 24 and 48 h postexposure discloses struc- tural damage in the RPE and numerous pyknotic nuclei in the outer nuclear layer. Minimal thermal lesions are always smaller than the image diameter on the retina because temperature is maximal at the center of the irradiated image. The damage is maximal at the center of the lesion, tapering off toward the periphery. This is in sharp contrast to minimal photochemical lesions, where damage is fairly uniform across the lesion with a definite border between injured and uninjured RPE cells at the periphery and only a few pyknotic nuclei in the outer nuclear layer.

The realization that extended exposure to short wavelength light at power levels well below the burn threshold can cause retinal injury has brought forth new theories concerning the etiology of retinal diseases (Young, 1981, 1982, 1986, 1987), has changed the ANSI safety standards for lasers (ANSI, Z 136.1, 1986), and has alerted the clinical ophthalmologist to the possibly damaging effects from the excessive use of indirect

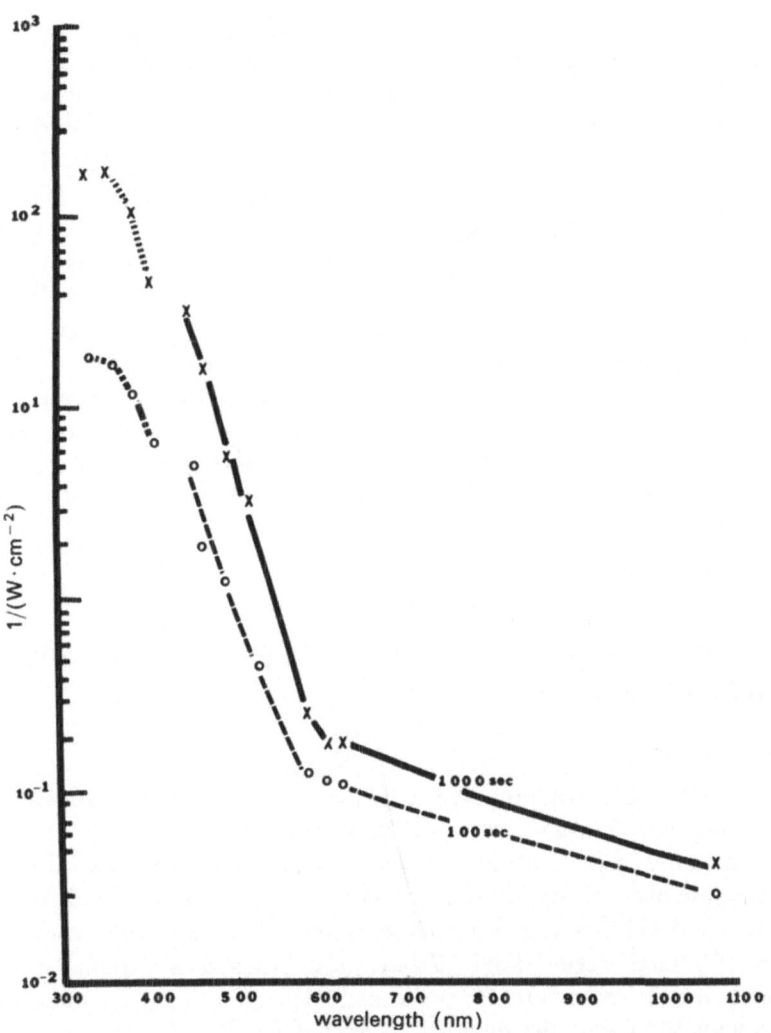

FIGURE 4. Action spectrum for sensitivity of rhesus retina to minimal photic damage from near-UV radiation for 10-nm bandwidths at 405, 380, 350, and 325 nm. The reciprocal of retinal irradiance in W cm^{-2} needed to produce a minimal lesion in 100 and 1000 s is plotted for comparison on the same graph with eight laser lines taken from Fig. 1. ○, 100 s; ×, 1000 s.

ophthalmoscopy, fluorescein angiogaphy, fundus photography, and the operating microscope during eye surgery, all of which procedures subject the retina to high levels of blue light. Several authors have emphasized the retinal hazard to aphakics from blue light and near ultraviolet radiation: Ham *et al.* (1976a, 1978, 1979, 1980a,b, 1981, 1982a,b, 1983), Mainster (1978a, 1978b, 1983), Blodi (1982), Zigman (1978), Lerman (1980), and Young (1981, 1982, 1986). Removal of the lens increases the amount of blue light and the near-UV radiation impinging on the retina.

In another study, Ham *et al.* (1981, 1982a) demonstrated that the rhesus retina is more sensitive to near UV than to 441-nm light by a factor of six. These authors exposed the aphakic eye in 3 monkeys to wavelengths of 405, 380, 350, and 325 nm as produced by a 2500-W xenon optical source with quartz optics through 10-nm bandpass interference filters. Exposure durations were 100 and 1000 s and the retinal spot size was 500 μm in diameter. Figure 4 is a replication of Fig. 1 where the near UV data for 100 and 1000 s are plotted on the same graph for comparison with the eight laser lines in the visible and near infrared regions of the spectrum. It can be seen that retinal sensitivity to photic damage continues to increase with decreasing wavelength. The radiant exposure required to inflict a minimal retinal lesion with blue light (441 nm) was $30 \, \text{J cm}^{-2}$ as compared to $5 \, \text{J cm}^{-2}$ for 350- and 325-nm UV radiation. The histology of the UV lesions will be discussed more thoroughly in Section 8. These findings support the previous forebodings of several authors as to the retinal hazard in the aphakic eye.

3. AGING AND DEGENERATIVE EFFECTS OF CHRONIC EXPOSURE TO LIGHT

Young (1981) has proposed a theory of central retinal disease based upon long-term chronic exposure of the fovea to light. In his words, "Perhaps some of the entities in the current .nosology of retinal disease are diseases provoked or aggravated by light..." and "Analysis reveals that it is possible to develop a theory which accounts for the pathogenesis of many forms of retinal degenerative disease, and provides a rational basis for prevention and treatment." He proceeds to develop a set of theoretical principles with empirical laws determined experimentally. He then cites a large number of empirical facts that are substantiated by numerous references to the literature. It is a convincing thesis, well documented and augmented by scholarly research. Particularly impressive is the relationship between long-term radiation exposure and degenerative or aging effects on the retina; these correlate well with the observations of Smith (1944), Ham *et al.* (1978), and Tso and Fine (1979) concerning hypopigmentation of the RPE, serous detachment of the RPE, macular edema, and age-related macular degeneration (AMD).

The concept that long-term, chronic exposure to sunlight is a contributing factor to aging of the retina and age-related macular degeneration (AMD), is not new, as Young (1986) has pointed out. Van der Hoeve (1920) supported a similar thesis. The aging of the retina, particularly in the macular region, is intimately associated with the gradual accumulation of debris (residual bodies, lipofuscin, melanolysosomes, and melanolipofuscin) in the RPE (Feeney-Burns, 1983, 1984; Ham, 1986a). Young (1987) has reviewed the clinical and histopathological features of AMD. In his words,

> Many of the features of AMD can be attributed to the progressive deterioration of the retinal pigment epithelium. It is proposed that this deterioration arises primarily from imperfections in metabolic processes concerned with intracellular renewal—the incessant replacement of the cells' molecular constituents. In particular, inefficiency of the cells' digestive apparatus seems to play a primary role in setting off a complex sequence of events involving residual bodies, alteration of Bruch's membrane, drusen, basal laminar deposits and pigmentary disturbances. Two major characteristics of AMD which remain unexplained by this hypothesis—the central location of the lesion and the protective effect of ocular pigmentation—can be accounted for by postulating a role for the damaging effects of radiant energy in the etiology of AMD.

The Division of Risk Assessment, National Center for Devices and Radiological Health of the Food and Drug Administration sponsored a Workshop on "Long-Term Visual Health and Optical Radiation" in September 1983. Seven working groups were established to identify and define specific long-term visual problems that could be induced by optical radiation. The Retinal Pigment Epithelial Working Group specifically considered aging and its possible relationship to light damage to the retina (Ham, 1986a). There is convincing evidence suggesting that both acute exposure (eclipse blindness and sun gazing) and long-term chronic exposure to sunlight causes a photochemical type of maculopathy in both humans and nonhuman primates that is closely related, if not identical, to the blue light lesion. Whether long-term exposure to short wavelength light and near ultraviolet radiation can be related to aging of the RPE and AMD remains questionable, but there are several reasons for believing that such a relationship exists. For example, a significant feature of blue light damage in the primate retina is the loss of melanin granules, suggesting an increase in the formation of complex granules (melanolysosomes and melanolipofuscin) and an acceleration of the depigmentation process that accompanies aging. It makes sense to postulate that light exposure increases the phagocytic role of the RPE, producing more lipofuscin and complex granules with extrusion of debris onto Bruch's membrane. For example, in the rhesus monkey each retinal rod produces 80 to 90 disks per day; the OS is replaced every 9 to 13 days. Thus, each RPE cell must phagocytize and digest approximately 3000 disks per day (Young, 1971). As Feeney (1973) has emphasized, "The pigment epithelial cell must have a highly developed phagocytic-lysomal system in order to digest these enormous amounts of exogenous material daily for 70 or more years." Again in Young's words,

> The senescent changes are centered on the macula, where the intensity of radiation appears to be the greatest. Accumulation of lipofuscin in the pigment epithelium begins in childhood. Several decades later, the cells are filled to overflowing with the remnants of failing molecular renewal, despite abortive attempts to clear the cytoplasm of debris by extruding it onto Bruch's membrane (Young, 1982).

While the evidence that chronic blue light exposure is a contributing agent in the aging and degeneration of the macula is not conclusive, it is suggestive enough to warrant recommending that the public wear protective yellow filters or sunglasses during exposure to bright light and near ultraviolet optical radiation. Such filters are harmless and may slow down the aging process in both the retina and the lens.

Wing *et al.* (1982) have demonstrated an inverse relationship between melanin and lipofuscin in the aging RPE of human subjects, as has also been shown by Feeny-Burns *et al.* (1980, 1984). The amount of melanin in the RPE appears to be constant in young eyes, and progressively decreases in old eyes, whereas lipofuscin accumulates with age. The amount of lipofuscin in the RPE of elderly persons may exceed that of melanin by a ratio 5–10 to 1 (Feeney-Burns *et al.*, 1984). These findings have implications as to the role of melanin in lipofuscin accumulation and its possible association with macular degeneration. Weiter *et al.* (1985, 1986) have investigated the relationship of senile macular degeneration* to ocular pigmentation in 650 white patients with senile macular degeneration to a control group of 363 patients without symptoms of macular degeneration. Significantly, 494 of the patients with senile macular degeneration (76%) had light-colored irides compared to 145 in the controls (40%). Fundus pigmentation corresponded closely with iris pigmentation. Furthermore, there was a tendency for individuals with lightly pigmented irides to have senile macular

---

* "Age-related" has been substituted for "senile" in the disorder commonly known as senile macular degeneration because of the pejorative connotation of the word "senile." Age-related macular degeneration (AMD) is used throughout this review except where other authors are quoted.

degeneration at an earlier age than those with dark irides. These authors conclude with the statement, "The strong association between ocular melanin and both senile macular degeneration and lipofuscin in the retinal pigment epithelium suggests a role for light damage in the eye and offers possibilities for research to prevent this important cause of blindness in our population."

A major thesis in Young's (1986) publication "Solar Radiation and Age-Related Macular Degeneration" is the protective role of melanin pigmentation against light damage to the retina. For example, blacks have more melanin in most ocular tissues than whites, especially in the iris and choroid, but not in the RPE (Geeraets et al., 1962; Weiter et al., 1985, 1986). Young points out that the prevalence of AMD is much higher in whites than in blacks. The same is true for whites versus Japanese. There is also epidemiological evidence for a relationship between macular degeneration and ocular pigmentation (Hyman et al., 1983). Blue eyes are more prone to AMD than those with dark pigmentation. The syndrome of "camp eyes" (Dekking, 1947; Churchill, 1945), previously described, is another example of the protective role of melanin pigmentation against damage. Weiter et al. (1986) have shown that while the melanin content in the RPE is similar for whites and blacks, the accumulation of lipofuscin in the RPE is much greater in whites than in blacks. This provides another example of the protective role of melanin against light damage. In an ever-increasing elderly world population, the loss of melanin (depigmentation of the RPE) with age increases the susceptibility of the retina to light damage. As will be shown, a major effect of blue light exposure is the depigmentation of the RPE. Thus, excessive light exposure can accelerate the aging process, and the aging retina becomes more and more susceptible to further light damage because of the loss of melanin.

In addition to Young's theory of radiation damage to the retina, a number of papers have emphasized the retinal hazard from extended or chronic exposure to white light sources containing appreciable amounts of blue light. A workshop of ocular safety and eye care was held at the Duke Eye Center in October 1978 under the auspices of the National Research Council, Committee on Vision, Wolbarsht (1980). Members of the workshop explored potential ocular hazards from radiation emitted by ophthalmic instruments currently in use. It was recommended that

> in general, instruments be designed to minimize the amount of ultraviolet and infrared radiation. Especially, chronic exposure to emission in the blue end of the spectrum should be reduced as far as possible to avoid photochemical damage to the retina. This "blue light" retinal hazard of any instrumentation may be evaluated by using guidelines as the proposed Threshold Limit Values (TLVs) of the American Conference of Governmental Industrial Hygienists (ACGIH, 1979).

A table is given that estimates the retinal hazard of chronic exposure to different wavelengths relative to 435–440 nm, judged to be the most dangerous visible wavelengths.

4. REPETITIVE EXPOSURES TO BLUE LIGHT

Greiss and Blankenstein (1981) have shown in the rhesus monkey that repeated subthreshold exposures to blue light produce cumulative retinal injury, which is countered by an exponential repair process. Retinae were exposed to 458-nm blue light from an argon-ion laser at a radiant exposure equivalent to one-half that required to produce a minimal lesion. At time intervals of 1, 2, 3, 4, and 6 days, the same sites were reexposed to determine the split radiant exposure threshold, which is related to the single radiant exposure threshold by an additivity constant A, which in turn depends upon the time interval between exposures. They found, for example, that the additivity A was 91% at 1 day, 57% at 2 days, and only 23% at 6 days. The parameter A could be expressed as an exponential $A = \exp(-\Delta t/\tau)$ where Δt is the interval between exposures and $1/\tau$ is the repair rate constant, tau (τ), which turned out to be almost exactly 4 days. The 4-day time constant helps to explain the split radiant exposures of Ham *et al.* (1979) and Lawwill *et al.* (1977). While the basic mechanisms producing photochemical or actinic damage as well as the repair mechanisms are unknown, the authors have described the molecular dynamics by a rate equation that under certain assumptions can be solved for repetitive exposures when the radiant threshold exposure for a single exposure is known. This research represents a valuable addition to our very limited knowledge concerning the cumulative nature of photochemical or actinic damage to the retina.

The cumulative or additive effect of repetitive light exposures to the same site on the macaque retina has also been investigated by Ham and Mueller (1986b). Their research protocol differed somewhat from that of Greiss and Blankenstein. Instead of using the "split-dose" technique, the retinae of 3 monkeys (3 eyes) were subjected to daily radiant exposures of 1000-s duration and 500-μm spot diameter for 21 consecutive days at each of three wavelengths, 440, 475, and 533 nm. The optical source was a 2500-W xenon lamp equipped with quartz optics and 10-nm interference filters peaked at 440, 475, and 533 nm. Initially a threshold radiant exposure in J cm^{-2} was determined in the other eye of each animal using the same parameters as above, i.e., exposure time, spot size, and wavelength. The criterion for a threshold from a single exposure was the appearance of a minimal lesion at 48 h postexposure as seen with the fundus camera. The other eye of each animal was used for repetitive daily exposures to the same site at 50%, 40%, 30%, 20%, and 10% of threshold for a single exposure at each wavelength. In a given eye, accordingly, there were 5 retinal sites for each wavelength or a total of 15 retinal sites. All retinal sites were in the paramacular area. They consisted of a parallel, horizontal row for each wavelength, either above of below the macula, and these were varied in each animal.

Results are shown in Table I, where the daily repetitive radiant exposures in J cm^{-2} are listed for each wavelength according to the number of exposures required to produce a minimal threshold lesion at 24 h postexposure. At 440 nm, three animals underwent the repetitive exposure protocol. The c monkey was removed from the experiments after 5 days because of a corneal opacity from an unfortunate blunder by the animal caretaker, but the other two monkeys received 20 and 21 repetitive exposures, respectively. In the a monkey 17 exposures to 6 J cm^{-2} produced

TABLE I. Radiant Exposure in J cm^{-2} per Exposure for Wavelengths 440, 475, and 533 nm versus Number of Exposures Required to Produce a Minimal Lesion in the Macaque Retina for Three Eyes in Animals Designated a, b, and c

No. of exposures	J cm^{-2}/Exp. 440 nm			J cm^{-2}/Exp. 475 nm		J cm^{-2}/Exp. 533 nm	
1	30a;	28b;	28c	94.6a;	92.4b	300a;	294b
2	15a;	14b;	14c				
3	12a;	11b;	11c	47.3a;	46.2b		
4				37.8a;	37.0b		147b
5	9a;	8.5b;	8.5c			150a	
6					27.7b		118b
7				28.4a		120a	
8							
9							
10							
11							
12							88.1b
13						90a	
14		5.6b					
15							
16							
17	6a						
18							
19							
20		2.8b					
21							

a lesion, but 21 exposures to $3 \, \mathrm{J \, cm^{-2}}$ did not. The b monkey showed threshold lesions after 14 exposures to $5.6 \, \mathrm{J \, cm^{-2}}$ and after 20 exposures to $2.8 \, \mathrm{J \, cm^{-2}}$. This demonstrates that the cumulative effect of daily exposures to subthreshold amounts of 440-nm light can damage the primate retina and suggests that some of the photochemical effects of light toxicity are irreversible even at the subthreshold level. The animals exposed to 475- and 533-nm light did not develop visible lesions for repetitive daily exposures at levels less than 30% of threshold of the single exposure, even though they received 21 daily exposures to approximately 20 and $60 \, \mathrm{J \, cm^{-2}}$ or total radiant exposures of 420 and $1260 \, \mathrm{J \, cm^{-2}}$, respectively. From Table I it can be seen that in monkey a, 3 exposures to $12 \, \mathrm{J \, cm^{-2}}$ of 440-nm light produced a lesion, while 3 exposures to $47.3 \, \mathrm{J \, cm^{-2}}$ of 475-nm light were required to produce a lesion; similarly, monkey b required 3 exposures to 11 and $46.2 \, \mathrm{J \, cm^{-2}}$, respectively, for these two wavelengths. Wavelength 440 nm is about 4 times more toxic than 475 nm. When 440-nm light is compared to 533 nm the toxicity ratio is about 17. However, this type of comparison is not entirely valid since different mechanisms are involved at different wavelengths. Nevertheless, these experiments illustrate the extreme sensitivity of the primate retina to the blue end of the visible spectrum.

5. CONTINUOUS EXPOSURE OF RHESUS MONKEY TO FLUORESCENT LIGHT

Sykes *et al.* (1981) exposed the eyes of rhesus monkeys (6 to 7 years of age and restrained in a chair) to daylight fluorescent (400–700 nm) irradiation ranging from 8.6 to $2 \, \mathrm{mW \, cm^{-2}}$ at the cornea. The monkey's head was completely surrounded by a triple bank of circular fluorescent lights. One eye of each animal was patched to serve as a control but both pupils were fully dilated (>8 mm) during exposure. After sacrifice at 15 h postexposure photic damage was evaluated by LM and EM. They found that a single 12-h exposure to $8.6 \, \mathrm{mW \, cm^{-2}}$ was well above the threshold for minimal damage which was restricted to the photoreceptor cells; the RPE was not damaged. Outer segments were twisted and swollen and disk membranes of both rods and cones were irregularly arranged and vesicularized. The paramacular areas were less sensitive to damage than macular areas of the same animal. Here again the entire rod and cone population was maintaned in a state of excessive bleach for 12 h and the initial damage was restricted to the photoreceptor cells. This experiment simulates in the primate the albino rat experiments of Noell, Kuwabara and

Gorn, and others and shows that under rather extraordinary conditions of exposure, even diurnal primates can undergo a similar type of damage. These results also clearly define the difference between the Noell syndrome and the blue light phenenomenon, i.e., the initial damage occurs in the photoreceptors in the former and in the RPE in the latter.

At first thought it would seem that the conditions employed to expose the monkey's eyes (complete close surround by fluorescent light, extreme pupillary dilation > 8 mm, and 12 h continuous exposure) in the Sykes experiment would result in an intense radiant exposure of the retina. Sliney (1980b) has derived a simple equation for calculating retinal irradiance when the radiance L in $W\ cm^{-2}\ sr^{-1}$ of the optical source is known:

$$E_r = \frac{\pi}{4} LT(d_p^2/f^2)$$

E_r is the irradiance on the retina in $W\ cm^{-2}$, T is the transmittance through the ocular media, d_p is the pupil diameter in centimeters, and f is the focal length of the emmetropic eye (1.7 cm humans, 1.35 cm rhesus monkey, 1.0 cm rabbit, and about 0.3 cm in the rat). The advantage of this equation is that the image size on the retina need not be known. On the assumption that the monkey's vision encompassed 90° horizontally and 90° vertically (rather maximal conditions) the maximum radiance of the source was about $2.74 \times 10^{-3}\ W\ cm^{-2}\ sr^{-1}$. For this radiance, a transmittance of 0.8, a pupil diameter of 0.8 cm, and a focal length of 1.35 cm, the retinal irradiance was $E_r = 9.43 \times 10^{-4}\ W\ cm^{-2}$, a value somewhat above that calculated by the authors. Multiplying this irradiance by $12 \times 3600\ s$ results in a radiant exposure of $41\ J\ cm^{-2}$, a value well below the threshold for blue light damage as determined for monochromatic wavelengths 514.5, 488, and 458 nm (Ham et al., 1976a). Actually, this irradiance on the retina $9.43 \times 10^{-4}\ W\ cm^{-2}$, is small in comparison to that resulting from gazing at the sun (18 W cm^{-2} at sea level with a 3-mm pupil) (Ham et al., 1973). The difference, of course, resides in the small image diameter of the sun on the human retina compared with the very large proportion of the retina irradiated in the experiments of Noell, Lawwill, Sykes, and others. Another interesting feature of this equation is that the retinal irradiance is inversely proportional to the square of the focal length of the eye. Thus, for example, under identical exposure conditions, a rat with a 0.3-cm focal length receives approximately 32 times more retinal irradiance than a human eye with a focal length of 1.7 cm and 20 times the retinal irradiance of the rhesus monkey with a focal length of 1.35 cm. This is one of the reasons why the retina of the rat is so sensitive to environmental light damage.

6. POTENTIAL EYE HAZARD FROM OPHTHALMIC INSTRUMENTATION

The workshop on ocular safety and eye care held at the Duke Eye Center in October 1978 has been followed by a number of important papers dealing with the potential eye hazards inherent in current ophthalmological clinical practice. Fuller *et al.* (1978) exposed the retinae of owl monkeys to a white light optic probe similar to those used for illumination during human pars plana vitrectomy. The light source was a 150-W halogenated tungsten projector bulb (3,300 K). At a 2-mm distance from the retina the spot size was 3.5 mm in diameter and they calculated the retinal irradiance to be 220 mW cm^{-2}. Thirty minutes of light exposure (396 J cm^{-2}) produced massive damage to both RPE and the photoreceptors. The threshold for ophthalmoscopically visible lesions was 15 min of exposure (198 J cm^{-2}), and microscopic changes were noted after 10 min (132 J cm^{-2}). Exposures less than 10 min damaged the neural retina but not the RPE. Removing wavelengths greater than 700 nm did not protect the retina. The earliest changes in threshold lesions (10-min exposures) occurred in the outer retina rather than in the RPE. This finding suggested that absorption of visible light by the photopigments in the outer segments (OS) is the initial event in the light injury cycle and that the damage is photochemical in nature. However, this tentative conclusion may not hold for diurnal primates, since the owl monkey is a nocturnal animal. Fluorescein angiography disclosed extensive leakage at the RPE and outer retina for animals exposed 30 and 20 min. However, there was no leakage in the retinal vasculature. This supports Tso's findings that the blood–retinal barrier is more vulnerable to photic damage at the RPE than at the retinal vasculature (Tso and Shih, 1977). Although the authors did not filter out wavelengths between 400 and 500 nm, they believe this would have been more helpful in protecting the retina than removing the infrared (>700 nm). In their opinion retinal damage caused by intraocular fiber optic light is primarily a photic mechanism and damage to the RPE may be secondary to neural retinal damage.

Tso and Shih (1977) have investigated the pathology of macular edema following lens extraction, the so-called Irvine–Gass–Norton syndrome (Irvine, 1953; Gass and Norton, 1966; Gass, 1977; Norton *et al.*, 1975). Tso and Shih developed an experimental model in the rhesus monkey. Lenses were extracted from 7 eyes of 4 adult monkeys. One eye underwent extracapsular lens extraction with vitreous loss, 2 eyes had intracapsular extraction with vitreous loss, and 4 eyes underwent uneventful intracapsular lens extraction. All fundi were examined by ophthalmoscopy, fundus photography, and fluorescein angiography before and after surgery and immediately before sacrifice. Eyes were enucleated at 6 h and 2, 3, 5, 10, 15, and 30 days postextraction. Thirty minutes prior to sacrifice each animal received an i.v. injection of horseradish peroxidase. Tissues were fixed and processed for localization of horseradish peroxidase. Both LM and EM sections were processed for analysis. In the macular region, the blood–retinal barrier at the RPE was broken in each of the 7 eyes while the blood–retinal barrier at the retinal vasculature was disrupted in only 3 eyes, all of which had vitreous loss during surgery. Fluorescein angiography detected no leakage for the first 30 days postsurgery in either the retinal or the choroidal circulation. However, horseradish peroxidase was found both intracellularly and extracellularly in the macular area but not in the peripheral retina. Since all eyes developed disruption of the blood–retinal barrier at the RPE, but only 3 showed similar disruption in the retinal vasculature, these results suggest that the blood–retinal barrier at the RPE is more vulnerable than that of the retinal vasculature. The authors also performed trabeculectomy on rhesus monkeys and found similar disruption of the blood–retinal barrier in both RPE and retinal vasculature. In their words, "These findings supported the concept that the occurrence of macula edema after intraocular surgery is not unique to lens extraction, but rather a more general phenomenon that occurs after various forms of intraocular surgery." While pointing out that the pathogenesis of the Irvine–Gass–Norton syndrome is unknown, the authors do not suggest that excessive exposure of the retina to light during cataract surgery might be a contributing cause.

However, Henry *et al.* (1977) noted an apparent increase in the incidence of chronic cystic maculopathy following cataract surgery in their practice, which they attributed to an increased amount and duration of exposure to operating room lights. They recommend placing a small opaque button on the cornea during that part of the operation that does not involve the extraction procedure and reducing the amount of illumination as much as practicable. Messner *et al.* (1978) exposed newborn stump-tail monkeys continuously to 400 ft-cd of cool, white, fluorescent light from a clinical phototherapy lamp for periods of 12 h, 24 h, 3 days, and 7 days. The monkeys were restrained in a supine position facing the lamp. One eye was covered as a control while the other eye was uncovered and allowed to blink at will. Irradiance at the cornea was 1.23 mW cm^{-2}. The animals were sacrificed immediately postexposure, eyes enucleated and prepared for LM and EM by standard techniques. No pathology was noted in the control eyes. In the exposed eyes, damage to the rod and cone receptors and the RPE was progressive with exposure time. This experiment, like that of Sykes *et al.* (1981), comes close to simulating in the primate the albino rat experiments of Noell (1966) and Kuwabara and Gorn (1968). When the entire retina, rather than a limited portion, is exposed for an extended time to short-wavelength light as in this experiment and in those of Lawwill (1976, 1977), the photopathology seems to involve primarily the photoreceptors rather than the RPE. It may be that when the entire rod and cone population is maintained in a constant state of bleach for long periods of time, damage appears first in the photoreceptor cells and particularly in the outer segments which contain the photopigments. For example, under these conditions of constant intensive bleach the concentration of retinal in the outer segments would increase. Delmelle (1978) has shown that when exposed to light retinal acts as a sensitizer to produce singlet oxygen, which in turn generates membrane damage in model systems (liposomes). Oxygen is very soluble in a lipid environment. The combination of retinal, oxygen, and light exposure in the outer segments would be expected to produce singlet oxygen, a very reactive molecule, especially for the lipid peroxidation of polyunsaturated fatty acids (PUFA), which are the major components of the outer segment disks and plasma membrane. This hypothesis receives additional support from the well-known fact that vitamin A deficiency is a protective factor from light damage caused by prolonged environmental exposure (Noell *et al.*, 1971).

Calkins *et al.* (1979, 1980a, b) and Hochheimer *et al.* (1979), in a series of studies, have explored the potential hazards to the retina of ophthalmological instrumentation employing white light sources. These include the direct and indirect ophthalmoscope, biomicroscopy of the fundus with the slit lamp using a plano contact lens, overhead surgical lamps, and the operating microscope. All of these sources provide white light with a substantial component of blue wavelengths. Although standards are yet to be developed for white light incoherent optical sources, the authors demonstrate that the established standards for coherent laser sources by the American National Standards Institute (ANSI Z 136.1, 1980) can be used conveniently and with considerable accuracy to define maximum permissible exposures (MPEs) of white light for the retina. They use ANSI laser guidelines to specify the time required to attain the MPE. For exposure times ranging from 10 s to a little less than 3 h the retinal MPE is $2.92 \, \text{J cm}^{-2}$, so that if the retinal irradiance in W cm^{-2} is known, the "safe" exposure time is defined. Retinal irradiances produced by a wide variety of ophthalmic instruments at the Wilmer Institute were measured. Direct ophthalmoscopy is the least hazardous procedure, the retinal irradiance averaging about $29 \, \text{mW cm}^{-2}$, which defines a safe exposure time of 100 s. There is little likelihood that a specific area of the retina would be viewed for such a long interval. The average retinal irradiance from the standard binocular indirect ophthalmoscope is about $69 \, \text{mW cm}^{-2}$ for lamps operating at their designed voltage (6.5 V) and 125 mW cm^{-2} for a maximum setting. This corresponds to safe viewing times of 42 and 23 s, respectively. By ANSI laser criteria examinations should be safe since normal fundus observations require less than 10 s for a given retinal site. However, viewing intervals of 40–60 s might be required to identify a pathological condition. At maximum intensity settings, the blue component in the spectrum is enhanced and safe viewing times can range between 15 and 23 s. It is evident that the indirect ophthalmoscope could become a hazardous instrument in inexperienced hands. Slit-lamp biomicroscopy can produce a retinal irradiance three times greater than the indirect ophthalmoscope. The authors caution that this is a matter of concern because such examinations are usually performed on a pathological fovea, which may be unusually susceptible to photic damage. Overhead surgical lamps illuminate an area of 5 disk diameters with a retinal irradiance of 24 mW cm^{-2}. This corresponds to a safe exposure time of 122 s, whereas various types of surgery could last for 45 min or longer. Moreover, the large area of illumination ensures that some parts of the fundus will be under constant irradiation. The retinal irradiance of several surgical microscopes was also measured. The average figure for myopes and emmetropes exposed to current models in usage was 460 mW cm^{-2}. A nonfiber optic model produced 970 mW cm^{-2}, which represents a safe exposure time of 1.8 s. Fortunately, surgical microscopes illuminate a smaller area of the fundus than overhead surgical lamps, but even so the exposed area is approximately 1.5 disk diameters. There can be little doubt that ophthalmic surgery represents a real hazard for the retina. This reintroduces the so-called Irvine–Gass–Norton syndrome, substantiates the conjectures of Henry *et al.*, reinforces Young's theory of retinal radiation damage, and supports Tso's findings about macula edema after cataract extraction. Calkins and Hochheimer urge manufacturers to design safer surgical illuminators by eliminating the most hazardous portion of the short-wavelength spectrum (400–500 nm) and the unnecessary long wavelengths greater than 700 nm. They also recommend similar filtration for diagnostic instruments. However, they conclude by emphasizing that the primary impetus for restricting non-essential light exposure should come from the ophthalmological profession.

A paper by Delori, Pomerantzeff, and Mainster (1980) examines the ocular safety of fundus photography and indirect ophthalmoscopy. These authors also use the ANSI laser standards (ANSI Z 136.1, 1986) to evaluate safety criteria for ophthalmic instruments. To characterize the potential hazards the following must be known: (1) size of retinal irradiated area, (2) exposure time, (3) power delivered to patients eye, (4) transmission through the ocular media, and (5) spectral distribution of the light source.

They investigated the Zeiss, Topcon, Nikon, and MIRA fundus cameras. For each instrument, they determined the percentage of total energy in the blue, green, red, and infrared regions of the spectrum. To describe the potential hazard of a particular exposure or setting, the weighted irradiance is compared to the maximum permissible exposure (MPE) as defined by the ANSI Z 136.1 standards for wavelengths between 400 and 550 nm. Most of the hazard ratios for different fundus cameras are below 50%, but it must be remembered that this is for a single exposure.

For continuous exposure, such as used in ophthalmoscopes, the maximum permissible time of exposure (MPT) is given. This is defined as the exposure duration that can be sustained before the MPE is reached for the short wavelength band 400–550 nm. Light levels produced by binocular indirect ophthalmoscopy using a 20 diopter lens were measured for (a) American Optical ophthalmoscopes with tungsten bulb, (b) MIRA indirect ophthalmoscopes with tungsten bulbs or fiber optic illuminators with tungsten-halogen lamps. The MPT for these systems was between 5 and 60 s. The authors point out that while an ophthalmoscopical examination may take about 15 min, the effective exposure time is considerably shorter because of frequent interruptions of the light beam by the observer, movement of patient's eye, movements of hand-held lens, and changes in the retinal field of view. Nevertheless, they note that it is clear that retinal exposure in indirect ophthalmoscopy may exceed the MPE recommended by the ANSI Z 136.1 standard.

The potential hazards of fluorescein angiography were also assessed. This diagnostic technique consists of serial fundus photography following injection of sodium fluorescein. Fluorescence in the choroidal and retinal vasculature is followed by flash photography during excitation by blue light. Normally, fundus pictures are taken at the rate of 0.5 to 2 frames per second. Two-millisecond flashes produce retinal radiant exposures ranging from 10 to 25 mJ cm^{-2}. At least 50%–80% of the flash energy resides in wavelengths beyond 700 nm; these are ineffective in exciting the fluorescein dye since it requires only wavelengths between 450 and 520 nm. Under routine conditions of fluorescein angiography the ANSI MPEs are not exceeded by exposures of 36 successive light flashes. The authors recommend elimination of useless infrared radiation by appropriate filtering, which would not affect the efficiency of the fluorescein angiography technique.

Light levels resulting from fundus observation with tungsten bulbs were measured also for the Zeiss and Topcon cameras. MPTs ranged from 5 to 43 min, indicating that normal observation of the fundus was not dangerous.

In conclusion, Delori, Pomerantzeff, and Mainster recommend the following: (1) filters should be used to remove wavelengths greater than

700 nm; (2) wavelengths below 470 nm could be eliminated to reduce the exposure hazard and would also improve image quality by reducing light scatter; (3) photography and ophthalmoscopy should use narrow bandpass spectral illumination (15–25 nm) whenever high-contrast imaging of fundus details is required. This would reduce the radiant exposure by a factor of at least 4 over that from white light. They recommend that manufacturers of ophthalmic instrumentation provide clinicians with information regarding the range of settings in which their devices could be potentially hazardous for eyes with clear ocular media; also give the variation in radiance that would be anticipated during the lifetime of a particular light source and some information on how the instrument could be used most effectively to minimize retinal irradiance for a given retinal luminance.

Sliney and Wolbarsht (1980) caution that any analysis of the retinal hazards of ophthalmic instrumentation based on criteria derived from current standards such as the ANSI Z 136.1 1980 laser standard requires some care in application because the assumptions of large eye movements and limited pupil size may not apply in many diagnostic and therapeutic situations. Simple measurements of the optical power emitted by a source and a calculation of total retinal irradiance are not sufficient to permit an adequate hazard evaluation for a specific ophthalmic instrument. Not only the spectral radiance of a lamp and an accurate assessment of the imaging conditions are needed, but also some consideration must be given to possible changes in sensitivity of the patient's eye as a result of pathological conditions. In the last analysis, only the clinician can assess and evaluate the photic risk to his patient's eye in terms of the diagnostic and/or therapeutic need.

The recent concern over the possible hazards to the patient of light sources used in ophthalmology has been brought about by the realization that extended exposure of the retina to short wavelengths in the visible spectrum produces photochemical or actinic effects, which may lead to macular degeneration. The hazards of extended exposure to blue light also apply to those who work or play continuously in a bright light environment, such as those engaged in sunbathing, skiing, mountain climbing, or yachting, as well as welders, surgeons and nurses in operating rooms, seamen, farmers, lifeguards, airplane pilots, and the like. In every case the hazard can be reduced or eliminated by using a proper filter to remove the near ultraviolet and blue light. Wavelengths beyond 700 nm should be removed, since they do not help vision and are a source of additional heat energy on the retina, which may enhance the photochemical effects. Fortunately, the large choroidal blood circulation tends to maintain retinal temperatures at near ambient even during exposure to intense environmental radiation.

7. PHOTOPATHOLOGY OF BLUE LIGHT

The pathological effects of blue light on the retina have been controversial to the extent that investigators differ in their opinions as to whether the initial insult involves the photoreceptors of the neural retina or the single layer of epithelial cells comprising the retinal pigment epithelium (RPE). The confusion has been augmented by the tendency to relate the blue light lesion in primates to the syndrome involving the loss or destruction of the photoreceptors in albino rats exposed to a continuous or constant white light environment. As previously pointed out, there are a number of important differences. Also, the long-term biological end points are completely different. In the albino rat, repeated exposures result in a complete loss of photoreceptors with resultant blindness (Noell, 1966); extended exposure of the primate retina to blue light leads to hypopigmentation of the RPE and an appearance that bears a close resemblance to age-related macular degeneration (Ham et al., 1978). Tso and Fine (1979) found in the rhesus that although the RPE may recover during the first few months, it develops serious complications at a later date, which may lead to serious detachment and foveal (macular) edema.

There is general agreement that photic damage to both photoreceptors and RPE occurs at exposure levels to blue light appreciably above those required to produce a minimal lesion. For example, Friedman and Kuwabara (1968) and Tso et al. (1972a, b, 1973) produced both photoreceptor and RPE damage at levels above threshold for blue light damage in the rhesus monkey, and Moon et al. (1978) demonstrated a loss in visual acuity in rhesus monkeys exposed to levels of blue light well above threshold. Most investigators agree that damage to the outer segments (OS) is reparable with time so long as the photoreceptor cells are not irreversibly damaged; and even the loss of a few photoreceptor cells following pyknosis of a few nuclei in the outer nuclear layer (ONL) is a relatively minor injury that does not affect visual function.

The distinction as to the initial or primary site of photic injury is extremely important because it can provide an insight into the long-term chronic effects of blue light. Only by investigating minimal or subminimal lesions after extended exposure (1000 s) can the events leading up to long-term effects be discovered. Overexposure obliterates in a badly damaged matrix of tissue the subtle effects of phototoxicity. Extended exposures beyond 1000–10,000 s are impractical in the laboratory, but repeated exposures on a daily basis to monkeys trained to sit in a chair and perform a visual task for a reward are feasible. Such experiments have been underway since 1979 at the Medical College of Virginia.

For example, a rhesus monkey was exposed monocularly over a period of 30 months (608 exposures) on a daily basis (1000 s per day, 5 days per week) to a short-wavelength spectrum (330–490 nm) provided by a 2500-W xenon lamp with quartz optics and suitable mirrors and filters (Ham *et al.*, 1982b). The corneal irradiance was 5 mW cm^{-2}; estimated radiant exposure to the retina was 8 J cm^{-2} per daily exposure, based on the assumption that the exposed eye remained fixed on the light source. In reality, both the exposed eye and the unexposed control eye were moving constantly. Nevertheless, the image size on the retina (1.2 mm diameter) was large enough to assure an appreciable overlap on the macular area. Periodic examinations with the fundus camera revealed no startling changes, though there was definite evidence of depigmentation in the temporal, superior area of the fundus. Fluorescein angiography was normal in both eyes. Histology on this animal was negative when sacrificed after 608 exposures. These results were inconclusive, especially as there was some doubt as to whether the depigmented area was included in the histological examination. It should be noted that only 27% of the energy in the 330- to 490-nm spectrum penetrated the ocular media to irradiate the retina. Wavelengths shorter than 400 nm were absorbed, primarily by the lens. When the lens is removed the effects on the retina are dramatic. In another experiment a monkey whose lens had been extracted surgically was exposed to a near ultraviolet spectrum, 330–420 nm, under similar conditions to those above except that the corneal irradiance was only 66.5 μW cm^{-2}. The radiant exposure to the retina was estimated to be 1.1 J cm^{-2} per daily exposure. This animal underwent 316 daily exposures before sacrifice for histology. Early on, after 80 exposures, fluorescein angiography had disclosed multiple focal areas of depigmentation in the RPE of the superior macula. Funduscopic examination before sacrifice showed a large lesion in the superior paramacular where an edematous area had been observed previously, as well as another large lesion in the temporal macula; also numerous small depigmented areas in the superior macula and what appeared to be a retinal hole in the periphery at about ten o'clock. Histological examination confirmed these *in vivo* findings. These results demonstrate the extreme sensitivity of the primate retina to repeated small exposures of near ultraviolet light. The histological findings after exposure to near ultraviolet radiation are described further on in this report.

In an attempt to investigate more thoroughly the basic mechanisms leading to the blue light lesion, Ham *et al.* (1978) prepared for light microscopy and electron ultrastructural analysis approximately 3000 sections from 20 eyes in 10 rhesus monkeys exposed to blue light (441 nm) at levels slightly above threshold. Sections were taken at postexposure times of 1 h and 1, 2, 5–6, 10–11, 30, 60, and 90 days. Experimental methods and

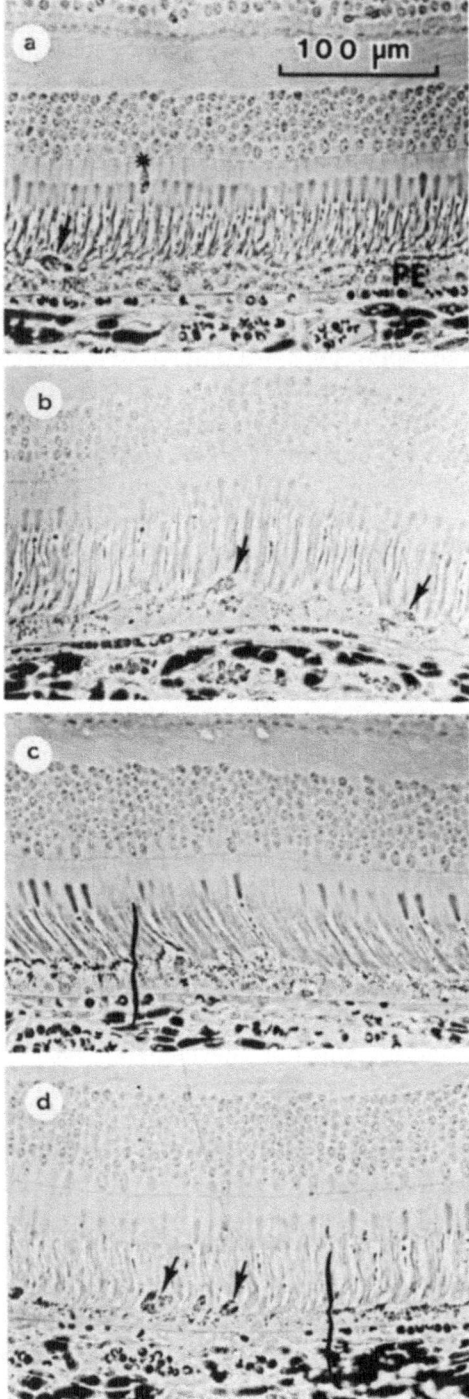

procedures are given in the publication. In what follows, the histological findings will be discussed in some detail.

In specimens examined 1 h postexposure the neural retina, RPE, and choroid appeared normal with the exception of a few pyknotic rod nuclei and a few dense cone ellipsoids. Findings were similar at one day postexposure, but at two days postexposure there were definite changes in the RPE, which was edematous in about 90% of the exposed area (1 mm diam.). The most characteristic feature of the lesion was a pigmentary change caused by the agglutination of melanin granules, which produced interstices in the curtain of melanin granules normally found in the apical region of the RPE. A few macrophages containing melanin granules were present in the subretinal space. This hypopigmentation of the RPE made the lesion visible funduscopically for the first time at two days postexposure. In more severe lesions the choroid was involved over the central 50% of the irradiated area. A mild choroidal response or none at all was a common finding for most of the lesions examined histologically at two days postexposure. In these two-day lesions the RPE was mildly inflamed but the OS of the photoreceptors were not grossly damaged. The initial lesion was localized predominantly in the RPE with widespread damage and possible necrosis of some cells. Figure 5a illustrates a two-day lesion in response to a radiant exposure of $35\,\mathrm{J\,cm^{-2}}$ of 441-nm light; exposure time was 1000 s. The injury is confined primarily to the RPE, while the overlying photoreceptor cells show little response. An arrow points to a macrophage in the subretinal space.

Lesions examined at 5 and 6 days postexposure usually showed a highly inflamed RPE, often with cellular proliferation (mitotic figures) and always with hypopigmentation. Several macrophages loaded with melanin granules were now visible in the subretinal space. For the first time the OS seemed to be disarranged and damaged. Figure 5b illustrates a lesion at 5 days postexposure. Cellular proliferation, hypopigmentation, and macrophages

FIGURE 5. Rhesus monkey paramacular retinal response to 441-nm light at a radiant exposure of $35\,\mathrm{J\,cm^{-2}}$ in 1000 s on a retinal spot size 1 mm in diameter. The figures are phase contrast micrographs of unstained plastic sections at the same magnification. (a) At two days postexposure the histological response to blue light injury is apparent in the retinal pigment epithelium, while the overlying photoreceptor cells show little response (damaged cone ellipsoid at asterisk). Macrophages are present in the subretinal space (arrow). (b) At 5 days postexposure the retinal pigment epithelium shows hypopigmentation, cell proliferation, and macrophages (arrows). (c) At 10 days postexposure the hypopigmented RPE tends to reform a single layer of cells. The border between lesion and undamaged adjacent tissue is marked by a bracket. (d) At 30 days postexposure the regenerated RPE remains hypopigmented, and large macrophages (arrows) are present in the subretinal space. A bracket marks the border between lesion and normal tissue.

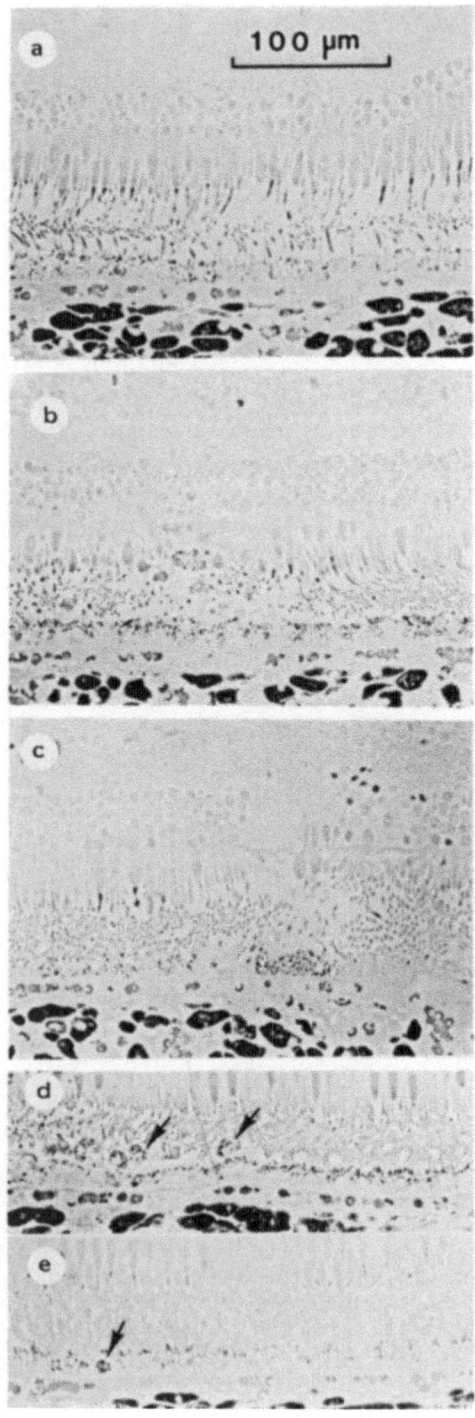

(arrows) in the subretinal space are clearly visible. The OS show mild disarrangement.

By 10 to 11 days postexposure lesions usually showed remarkable recovery. The RPE, while hypopigmented, had returned to a single layer of cells; macrophages loaded with melanin granules persisted in the subretinal space, but the OS of the photoreceptors appeared fairly normal. Figure 5c shows the edge of a 10-day lesion. Unexposed tissue is shown to the left of the bracket where a normal curtain of melanin granules can be seen. The exposed area is hypopigmented and there are still macrophages in the subretinal space.

At 30 days postexposure, most lesions showed a normal RPE except for hypopigmentation and the continued presence of macrophages in the subretinal space. Figure 5d illustrates a typical 30-day lesion. A bracket marks the edge of the lesion. Macrophages (arrows) and hypopigmentation are clearly evident.

In lesions examined at 60 days postexposure the macrophages had disappeared from the subretinal space. Except for hypopigmentation the RPE and neural retina appeared normal. The same was true at 90 days postexposure except that the hypopigmentation seemed less evident. From these observations it would appear that partial to almost complete recovery had occurred by 90 days postexposure. It must be remembered, however, that these were very mild lesions, slightly over threshold. Also, Tso and Fine (1979), as pointed out in Section 1, found that initial recovery from a blue light lesion in the retina may be followed several years later by functional incompetence leading to serious detachment of the RPE and foveal (macular) edema.

The characteristic features of the blue light lesion as outlined above closely resemble the clinical events leading to solar retinitis and eclipse blindness as well as to the syndrome of foveomacular retinitis reviewed by

─────

FIGURE 6. Retinal response to near-UV radiation (350 nm) at a radiant exposure of 5.5 J cm^{-2} in 100 s on a retinal spot size of 500 μm in diameter in one aphakic eye of a rhesus monkey. The figures are phase contrast micrographs of unstained plastic sections taken at the same magnification. (a) Parafoveal response at 2 days postexposure, showing RPE pigmentary changes and apparently moderate direct damage to photoreceptors. (b) Paramacular response at 5 days postexposure showing a moderate RPE response and severe damage to photoreceptor cells. (c) Response at 5 days in a second paramacular location. Note the clumped melanin granules in the RPE beneath a region of destroyed photoreceptor cells. (d) At 10 days postexposure the paramacular lesion showed RPE hypopigmentation and macrophages (arrows) in the subretinal space. The neural retina did not show significant damage. (e) At 30 days postexposure the paramacular lesion appeared well healed, with slight depigmentation of the RPE, an unknown inclusion (arrow), and no apparent macrophages in the subretinal space.

Marlor (1973) and the pigment degeneration of the macula in 150 servicemen stationed on a tropical island in the Pacific (Smith, 1944). The recovery phase after blue light exposure is similar to the clinical data on eclipse gazing reported by Penner and McNair (1966), who reported recovery to preexposure vision in 59% of their patients at six months postexposure. Hatfield (1970) reported 145 cases of solar retinopathy during the 1970 eclipse; 45% of those afflicted returned to normal vision. Tso and La Piana (1975) exposed three patients scheduled for enucleation because of melanoma to direct sungazing for a period of one hour. The eyes were removed 38 to 48 hours after exposure and examined histologically. They found varying degrees of damage to the RPE including irregular pigmentation, necrotic RPE cells, and edema, but the photoreceptor cells appeared normal. The vision of two patients had returned to preexposure levels before their eyes were enucleated.

Henkes (1977) has presented two case histories of patients suffering from a condition resembling acute posterior multifocal placoid pigment epitheliopathy after exposure in one case to a halogen bumper light on a truck and in the other case to bright sunshine. In neither case could the injury be attributed to thermal effects. He discusses a possible relationship between the appearance of fundus changes (unclassifiable edematous maculopathies) and the direct impact of light.

Moon et al. (1978) were able to demonstrate that the photopathology of the blue light lesion correlated well with monocular visual acuity tests in the rhesus monkey as defined by the Landolt ring technique. Exposure of the fovea to paramacular threshold levels of 441-nm light ($30 \, J \, cm^{-2}$ in 1000 s) did not impair vision, but $60 \, J \, cm^{-2}$ in the fovea produced a decline in 20/20 vision on about the fifth or sixth day postexposure; the visual acuity gradually returned to normal at 30 days postexposure. An animal exposed to $90 \, J \, cm^{-2}$ in the fovea lost 20/20 vision permanently. When this animal was sacrificed $4\frac{1}{2}$ years later, histological examination disclosed plaque formation in the RPE similar to that reported by Tso and Fine (1979). It has been demonstrated conclusively in the rhesus monkey that solar retinitis is caused by exposure to the blue component in the solar spectrum (Ham et al., 1980b). All these data support the thesis of Young (1981) that "some of the entities in the current nosology of retinal disease are *radiation diseases* provoked or aggravated by light...." In particular, there is ample reason to postulate that exposure to blue light plays a role in the Irvine–Gass–Norton syndrome, especially after the investigation of Tso and Shih (1977) regarding the pathology of macular edema following lens extraction in the rhesus monkey and the recent measurements of retinal light exposure from operation microscopes and surgical overhead lamps made by Calkins and Hochheimer (1979).

8. PHOTOPATHOLOGY OF NEAR-ULTRAVIOLET LESION

The ocular media, especially the lens, protect the primate retina from short-wavelength light and near ultraviolet (UV) radiation, but this protection is largely forfeited when the lens is removed. Since over 575,000 cataract operations were performed in the United States in 1981 (Maumenee, 1981), the question arises as to whether blue light and near-UV radiation constitute a hazard for the aphakic eye. As mentioned in Section 2, Ham *et al.* (1981, 1982a, b) have shown that the rhesus retina is six times more sensitive to near-UV radiation than to blue light. The radiant exposure required to produce a minimal lesion was approximately $5\,\mathrm{J\,cm^{-2}}$ for a 100 or 1000 s exposure to 350 and 325 nm, as contrasted with $30\,\mathrm{J\,cm^{-2}}$ for 441-nm blue light. Enough histological data are available now to demonstrate that near-UV lesions differ from blue light lesions in several important respects (Ham *et al.*, 1987). The UV lesion is funduscopically visible immediately after exposure as contrasted to a latent period of 48 h before a minimal blue light lesion appears, and the photoreceptors as well as the RPE are damaged, particularly the cone ellipsoids, which appear to be especially vulnerable, probably because of absorption by the metallo-flavoproteins and cytochromes in the mitochondria. Rhodopsin and the cone photopigments have strong absorption peaks in the near UV that also may account for the sensitivity of the photoreceptors. In both types of exposure (blue light and near UV) damage to the RPE plays an important and similar role.

Only one aphakic eye in a rhesus monkey was available for histological analysis. This eye was exposed to $5.5\,\mathrm{J\,cm^{-2}}$ of 350-nm radiation in 100 s on a retinal spot size of about $500\,\mu\mathrm{m}$. All exposures were paramacular and scheduled so that at sacrifice lesions could be examined at postexposure times of 2, 5, 10, and 30 days.

Figure 6a displays a two-day lesion as a phase contrast micrograph, unstained. It is immediately apparent that the OS of the photoreceptors are damaged, a finding in direct contrast with a blue light lesion of two days postexposure where the photoreceptors are virtually intact. There is mild damage and some depigmentation in the RPE. The primary focus of damage seems to be the OS of the photoreceptors and possibly just perceptible thinning of the nuclei in the outer nuclear layer (ONL).

Figure 6b illustrates paramacular response 5 days postexposure. Again there is obvious damage to the photoreceptors, especially the OS, and some damage to the nuclei in the ONL. The RPE is also involved with mild derangement among the melanin granules. Damage is confined mainly, however, to the photoreceptor cells of the neural retina.

The next micrograph, Fig. 6c, is also 5 days postexposure and shows another paramacular lesion where the photoreceptors are severely damaged; within a small area the entire photoreceptor population including the ONL has disappeared. Below the damaged area and resting on Bruch's membrane is a clump of melanin granules. Toward the left of the damaged area the RPE and neural retina appear reasonably normal. It is difficult to explain this highly localized damage. Perhaps it is due to refractive errors producing focal hot spots at these short wavelengths. In any event, it demonstrates that near-UV radiation is lethal to photoreceptor cells, a phenomenon which is not apparent with blue light at threshold levels.

Figure 6d represents another paramacular lesion at 10 days post-exposure. Damage to the photoreceptors is mild if present at all, but the RPE shows hypopigmentation and macrophages in the subretinal space. The lesion is reminiscent of the blue light lesion at 10 days postexposure. It is as though the photoreceptors had "closed in" to fill the spaces formerly occupied by the destroyed photoreceptor cells seen at 5 days postexposure. This, of course, is speculation.

The last micrograph, Fig. 6e, depicts another paramacular lesion in the same eye at 30 days postexposure. Here the lesion appears almost completely healed except for a slight depigmentation of the RPE and an unknown inclusion (arrow) just above Bruch's membrane. No macrophages are seen in the subretinal space.

The histological data to date would seem to indicate that near-UV radiation attacks both the photoreceptors cells of the neural retina and the RPE, but that the primary effect is the destruction of the photoreceptor cells, especially the cones. Presumably, radiant exposures well above threshold would result in a massive loss of photoreceptor cells and irreparable damage to the retina.

Calculations based on an estimated radiance from sun and sky of $1 \text{ mW cm}^{-2} \text{ ster}^{-1}$ for the near UV, a pupillary diameter of 2.5 mm, and a transmittance for the aphakic eye of 0.5 yield a retinal irradiance of about $8.5 \text{ } \mu\text{W cm}^{-2}$. Exposure for three hours on a bright sunny day at sea level would give a radiant exposure of 92 mJ cm^{-2}, which is far below the threshold of 5 J cm^{-2} for retinal damage in the rhesus monkey. Even if no repair processes were operative in the retina, it would require about 54 repetitive daily exposures to accumulate a radiant exposure of 5 J cm^{-2}. While these calculations may be reassuring, it is also prudent to assume that near-UV exposures may be partly cumulative. For example, the loss of a few. photoreceptors daily would go unnoticed, but over a period of years the depletion in photoreceptor population would lead to serious consequences. Also, it is important to realize that removal of the lens increases the retinal exposure to blue light. The tremendous preponderance of blue light over near-UV radiation in the solar spectrum more than compensates for the toxicity ratio of 6 to 1 for near-UV versus blue light. Transmittance of the aphakic eye to the bandwidth 400–500 nm is about 0.7. Nothing is known about possible synergistic effects of short-wavelength light and the near-UV radiation.

9. VALIDITY OF ANIMAL DATA AS RELATED TO MAN

Nearly all the data relating to the effects of blue light and near-UV radiation on the retina are based on animal experiments, particularly the rhesus monkey. It may be asked whether it is appropriate to extrapolate from monkey to man. Comparative data on retinal damage from light in man, monkey, and rabbit are scarce, and those that are available are based on thermal damage rather than photochemical effects. To the authors' knowledge no human volunteers have been exposed to strictly blue light and no aphakic patients have been deliberately exposed to near-UV radiation, with the possible exception of Wald's experiments (1952). He exposed both normal and aphakic human volunteers to a 250-W mercury-arc lamp placed 50 cm from a ground Pyrex glass screen, which in turn was viewed at a distance of 50 cm for periods up to 10 min. No instrumentation was available at that time to measure radiant exposure in the ultraviolet. He found that his aphakic volunteers had 20/20 vision in the near UV and concluded "that ultraviolet radiations which might harm the retina do not reach it; those which reach the retina do it no harm, and at most can be seen."

Most comparative data on the thermal effects of light on the retinae of man and monkey indicate that the latter is more sensitive by a factor of about two. Presumably this is because the monkey fundus is more uniformily and densely pigmented than that of man. Except for pigmentation, the morphology of the rhesus retina is remarkably similar to that of man (Kuwabara, 1970). Table II, taken from Ham et al. (1973), compares retinal burn thresholds in man, monkey, and rabbit.

Although these data support the conclusion that the rhesus monkey is the experimental animal of choice for extrapolating to man, it must be remembered that no human subject has been exposed to a strictly blue light source for an extended period of time. Until such data are forthcoming it would be wise to consider with caution the validity of extrapolating from monkey to man. However, it can be said that most investigators in the field of safety standards are of the opinion that the rhesus monkey provides a factor of safety in estimating human thresholds.

TABLE II. Comparison of Retinal Burn Thresholds in Man, Monkey, and Rabbit[a]

	Area exposed		
Species	Paramacula	Fovea	Comments
Man (A.E.G.)	9.0–12.2	13.8	Only temporary afterimage
Man (M.Y.)	9.5– 9.9	9.7	Absolute central scotoma
Man (M.V.)	9.7	—	Foveal detachment
Man (White)	9.3 ± 1.56		18 patients
Man (Black)	7.9 ± 186		10 patients
Monkey	5.9 ± 1.5	5.7 ± 0.35	22 rhesus eyes
Rabbit	4.1 ± 0.4		100 rabbits

[a] Human volunteers and patients treated for diabetic retinopathy are compared to monkeys and rabbits under identical conditions of exposure, namely, retinal image diameter 1 mm; exposure time 135 ms; optical source, filtered Osram lamp (XBO 2500 W) producing spectrum 400–800 nm. Radiant exposure on retina given in $J\ cm^{-2}$.

10. PROTECTIVE FILTRATION

A final word about protective measures that can be taken to avoid the ocular hazards of exposure to blue light and near-UV radiation. Ideally, all spectacles, contact lenses, diagnostic and therapeutic ophthalmic instrumentation, and environmental light sources should transmit and/or emit only wavelengths between 400 and 700 nm. Wavelengths beyond 700 nm could be eliminated in optical sources used in ophthalmic instrumentation by the simple and inexpensive expedient of inserting a "hot mirror" (reflects wavelengths > 700 nm and transmits wavelengths between 435 and 700 nm) between the source and the optical system. Near-infrared radiation contributes nothing to visual image quality; its removal reduces light exposure and, more importantly, reduces thermal enhancement of the photochemical effects. Again, wavelengths below 400 nm contribute nothing to visual quality and are extremely toxic for both the retina and the lens. In the case of the aphakic eye it is imperative to remove the near-ultraviolet radiation and the visible short wavelengths below 450 nm. Intraocular lenses are now available that transmit only wavelengths greater than 400 nm. The mature human lens filters out most of the toxic blue wavelengths and all of the violet and near-ultraviolet radiation. An intraocular lens that simulated the transmission characteristics of the mature human lens would be the ideal solution for the aphakic eye. Filtering out wavelengths below 450 nm would eliminate the near ultraviolet radiation and those blue wavelengths that are potentially most hazardous, while improving image quality by reducing light scatter and chromatic aberration and only minimally affecting image color balance. Filters are commercially available that strongly attenuate wavelengths below 450, 480, 511, and 550 nm. The latter, 550 nm, is designed specifically for retinitis pigmentosa patients. The consumer can select that filter for his or her sunglasses that best suits his or her aesthetic color sense. Any of them will provide partial to complete protection from the toxic and aging effects of solar radiation. Until more conclusive data regarding the ocular hazards of blue light and near UV become available it would seem advisable for those who work or play in bright light environments, and certainly for those who have had their lenses removed, to wear protective filters. These filters do not harm the eye and may help to delay the onset of future ocular problems concerned with aging of the macula and the lens.

11. POSSIBLE MECHANISMS LEADING TO LIGHT DAMAGE IN THE PRIMATE RETINA

Actinic or photochemical injury to the retina begins at wavelengths below approximately 550 nm at retinal irradiance levels too low to produce appreciable temperature rises (Ham *et al.*, 1980a). The transition from thermal to actinic damage as wavelength decreases is gradual with an ill-defined mixture of both types of insult through the range 550–500 nm. Below 500 nm, photochemical effects predominate. Because chemical reactions are a strong function of temperature there is a region of thermally enhanced photochemical reactions bridging the gap between predominately thermal and predominately photochemical events.

The basic mechanisms leading to photochemical effects in the retina are not known, yet there is no scarcity of deleterious reactions in photobiology that might play a role and there is little experimental evidence to single out a specific reaction to the exclusion of others. The mammalian retina is unique among body tissues in that light is focused directly on a group of cells that are highly oxygenated. According to Parver *et al.* (1980), the choroidal circulation accounts for 85% of all ocular blood flow. Per gram of tissue the choroid has four times the volume of blood found in the renal cortex and is structured so that a dense matrix of small blood vessels with a large surface area is immediately adjacent to the RPE and the outer layers of the neural retina. Parver has shown that the choroidal circulation plays a key role in dissipating the heat generated by the absorption of light in the RPE and the choroid. Choroidal circulation is unusual in having a low arteriovenous oxygen differential (approximately 5%), suggesting that this high flow characterstic may serve purposes above and beyond supplying metabolites and oxygen, i.e., a heat-dissipating mechanism for the macula. The presence of numerous large mitochondria in the ellipsoid of the photo-receptor cell demonstrates how dependent the retina is on oxygen. Indeed, the photoreceptor and RPE cells are among the most metabolically active cells in the body.

Either light or oxygen individually can damage cells. The retina would be subject to oxygen toxicity even without light, but the combination of the two greatly enhances the probability of deleterious reactions. Thus, nature's dilemma: light is essential for vision but light is toxic; oxygen is essential for life but oxygen is also toxic. Fridovich (1976) has pointed out that all respiring organisms survive by virtue of maintaining a delicate balance between their energy requirements as obtained by the reduction of oxygen to water and the toxic effects engendered in tissue by the free radicals produced during these catabolic processes.

Photochemical reactions are initiated by photons of light (hv) exciting a molecular sensitizer, (S), to form an initially excited electronic state, the singlet state, ^1S, which has a very short lifetime ($< 10^{-8}$ s). There are three major ways in which the molecule ^1S can dissipate its quantum of absorbed energy: by reaction with a solvent, usually water; by emission of a photon (fluorescence); or by a radiationless transition or crossing-over to a triplet or metastable state ^3S, which has a much longer lifetime than ^1S and therefore has more time to react with other molecules. The triplet state ^3S is believed to be the pathway leading to most photochemical reactions. The most effec-tive sensitizers are those that yield a long-life triplet state in high quantum yield (Foote, 1976). The retina is replete with molecular species which could serve as photosensitizers. Examples are hematoporphyrins, flavins, and aromatic hydrocarbons, which are distributed ubiquitously throughout mammalian tissue. These are among the many chromophores that can absorb visible or near-UV radiation to become sensitizers leading to photochemical reactions. In addition to endogenous sensitizers account must also be taken of exogenous substances that can also act as chromophores, e.g., certain drugs, foods, dyes, etc.

There are two major types of photochemical reactions designated as Type I and Type II. In Type I, the redox reactions do not involve oxygen and ^3S reacts directly with the substrate to produce cellular damage. In Type II reactions, usually called photodynamic reactions, ^3S reacts directly with molecular oxygen to produce either excited singlet oxygen, $^1O_2$, or the superoxide anion radical O_2^-. In either Type I or Type II, the net result is the production of active free radicals which can attack other molecules (Krinsky, 1976). According to Foote (1976), the majority of Type II processes involve singlet oxygen as the primary reactive species.

Molecular oxygen has two unpaired electrons with parallel spins in the triplet ground state, $^3O_2$. To react with other molecules one electron spin requires inversion. Spin inversion is a slow process in comparison to the lifetime between molecular collisions. Because of spin restriction, $^3O_2$ is not as highly reactive a molecule as singlet oxygen. Excitation of $^3O_2$ to singlet oxygen $^1O_2$ results in a spin inversion so that $^1O_2$ becomes a very reactive molecule, especially for the lipid peroxidation of polyunsaturated fatty acids, PUFA. Singlet oxygen has a half-life in water of 3.3 μs (Rodgers, 1983) and an excitation energy of 22 kcal or 0.98 eV corresponding to a photon wavelength of 1270 nm (Fridovich, 1975, 1976, 1977; Foote, 1984).

The reduction of oxygen to water requires the removal of four electrons. Oxygen is toxic, not because of its own reactivity, but because its reduction to water tends to favor (because of the spin restriction) a series of univalent single-electron transfers which generate superoxide radical, O_2^-, hydrogen peroxide, H_2O_2, and hydroxyl radical, $OH\cdot$. The latter is the most potent oxidant known. It is mainly these intermediates that cause oxygen toxicity. Most of the oxygen reduction in respiring cells proceeds by pathways that are directly multivalent. Thus, cytochrome C oxidase, which accounts for most of the oxygen consumption by aerobes, produces H_2O tetravalently without the intermediate production of radicals. This enzyme represents the cell's first line of defense again oxygen toxicity. There are also flavin-containing enzymes that perform the divalent reduction of O_2 to H_2O. Such enzymes also represent part of the cell's defense against oxygen toxicity since they skip the univalent reduction of oxygen, which produces O_2^-.

However, recent research has shown that there are numerous spontaneous oxidations as well as enzymatic oxidations in biological systems that can generate free radicals (Freeman, 1982). For example, the autooxidation of epinephrine, leucoflavin, hydroquinones, and hemoglobin are known to generate superoxide, O_2^-. Mitochondria and phagocytic cells have been shown to produce O_2^-. Babior *et al.* (1973) have demonstrated that during phagocytosis granulocytes show an increased production of O_2^-. They were among the first to suggest that superoxide may be a bactericidal agent. There is no cell in the body with greater powers of phagocytosis than the RPE cell, one of whose major functions is the digestion of outer segments that have been discarded by the rods and cones. Part of the digestive process may involve the oxygen radicals and singlet oxygen. The spontaneous generation of free radicals by metabolic processes in living systems is thought to be a major cause of aging (Tappel, 1968; Feeney and Berman, 1976; Harman, 1982; Cutler, 1984).

Whenever superoxide anion radical, O_2^-, is generated in aqueous media, hydrogen peroxide, H_2O_2, is also produced. This is because superoxide is not stable and dismutates spontaneously to O_2 and H_2O_2. The presence in solution of both O_2^- and H_2O_2 can also result in the production of the powerful hydroxyl radical OH· under special circumstances when iron salts are present (Fridovitch, 1981). The OH· radical indiscriminately attacks all organic compounds, while singlet oxygen preferentially attacks carbon–carbon double bonds. Singlet oxygen is the major product of Type II photodynamical reactions. Thus the combination of oxygen and light in the retina can produce a quartet of toxic poisons, viz., O_2^-, H_2O_2, OH·, and $^1O_2$.

12. PROTECTIVE MECHANISMS AGAINST LIGHT DAMAGE

In view of the above it becomes obvious that cells that utilize both light and oxygen must have protective mechanisms to minimize the production of these toxic substances and also scavenge effectively those whose production cannot be avoided. Those enzyme systems that reduce oxygen to water by tetravalent or divalent pathways represent one type of defense system. The recent recognition that cells, particularly the RPE cell, are constantly digesting their own cytoplasmic constituents and synthesizing new molecules to replace them represents another defence system (Young and Bok, 1979). This process of molecular turnover or renewal provides the cell with a powerful tool for combating the damage caused by toxic substances. Enzymatic repair systems like the well-known DNA repair enzymes constitute still another method of defense against damage from free radicals. The photoreceptor cells have developed a unique defense against the peroxidation of the PUFA, which are the major fatty acid constituents of the outer segments. Both rods and cones renew their outer segments, rods at a daily rate of 10% while cones have a lower but appreciable turnover rate. The tips of the rod outer segments represent those disks that have been exposed the longest to the deleterious effects of light and oxygen, and it is those tips (sometimes 100 disks or more) that are pinched off and phagocytized by the RPE.

In addition to the defense systems listed above, cells have very specific mechanisms, enzymes and antioxidants, to scavenge radicals and/or inhibit their action on susceptible structures, notably membranes. One such mechanism involves superoxide dismutase (SOD), an enzyme that catalyzes by dismutation the conversion of O_2^- to H_2O_2 and O_2. Altogether, there are four different kinds of SOD. One of these is found in the cytosol of mammalian cells. It contains both copper and zinc and has a molecular weight of 32,000 (Fridovich, 1975). This enzyme has been isolated from a wide variety of eukaryotic cells including those from humans, cows, chickens, yeast, and bread mold. Another superoxide dismutase containing manganese is found in mitochondria. The other two types are bacterial in origin and do not occur in eukaryotic cells. It is interesting to note that the striking similarities in the amino acid sequences between bacterial and mitochondrial superoxide dismutases provide biochemical evidence on the evolution of unique dismutases by prokaryotes and protoeukaryotes during the period when the blue-green algae transformed the earth's atmosphere from anaerobic to aerobic. Biochemically, man bears close kin to *E. coli*!

While SOD does protect the cell from superoxide anion radical, in so doing it produces H_2O_2. Hydrogen peroxide is also generated by the divalent reduction of oxygen to water and by some photochemical reactions; it also is toxic to the cell, and in the presence of Fe^{2+} and O_2 can generate the extremely reactive hydroxyl radical (Fridovich, 1981). There are two classes of related enzymes, the catalases and the peroxidases, that catalyze the divalent reduction of H_2O_2 to water. The catalases are found predominately in liver, kidney, and red blood cells. They can reduce H_2O_2 to water directly without the aid of an electron donor. Most tissues need little catalase because the circulating blood can remove and decompose the H_2O_2 excreted by those tissues. An important function of the choriocapillaris may be the removal of H_2O_2 generated in the outer retina, RPE, and choroid by the combined actions of light and oxygen. Peroxidases acting on H_2O_2 require a cosubstrate or hydrogen donor such as glutathione or ascorbic acid. Glutathione peroxidase is a seleno-enzyme that is widely distributed in mammalian cells, e.g., leucocytes, mammary, thyroid, salivary glands, and RPE (Feeney and Berman, 1976). Glutathione peroxidase is effective at low concentrations of H_2O_2 and can act also upon a wide range of hydroperoxides by converting them to harmless hydroxy fatty acids.

The dismutases, catalases, and peroxidases protect the cell from superoxide anion radical and hydrogen peroxide but do little to inhibit singlet oxygen, the major product of Type II photochemical reactions. Delmelle (1977, 1978) has proposed that light damage to the retina could be due in part to photosensitized reactions involving singlet oxygen, and Lion *et al.* (1976) have been able to detect singlet oxygen when the retina is illuminated. Direct proof of the photosensitized formation of singlet oxygen in aqueous media is lacking, but indirect evidence definitively supports the role of singlet oxygen in many solution photooxidations. Proteins, polypeptides, and individual amino acids affected are methionine, histidine, tryptophane, tyrosine, and cysteine either in the free state or in peptides. There is no breaking of peptide or disulfide bonds, but there is loss of conformation which can lead to inactivation of many enzymes. However, the most destructive role of singlet oxygen is the peroxidation of the polyunsaturated lipids which represent the main constituents of membrane structure. Antioxidants supply the cell's main line of defense against lipid peroxidation of membranes. Foremost among naturally occurring antioxidants is vitamin E, which is distributed throughout mammalian cells. Lipid peroxidation can be a chain reaction in membranes (Barber and Bernheim, 1976). It is the nature of chain reactions that a single initiating event propagates itself to adjacent molecules in the membrane in a domino process, which is called autooxidation. Vitamin E or alpha-tocopherol is able to intercept or terminate the autooxidation chain reaction and protect the membrane. Vitamin E is especially concentrated in the outer segments of the photoreceptors. Vitamine E is also thought to be a scavenger of singlet oxygen and to act synergistically with selenium to protect cells from oxygen damage, but the inhibition of membrane autooxidation is probably the most important function of alpha-tocopherol in the retina.

Hayes (1974) has assessed the effects of vitamin E deficiency on the retina in two species of monkeys over a period of $2\frac{3}{4}$ years. Macular degeneration developed after two years on a vitamin E-deficient diet. The lesion was characterized by focal, massive disruption of photoreceptor outer segments, which was attributed to lipid peroxidation of those lipoprotein structures containing highly unsaturated fatty acids. The remarkable accumulation of lipofuscin pigment in the RPE was identical to that previously described in dogs (Hayes et al., 1970) and demonstrated that the RPE is capable of extreme phagocytic activity and lysosomal digestion. Robison et al. (1979, 1980) and Katz et al. (1978) have shown that vitamin E-deprived rat retinas show massive accumulations of lipofuscin in the RPE, disorganization of rod outer segment membranes, and loss of photoreceptor cells. The role of vitamin E in protecting the outer segments of photoreceptor cells from lipid peroxidation is unequivocally demonstrated by these experiments.

If singlet oxygen is indeed the reactive species in most Type II photodynamic effects, what protection does the cell have beyond those provided by alpha-tocopherol? One of the major protective devices in bacterial systems consists of colored carotenoid pigments (Krinsky, 1976). There is convincing experimental evidence that beta-carotene is an effective quencher of $^1O_2$ in mammalian cells. That singlet oxygen quenching is involved in the protective action of carotenes comes from the fact that the rate of $^1O_2$ quenching is a function of the number of conjugated double bonds in the polyene chain. Carotenes with nine or more conjugated double bonds are efficient quenchers (Foote, 1976). Quenching of $^1O_2$ by beta-carotene is due to energy transfer with the resulting production of the triplet state of beta-carotene; the latter dissipates this energy directly to the solvent without damage and therefore can react again in cyclic fashion. The energy of the transition from the ground state to the first excited triplet state must be equal to or less than that of the singlet–triplet transition of the excited oxygen, which is 22 kcal mol^{-1} or 0.98 eV. One of the major functions of carotenoids in nature is to protect cells from harmful or lethal photo-dynamic effects. In addition to quenching $^1O_2$ directly, carotenoids can intercept the photosensitization reaction at an earlier stage by effectively quenching the sensitizer ^3S, thereby preventing the formation of $^1O_2$ and thus further decreasing the $^1O_2$ available for initiating photodynamic damage.

Still another potent defense against the toxic effects of oxygen and light is melanin, the major ingredient of the melanin granules situated primarily in the apical portion of the RPE where they are in close apposition to the outer segments of the photoreceptor cells and the major constituent of the melanocytes in the choroid. Until recently, it was generally assumed that the major role or function of melanin was to shield the outer segments of the photoreceptors from scattered light and to convert absorbed photons into harmless heat. There is little doubt that this concept of the role of melanin is generally valid. However, melanin may have other functions above and beyond the mere conversion of light to heat.

Photoprotection of the skin is a major function of melanin (Pathak *et al.*, 1976; McGinness *et al.*, 1979), and it may well play a similar role in the RPE. One hypothesis (Proctor *et al.*, 1974) is that melanin plays a protective role at low rates of energy input by a conversion to innocuous phonovibrational modes (heat) but that at a high rate of energy transfer via photon absorption, melanin becomes cytotoxic. Melanin may be equally as protective by absorbing the energy of potentially disruptive excited state species or free radicals as in absorbing blue light or UV radiation. While at lower doses or UV or blue light, melanin may thus have a protective effect, there is some evidence that at high radiant exposures the melanin itself becomes cytotoxic. Data to support this thesis come from the histological effects of the blue light lesion (Ham *et al.*, 1978), where 48 h after exposure the RPE undergoes an inflammatory reaction accompanied by agglutination of melanin granules and some phagocytosis of melanin granules by macrophages. Presumably, the melanin granules have undergone some type of damage from overexposure to blue light.

Feeney and Berman (1976) suggest that biochemical damage to the RPE by light and/or oxygen should be reexamined in view of the free radical character of melanin and its possible role as an electron-transfer agent. Melanin is a heterogeneous random polymer comprising several different monomers coupled by various bond types into an amorphous substance containing stable free radicals and semiconductor properties that ensure efficient electron transfer for redox systems. Recent research has shown that melanin is a complex substance with a number of interesting features. Cope *et al.* (1963) demonstrated by electron spin resonance measurements that the melanin granules of the mammalian eye generate free radicals when irradiated with visible light. Melanin was for a long period considered to be an inert substance, but Gan *et al.* (1976) present data to demonstrate that melanin functions as an efficient electron transfer agent in redox systems. Blois and associates (1964) conclude that melanin is a highly irregular, three-dimensional polymer whose optical absorption spectroscopy in the UV and visible regions of the spectrum reveals a lack of structure. This is in accord with Wolbarsht *et al.* (1981), who believe that melanin has little biological significance other than its absorption of light. However, Menon and Haberman (1977) have presented data to indicate that the protective effect of melanin is not entirely due to the absorption of light. They suggest that more attention be paid to the protective and deleterious effects of melanin and that pigment biologists keep an open mind for other possible biological effects that are not presently recognized.

Felix *et al.* (1978) have reported that there is rapid scavenging of oxygen by melanin in the presence of light with saturation of the electron spin resonances of free radicals and reduction of the scavenged oxygen to hydrogen peroxide, accompanied by some production of superoxide. Chedekel *et al.* (1980) provide evidence that the absorption of light by pheomelanin (polymeric pigments found in the hair of red-headed individuals) in aerated aqueous media produces superoxide and hydroxyl radicals as well as solvated electrons. The action spectrum for superoxide production is greatest in the UV spectral region but continues well into the visible wavelengths.

Another interesting property of melanin is its ability to bind metals and certain drugs. This could be a mixed blessing—a storehouse either for needed materials or for toxic substances that poison the RPE. Sarna *et al.* (1976) found several specific types of metal binding sites on melanin. The interaction of Cu^{2+} with melanin was studied in some detail. Other metals that bind to melanin include Mn^{2+}, Ni^{2+}, and Zn^{2+}. Since melanin binds metals very tightly, these authors compared the binding strength with that of ethylenediaminetetraacetic acid (EDTA). Melanin has sites that bind metals more tightly than EDTA and some that bind them less tightly. Lindquist (1973) has reviewed the literature on the affinity of drugs for melanin. Chloroquine, quinine, and antibiotics of the streptomycin group are rapidly localized in the melanin-containing tissues of the eye. Adrenaline, dopamine, and noradrenaline bind reversibly to melanin while tyrosine and DOPA lack melanin affinity; this affinity appears to be an important factor in drug-induced lesions. The authors recommend that new drugs should be tested for melanin affinity before clinical use.

Mainster (1978b) cites melanin phagocytosis and pigment clumping in photic and senile maculopathies as examples of the part played by melanin in the transfer of free radicals in the RPE, and Feeney (1978) points out that the RPE of the eyes of elderly humans often contain complex granules consisting of both melanin and lipofuscin, suggesting an interrelated biological history for these two substances. Feeney studied the history of melanin and lipofuscin granules in 30 human RPEs, spanning a lifetime of 90 years. Her data indicate that RPE melanin undergoes autophagic remodeling and degradation during lifespan. She postulates that melanin plays a key role in protecting cells from light-generated free radicals and suggests that the loss or degeneration of melanin can lead to senile changes in the RPE. As mentioned earlier, Young (1986) attributes a major role to melanin as a protective agent against AMD.

In view of the many interesting properties of melanin as listed above—i.e., as a semiconductor, electron transfer, scavenging of singlet oxygen and free radicals, production of superoxide and other free radicals when irradiated with light, affinity for drugs and metals, liason with aging phenomena involving lipofuscin and drusen—it seems plausible to propose that melanin plays both a protective and a cytotoxic role in retinal photopathology. The association of pigmentary disturbances with specific disease symptomology, such as deafness, inflammatory lesions, neurological disorders, or pigment retinopathies, suggests an active rather than a passive role for melanin in biological systems (McGinness and Proctor, 1973; Barr et al., 1983; Proctor, 1976).

Thus, nature has evolved an impressive array of defense mechanisms to protect the retina (and also the lens) from the toxic effects of light and oxygen. Modern science and medicine, however, have doubled the life span of man, thereby subjecting the eye to aging effects like cataract and macular degeneration. There can be little doubt that long-term, chronic exposure to light and oxygen accelerates the aging process, despite the body's defense mechanisms. It is interesting to note that rhesus monkeys, like man, are also subject to retinal degeneration (El-Mofty et al., 1980; Bellhorn et al., 1981). It is possible to overwhelm the retina's protective mechanisms as is done when albino rats (Noell, 1966) are exposed continuously to light. Under these conditions of continuous bleach, the photoreceptor cells are destroyed first, followed by damage to the RPE. This is not a surprising result considering the lack of both a protective pigment and a nocturnal environment. The same result can be achieved in diurnal pigmented primates (Sykes et al.,

1981) but only by exposing the eye continuously for 12 h to bright light levels through a dilated pupil. The blue light lesion (Ham *et al.*, 1978) is another example of overwhelming the defense mechanisms of the retina, but the damage first appears in the RPE and not in the photoreceptors and the biological end point is a depigmented RPE which bears a close resemblance to the early stages of age-related macular degeneration. The retina is also extremely sensitive to near-UV radiation. Here the damage appears in both the photoreceptors and the RPE. Retinal defense mechanisms against near-UV radiation are probably minimal because the ocular media of the normal eye with intact lens transmits very little UV radiation.

13. THE EFFECT OF OXYGEN ON LIGHT TOXICITY

Although definitive proof is lacking, there is suggestive evidence that oxygen free radicals and reactive molecules, superoxide (O_2^-), hydrogen peroxide (H_2O_2), hydroxyl radical (OH·), and singlet oxygen ($^1O_2$) play an important role in photochemical damage to the retina. To further test this hypothesis, Ham *et al.* (1984a) exposed macaque monkeys under oxygenation to blue light (435–445 nm) peaked at 440 nm and compared the threshold for retinal damage to that determined under normal conditions (breathing air). Monkeys under anesthesia respired through an endotracheal tube with attached gas bag. They breathed various ratios of oxygen/nitrogen ranging from 20/80 (air) to 80/20 and 100% oxygen. Arterial blood samples were taken before and after 30 min of breathing a specific mixture and analyzed for PO_2 in mm of Hg. The decrease in radiant exposure ($J\ cm^{-2}$) was exponential with increase in PO_2. An empirical equation, $H = 39.3\ \exp(-0.0049\ PO_2)$ was established from the data on 8 eyes in 4 monkeys breathing various mixtures of oxygen/nitrogen; e.g., at a PO_2 of 270 mm Hg, the threshold was $10.5\ J\ cm^{-2}$ as contrasted with $30\ J\ cm^{-2}$ for monkeys breathing 20/80 (air). Histopathology disclosed more severe damage to the RPE than that observed under normal conditions (Ruffolo *et al.*, 1984). RPE cells were swollen and distorted at 24 h postexposure rather than at 48 h as normally found for threshold blue light lesions.

Continuing their research, Ham *et al.* (1987) exposed the aphakic eye (lens surgically removed) of a rhesus monkey to 325 nm ultraviolet radiation while elevating the arterial blood oxygen level to various PO_2's (398, 389, 278, and 139 mm Hg). Radiant exposures for threshold damage were reduced from $5.5\ J\ cm^{-2}$ at normal PO_2's (75–100 mm Hg) to less than $2\ J\ cm^{-2}$ for PO_2's greater than 300 mm Hg. The actual threshold was not determined but estimated to be less than $1 \, J \, cm^{-2}$. The histological appearance of these lesions was dramatic. There was severe damage to cone ellipsoids and in some histologic specimens the effect of oxygen plus near ultraviolet radiation was devastating. These findings certainly demonstrate that increased arterial blood-oxygen tension increases the sensitivity of the primate retina to radiation damage but they do not prove that oxygen radicals and reactive molecules are involved. The oxygen effect, while suggestives, does not exclude many other reactions from taking place. Meanwhile, lack of understanding of the basic mechanisms underlying photochemical light damage should not obscure the practical and clinical significance of the oxygen effect.

The protective features of beta-carotene have been shown in one rhesus monkey at PO_2 levels of 226 and 316 mm Hg. The radiant exposure needed to produce a minimal blue light lesion was increased by 60% and 44%, respectively. This experiment can be interpreted as presumptive evidence for singlet oxygen toxicity in the retina, but definite proof is lacking since beta-carotene can desensitize other excited or reactive molecules as well as singlet oxygen. Other experiments with the steroid methylprednisolone and the enzymes superoxide dismutase (SOD) and catalase were inconclusive, so that solid proof that oxygen radicals are responsible for photochemical light toxicity is still lacking.

14. SUMMARY

For a long time it was believed that light damage to the retina was thermal in nature, but research during the past two decades has shown that actinic or photochemical effects are also important, especially for extended exposure to the short wavelengths in the visible spectrum. These findings have suggested that long-term chronic exposure to light may be a contributing factor in age-related macular degeneration and in several other retinal diseases. In consequence of recent research, diagnostic and surgical procedures in ophthalmology that expose the retina unduly to blue light should be evaluated clinically on a risk–benefit basis. The retina is six times more sensitive to near-ultraviolet radiation than to blue light. When the lens is removed the retina is exposed to both blue light and the near-ultraviolet radiation, so that the aphakic patient should be protected by suitable filters. The basic mechanisms underlying photochemical effects on the retina are unknown, but the possible effects of light and oxygen toxicity are discussed in terms of the known protective mechanisms that the eye has evolved.

ACKNOWLEDGMENTS. The authors wish to acknowledge the invaluable assistance of their colleagues John J. Ruffolo, Jr., Dupont Guerry, III, and R. Kennon Guerry. They provided major support for the research reported here. Support for research came from the following sources: U.S. Army Medical Research and Development Command, under contract No. DADA-17-72-C-2; NIH, NEI grant No. EY-02324; Corning Glass Works; Polaroid Corporation; The Bell Laboratories; NCR Corporation, The Xerox Corporation, and the Illuminating Engineers Society.

REFERENCES

American Conference Governmental Industrial and Hygienista (ACGIH), 1979, TLVs, threshold limit values, for chemical substances and physical agents in the workroom environment with intended changes for 1979, Pub. Office, ACGIH, P.O. Box 1937 Cincinnati, OH 45201, $20.00.

American National Standard for the Safe Use of Lasers, ANSI Z 136.1, 1986, American National Standards Institute, Inc., 1430 Broadway, New York, NY 10018.

Babior, B. M., Kipnes, R. S., and Curnute, J. T., 1973, The production by leucocytes of superoxide: A potential bactericidal agent, *J. Clin. Invest.* **52**:741.

Barber, A. A., and Bernheim, F., 1976, Lipid peroxidation: Its measurement, occurrence, and significance in animal tissues, *Adv. Gerontol. Res.* **2**:355–403.

Barr, F. E., Saloma, J. S., and Buchele, M. J., 1983, Melanin: The organizing molecule, *Med. Hypothesis* **11**:1–140.

Bellhorn, R. W., King, C. D., Aguirre, G. D., Ripps, H., Siegel, I. M., and Tsai, H. C., 1981, Pigmentary abnormalities of the macula in rhesus monkeys: Clinical observations, *Invest. Ophthalmol. Vis. Sci.* **21**:771–781.

Birch-Hirschfeld, A., 1912, Zum Kapitel der Sonnenblendung des Auges, *Z. Augen.* **28**:324.

Blodi, F. C., 1982, Ophthalmology, *Am. Med. Assoc.* **247**:2970–2971.

Blois, M. S., Zahlan, A. B., and Mailing, J. E., 1964, Electron spin resonance studies on melanin, *Biophys. J.* **4**:471–490.

Cain, C. P., and Welch, A. J., 1974, Measured and predicted laser induced temperature rise in the rabbit fundus, *Invest. Ophthalmol.* **13**:60–65.

Calkins, J. L., and Hochheimer, B. F., 1979, Retinal light exposure from operation microscopes, *Arch. Ophthalmol.* **97**:2363–2367.

Calkins, J. L., and Hochheimer, B. F., 1980, Retinal light exposure from ophthalmoscopes, slit lamps and overhead surgical lamps. An analysis of potential hazards, *Invest. Ophthalmol. Vis. Sci.* **19**:1009–1015.

Calkins, J. L., Hochheimer, B. F., and D'Anna, S. A., 1980, Potential hazards from specific ophthalmic devices, *Vis. Res.* **20**:1039–1053.

Chedekel, M. R., Agin, P. P., and Sayre, R. M., 1980, Photochemistry of pheomelanin: Action spectrum for superoxide production, *Photochem. Photobiol.* **31**:553–555.

Churchill, M. H., 1945, Dietary deficiency diseases among prisoners of war, *J. R. Army Med. Corps.* **35**:294–298.

Cope, F. W., Sever, R. J., and Polis, B. D., 1963, Reversible free radical generation in the melanin granules of the eye by visible light, *Arch. Biochem. Biophys.* **100**:171–177.

Cordes, F. C., 1944, A type of fovea-macular retinitis observed in the U.S. Navy, *Am. J. Ophthalmol.* **27**:803–816.

Cutler, R. G., 1984, Antioxidants, aging and longevity, in *Free Radicals in Biology*, Vol. VI (W. A. Pryor, ed.), Academic, New York.

Dekking, H. M., 1947, Tropical nutritional amblyopia ("camp eyes"), *Ophthalmologica* **113**:63–92.

Delmelle, M., 1977, Retinal damage by light: Possible implication of singlet oxygen, *Biophys. Struct. Mechanism* **3**:195–198.

Delmelle, M., 1978, Retinal sensitized photodynamic damage to liposomes, *Photochem. Photobiol.* **28**:357–360.

Delori, F. C., Pomerantzeff, O., and Mainster, M. A., 1980, Light levels in ophthalmic diagnostic instruments, *Proc. Soc. Photo-Optical Instr. Eng. (SPIE)* **229**:154–160.

El-Mofty, A. A. M., Eisner, G., Balazs, E. A., Denlinger, J. L., and Gouras, P., 1980, Retinal degenration in rhesus monkeys, macaca mulatta. Survey of three seminatural free-breeding colonies, *Exp. Eye Res.* **31**:147–166.

Feeney, L., 1973, The phagolysosomal system of the pigment epithelium. A key to retinal disease, *Invest. Ophthalmol.* **12/9**:635–638.

Feeney, L., and Berman, E. R., 1976, Oxygen toxicity: Membrane damage by free radicals, *Invest. Ophthalmol.* **15/10**:789–792.

Feeney, L., 1978, Lipofuscin and melanin of human retinal pigment epithelium, *Invest. Ophthalmol. Vis. Sci.* **17**:583–600.

Feeney-Burns, L., 1980, The pigments of the retinal pigment epithelium, in *Current Topics in Eye Research* (J. A. Zadunaisky and H. Dawson, eds.), Academic, New York, Vol. 2, pp. 120–173.

Feeney-Burns, L., and Eldred, G. E., 1983, The fate of the phagosome: Conversion to "age pigment" and impact in human retinal pigment epithelium, *Trans. Ophthal. Soc. U.K.* **103**:416–421.

Feeney-Burns, L., Hildebrand, E. S., and Eldridge, S., 1984, Aging human RPE: Morphometric analysis of macular, equatorial and peripheral cells, *Invest. Ophthalmol. Vis. Sci.* **25**:71–76.

Felix, C. C., Hyde, J. S., Sarna, T., and Sealy, R. C., 1978, Melanin photoreactions in aerated media: electron spin resonance evidence for production of superoxide and hydrogen peroxide, *Biochem. Biophys. Res. Commun.* **84**:335–341.

Foote, C. S., 1976, Photosensitized oxidation and singlet oxygen: Consequences in biological systems, in *Free Radicals in Biology* (W. A. Pryor, ed.), Academic, New York, pp. 85–133.

Foote, C. S., Shook, F. C., and Abakerli, R. B., 1984, Characterization of singlet oxygen, in *Methods in Enzymology*, Vol. 105, *Oxygen Radicals in Biological Systems* (L. Packer, ed.), Academic, New York.

Freeman, B. A., and Crapo, J. D., 1982, Biology of disease. Free radicals and tissue injury, *Lab. Invest.* **47**:412–426.

Fridovich, I., 1975, Oxygen: Boon or bane, *Am. Sci.* **63**:54–59.

Fridovich, I., 1976, Oxygen radicals, hydrogen peroxide and oxygen toxicity, in *Free Radicals in Biology* (W. A. Pryor, ed.), Academic, New York, Vol. 1, pp. 239–277.

Fridovich, I., 1977, Oxygen is toxic, *Bioscience* **27**:462–466.

Fridovich, I., 1981, The biology of superoxide and of superoxide dismutases—In brief, in *Oxygen and Oxy-radicals in Chemistry and Biology* (M. A. Rogers and E. L. Powers, eds.), Academic, New York, pp. 197–205.

Friedman, E., and Kuwabara, T., 1968, The retinal pigment epithelium. IV. The damaging effects of radiant energy, *Arch. Ophthalmol.* **80**:265–279.

Fuller, D., Machemer, R., and Knighton, R. W., 1978, Retinal damage produced by intraocular fiber optic light, *Am. J. Ophthalmol.* **85**:519–537.

Gan, E. V., Haberman, H. F., and Menon, I. A., 1976, Electron transfer properties of melanin, *Arch. Biochem. Biophys.* **173**:666–672.

Gass, J. D. M., and Norton, E. W. D., 1966, Cystoid macular edema and papilledema following cataract extraction, *Arch. Ophthalmol.* **76**:646–661.

Gass, J. D. M., 1977, *Stereoscopic Atlas of Macular Diseases*, 2nd edition, C. V. Mosby Co., St. Louis, Missouri, pp. 254–256.

Geeraets, W. J., Williams, R. C., Chan, G., Ham, W. T., Jr., Guerry, D., III, and Schmidt, F. H., 1962, The relative absorption of thermal energy in retina and choroid, *Invest. Ophthalmol.* **1**:340–347.

Griess, G. A., and Blankenstein, M. F., 1981, Additivity and repair of actinic retinal lesions, *Invest. Ophthalmol. Vis. Sci.* **20**:803–807.

Ham, W. T., Jr., Mueller, H. A., Williams, R. C., and Geeraets, W. J., 1973, Ocular hazard from viewing the sun unprotected and through various windows and filters, *Appl. Opt.* **12**:2122–2129.

Ham, W. T., Jr., Mueller, H. A., Goldman, A. I., Newman, B. E., Holland, L. M., and Kuwabara, T., 1974, Ocular hazard from picosecond pulses of Nd:YAG laser radiation, *Science* **185**:362–363.

Ham, W. T., Jr., Mueller, H. A., and Sliney, D. H., 1976, Retinal sensitivity to damage from short wavelength light, *Nature* **260**:153–155.

Ham, W. T., Jr., 1976, Retinal sensitivity to short wavelength light, *Nature* **262**:629–630.

Ham, W. T., Jr., Ruffolo, J. J., Jr., Mueller, H. A., Clarke, A. M., and Moon, M. E., 1978, Histological analysis of photochemical lesions produced in rhesus retina by short wavelength light, *Invest. Ophthalmol. Vis. Sci.* **17**:1029–1035.

Ham, W. T., Jr., Mueller, H. A., Ruffolo, J. J., Jr., and Clarke, A. M., 1979, Sensitivity of the retina to radiation damage as a function of wavelength, *Photochem. Photobiol.* **29**:735–743.

Ham, W. T., Jr., Ruffolo, J. J., Jr., Mueller, H. A., and Guerry, D., III, 1980a, The nature of retinal radiation damage: Dependence on wavelength, power level and exposure time, *Vis. Res.* **20**:1105–1111.

Ham, W. T., Jr., Mueller, H. A., Ruffolo, J. J., Jr., and Guerry, D., III, 1980b, Solar retinopathy as a function of wavelength: Its significance for protective eyewear, in *The Effects of Constant Light on Visual Processes* (T. P. Williams and B. N. Baker, eds.), Plenum Press, New York, pp. 319–346.

Ham, W. T., Jr., Mueller, H. A., and Ruffolo, J. J., Jr., 1981, Near UV action spectrum for retinal damage, *ARVO Abs. Invest. Ophthalmol. Vis. Sci. Suppl.* **20**:164.

Ham, W. T., Jr., Mueller, H. A., Ruffolo, J. J., Jr., Guerry, D., III, and Guerry, R. K., 1982a, Action spectrum for retinal injury from near UV radiation in the aphakic monkey, *Am. J. Ophthalmol.* **93**:299–306.

Ham, W. T., Jr., Mueller, H. A., Ruffolo, J. J., Jr., Guerry, D, III, and Guerry, R. K., 1982b, Effects from repetitive exposures of rhesus eye to near UV and blue light, *ARVO Abs. Invest. Ophthalmol. Vis. Sci. Suppl.* **22**:198.

Ham, W. T., Jr., 1983, Ocular hazards of light sources: Review of current knowledge, *J. Ocup. Med.* **25**:101–103.

Ham, W. T., Jr., Mueller, H. A., Ruffolo, J. J., Jr., Millen, J. E., Cleary, S. F., Guerry, R. K., and Guerry, D., III, 1984a, Basic mechanisms underlying the production of photochemical lesions in the mammalian retina, *Curr. Eye. Res.* **3**:165–174.

Ham, W. T., Jr., Mueller, H. A., Ruffolo, J. J., Jr., Guerry, R. K., and Clarke, A. M., 1984b, Ocular effects of GaAs lasers and near infrared radiation, *Appl. Opt.* **23**:2181–2186.

Ham, W. T., Jr., Allen, R. G., Feeney-Burns, L., Marmor, M. F., Parver, L. M., Proctor, P. H., Sliney, D. H., and Wolbarsht, M. L., 1986, The involvement of the retinal pigment epithelium (RPE), in *Optical Radiation and Visual Health*, Chap. 3 (M. Waxler and V. M. Hitchins, eds.), CRC Press, Boca Raton, Florida.

Ham, W. T., Jr., and Mueller, H. A., 1986, Biological Applications and Effects of Optical Masers, Quarterly Progress Report (April 8, 1986), Contract No. DAMD 17-82-C-2083, Letterman Army Institute of Research, ATTN: SGRD-ULZ-RCM, Presidio of San Francisco, California.

Ham, W. T., Jr., Mueller, H. A., and Guerry, R. K., 1987, Light damage, in 10th Symposium on Ocular and Visual Development, *Cell and Developmental Biology of the Eye*, Vol. VI (J. B. Sheffield and S. R. Hilfer, eds.), Springer-Verlag, New York.

Harman, D., 1982, The free-radical theory of aging, in *Free Radicals In Biology*, Vol. V (W. A. Pryor, ed.), Academic Press, New York.

Harwerth, R. S. and Sperling, H. G., 1971, Prolonged color blindness induced by intense spectral lights in rhesus monkeys, *Science* **174**:520–523.

Hatfield, E. M., 1970, Eye injuries and the solar eclipse, *Sight Saving Rev.* **40**:79.

Hayes, K. C., Rouseau, J. E., Jr., and Hegstead, D. M., 1970, Plasma tocopherol concentrations and vitamin E deficiency in dogs, *J. Am. Vet. Med. Assoc.* **157**:64–68.

Hayes, K. C., 1974, Retinal degeneration in monkeys induced by deficiencies of vitamin E or A, *Invest. Ophthalmol.* **13**:499–510.

Henkes, H. E., 1977, Photic injury to the retina and the manifestation of acute posterior multifocal placoid pigment epitheliopathy, *Doc. Ophthalmol.* **44**:113–120.

Henry, M. M., Henry, L. M., and Henry, L. M., 1977, A possible cause of chronic cystic maculopathy, *Ann. Ophthalmol.* **9**:455–457.

Hochheimer, B. F., D'Anna, S. A., and Calkins, J. L., 1979, Retinal damage from light, *Am. J. Ophthalmol.* **88**:1039–1044.

Hyman, L. A., Lilienfeld, A. M., Firris, F. L., III, and Fine, S. L., 1983, Senile macular degeneration, A case control study, *Am. J. Epidemiol.* **118**:213–227.

Irvine, S. R., 1953, A newly defined vitreous syndrome following cataract surgery, *Am. J. Ophthalmol.* **36**:599–601.

Kaitz, M., and Auerbach, E., 1979, Action spectrum for light-induced retinal degeneration in dystrophic rats, *Vis. Res.* **17**:1041–1044.

Katz, M. L., Stone, W. L., and Dratz, E. A., 1978, Fluorescent pigment accumulation in retinal pigment epithelium of antioxidant-deficient rats, *Invest. Ophthalmol. Vis. Sci.* **18**:437–445.

Kremers, J. J. M., and van Norren, D., 1988, Two classes of photochemical damage of the retina, *Lasers Light Ophthalmol.* **2**:41–52.

Krinsky, N. I., 1976, Cellular damage initiated by visible light, *Symp. Soc. Gen. Microbiol.* **26**:209–239.

Kuwabara, T., and Gorn, R. A., 1968, Retinal damage by visible light: An electron microscopic study, *Arch. Ophthalmol.* **79**:69–78.

Kuwabara, T., 1970, Structure of the retina, Howe Laboratory of Ophthalmology, Harvard University Medical School, Boston.

Lanum, J., 1978, The damaging effects of light on the retina. Empirical findings, theoretical and practical implications, *Surv. Ophthalmol.* **22**:221–249.

Lawwill, T., Sharp, F., and Speed, N., 1973, Study of occular effects of chronic exposure to laser radiation, Report #111, Army contract No. DADA 17-68-C-8105.

Lawwill, T., Crockett, S., and Currier, G., 1977, Retinal damage secondary to chronic light exposure, *Doc. Ophthalmol.* **44**:379–402.

Lawwill, T., 1982, Three major pathologic processes caused by light in the primate retina: A search for mechanisms, *Trans. Am. Ophthalmol. Soc.* **LXXX**:517–579.

Lerman, S., 1980, *Radiant Energy and the Eye*, Macmillan, New York, pp. 210.

Lindquist, N. B., 1973, Accumulation of drugs on melanin, *Acta Radiol. Suppl.* **325**:5–84.

Lion, Y., Delmelle, M. and van de Vorst, A., 1976, New method of detecting singlet oxygen production, *Nature* **263**:442–443.

Mainster, M. A., White, T. J., and Allen, R. G., 1970a, Spectral dependence of retinal damage produced by intense light sources, *J. Opt. Soc. Am.* **60**:848–858.

Mainster, M. A., White, T. J. Tips, J. H., and Wilson, P. W., 1970b, Retinal temperature increases produced by intense light sources, *J. Opt. Soc. Am.* **60**:262–270.

Mainster, M. A., 1978a, Spectral transmittance of intraocular lenses and retinal damage from intense light sources, *Am. J. Ophthalmol*, **85**:167–171.

Mainster, M. A., 1978b, Solar retinitis, photic maculopathy and the pseudophakic eye, *Am. Intra-Ocular Implant Soc. J.* **IV**:84–86.

Mainster, M. A., Ham, W. T., Jr., and Delori, F. C., 1983, Potential retinal hazards. Instrument and environmental light sources, *Ophthalmology* **90**:927–932.

Marlor, R. L., Blair, B. R., Preston, F. R., and Boyden, D. G., 1973, Foveomacular retinitis, an important problem in military medicine: Epidemiology, *Invest. Ophthalmol.* **12**:5–16.

Maumenee, A. E., 1983, Congressional testimony presented to the U.S. Senate and House of Representatives, Appropriations Subcommittee for Labor, Health and Human Services.

McGinness, J. E., Keno, R., and Moorhead, W. D., 1979, The melanosome: Cytoprotective or cytotoxic, in *Pigment Cell*, Vol. 4, Karger, Basel, Switzerland, pp. 270–276.

McGinness, J. E., and Proctor, P., 1973, The importance of the fact that melanin is black, *J. Theoret. Biol.* **39**:677–678.

Menon, I. A., and Haberman, H. F., 1977, Mechanisms of action of melanins, *Brit. J. Dermatol.* **97**:109–112.

Messner, K. H., Maisels, M. J., and Leure-duPree, A. E., 1978, Phototoxicity to the newborn primate retina, *Invest. Ophthalmol. Vis. Sci.* **17**:178–182.

Moon, M. E., Clarke, A. M., Ruffolo, J. J., Jr., Mueller, H. A., and Ham, W. T., Jr., 1978, Visual performance in the rhesus monkey after exposure to blue light, *Vis. Res.* **18**:1573–1577.

Noell, W. K., Walker, V. S., Kang, B. S., and Berman, S., 1966, Retinal damage by light in rats, *Invest. Ophthalmol.* **5**:450–473.

Noell, W. K., Delmelle, M. C., and Albrect, R., 1971, Vitamin A deficiency effect on retina: Dependence on light, *Science* **172**:72–75.

Norton, A. L., Brown, W. J., Carlson, M., Pilger, I. S., and Riffenburg, R. S., 1975, Pathogenesis of aphakic macular edema, *Am. J. Ophthalmol.* **80**:101.

Parver, L. M., Auker, C., and Carpenter, D. O., 1980, Choroidal blood flow as a heat dissipating mechanism in the macula, *Am. J. Ophthalmol.* **89**:641–646.

Pathak, M. A., Jimbow, K., Szabo, G., and Fitzpatrick, T. B., 1976, Sunlight and melanin pigmentation, *Photochem. Photobiol. Rev.* **1**:211–239.

Penner, R., and McNair, J. N., 1966, Eclipse blindness, *Am. J. Ophthalmol.* **61**:1452.

Priebe, L. A., Cain, C. P., and Welch, A. J., 1975, Temperature rise required for the production of minimal lesions in macaca mulatta retina, *Am. J. Ophthalmol.* **79**:405.

Proctor, P., McGinness, J., and Corry, P., 1974, A hypothesis on the preferential destruction of melanized tissues, *J. Theor. Biol.* **48**:19–22.

Proctor, P., 1976, The role of melanin in human neurological disorders, in *Pigment Cell*, Vol. 3, Karger, Basel, Switzerland, pp. 378–383.

Robison, W. G., Jr., Kuwabara, T., and Bieri, J. G., 1979, Vitamin E deficiency and the retina: Photoreceptor and pigment epithelial changes, *Invest. Ophthalmol. Vis. Sci.* **18**:683–690.

Robison, W. G., Kuwabara, T., and Bieri, J. G., 1980, Deficiencies of vitamin E and A in the rat, *Invest. Ophthalmol. Vis. Sci.* **19**:1030–1037.

Rodgers, M. A. J., 1983, Time resolved studies of 1.27 m luminescence from singlet oxygen generated in homogeneous and microheterogeneous fluids, *Photochem. Photobiol.* **37**:99–103.

Ruffolo, J. J. Jr., Ham, W. T. Jr., and Mueller, H. A., 1981, Retinal photopathology; Distinction between photochemical and thermal lesions, *ARVO Abs. Invest. Ophthalmol. Vis. Sci. Suppl.* **20**:162.

Ruffolo, J. J. Jr., Ham, W. T. Jr., Mueller, H. A., and Millen, J. E., 1984, Photochemical lesions in the primate retina under conditions of elevated blood oxygen, *Invest. Ophthalmol. Vis. Sci.* **25**:893–898.

Sarna, T., Hyde, J. S., and Swartz, H. M., 1976, Ion-exchange in melanin: An electron spin resonance study with lanthanide probes, *Science* **192**:1132–1134.

Sliney, D. H., and Wolbarsht, M. L., 1980a, Safety standard and measurement techniques for high intensity light sources, *Vis. Res.* **20**:1133–1141.

Sliney, D. H., and Wolbarsht, M. L., 1980b, *Safety with Lasers and Other Optical Sources: A Comprehensive Handbook*, Plenum Press, New York, p. 123.

Smith, Homer E., 1944, Actinic macular retinal pigment degeneration, *U.S. Naval Med. Bull.* **42**:675–680.

Sperling, H. G., 1979, Prolonged intense spectral light effects on rhesus retina, in *The Effects of Constant Light on Visual Processes* (T. P. Williams and B. N. Baker, eds.), Plenum Press, New York, pp. 195–214.

Sykes, S. M., Robison, W. G. Jr., Waxler, M., and Kuwabara, T., 1981, Damage to the monkey retina by broad-spectrum fluorescent light, *Invest. Ophthalmol. Vis. Sci.* **20**:425–434.

Tappel, A. L., 1968, Will antioxidant nutrients slow aging processes? *Geriatics* **23**:97–105.

Tso, M. O. M., Fine, B. S., and Zimmerman, L. E., 1972a, Photic maculopathy produced by the indirect ophthalmoscope. 1. Clinical and histopathologic study, *Am. J. Ophthalmol.* **73**:686–699.

Tso, M. O. M., Wallow, I. H. L., Powell, J. O., and Zimmerman, L. E., 1972b, Recovery of the rod and cone cells after photic injury, *Trans. Am. Acad. Ophthalmol. Otol.* **76**:1247–1262.

Tso, M. O. M., 1973, Photic maculopathy in rhesus monkey, *Invest. Ophthalmol.* **12**:17–34.

Tso, M. O. M., and LaPiana, F. G., 1975, The human fovea after sungazing, *Trans. Am. Acad. Ophthalmol. Otol.* **79**:788.

Tso, M. O. M., and Shih, C. Y., 1977, Experimental macular edema after lens extraction, *Invest. Ophthalmol. Vis. Sci.* **16**:381–392.

Tso, M. O. M., and Fine, B. S., 1979, Repair and late regeneration of the primate foveola after injury by argon laser, *Invest. Ophthalmol. Vis. Sci.* **18**:447–461.

van der Hoeve, J., 1920, Eye lesions produced by light rich in ultraviolet rays. Senile cataract, senile degeneration of the macula, *Am. J. Ophthalmol.* **3**:178–194.

Vassiliadis, A., 1972, Hazard evaluation for scanning laser beam, Prepared for International Business Machines Corp., P.O. Box 2195, Research Triangle Park, North Carolina.

Verhoeff, F. H., and Bell, L., 1916, The pathological effects of radiant energy on the eye, *Proc. Am. Acad. Arts Sci.* **51**:630–759.

Vos, J. J., 1962, A theory of retinal burns, *Bull. Math. Biophys.* **24**:115–128.

Wald, G., 1952, Alleged effects of the near ultraviolet on human vision, *J. Opt. Soc. Am.* **42**:171–176.

Walker, C. B., 1916, Systematic review of the literature relating to the effects of radiant energy upon the eye, *Proc. Am. Acad. Arts Sci.* **51**:760–818.

Weiter, J. J., Delori, F. C., Wing, G. L., and Fitch, K. A., 1985, Relationship of senile macular degeneration to ocular pigmentation, *Am. J. Ophthalmol.* **99**:185–187.

Weiter, J. J., Delori, F. C., Wing, G. L., and Fitch, K. A., 1986, Retinal pigment epithelial lipofuscin and melanin and choroidal melanin in human eyes, *Invest. Ophthalmol. Vis. Sci.* **27**:145–152.

White, T. J., Mainster, M. A., Wilson, P. W., and Tips, J. H., 1971, Chorioretinal temperature increases from solar observation, *Bull. Math. Biophys.* **33**:1–17.

Wing, G. L., Delori, F. C., Weiter, J. J., and Kunis, K., 1982, The relation between melanin and lipofuscin in human RPE, *ARVO Abs. Invest. Ophthalmol. Vis. Sci. Suppl.* **22**:173.

Wolbarsht, M. L., 1980, Workshop Chairman: Ocular safety and eye care, NRC Comm. on Vision, *Invest. Ophthalmol. Vis. Sci.* **19**:1124.

Wolbarsht, M. L., Walsh, A. W., and George, G., 1981, Melanin, a unique biological absorber, *Appl. Opt.* **20**:2184–2186.

Young, R. W., 1971, The renewal of rod and cone outer segments in the rhesus monkey, *J. Cell Biol.* **49**:303–315.

Young, R. W., and Bok, D., 1979, Metabolism of the retinal pigment epithelium, in *The Retinal Pigment Epithelium* (K. M. Zinn and M. F. Marmor, eds.), Harvard University Press, Cambridge, Massachusetts, pp. 103–123.

Young, R. W., 1981, A theory of central retinal disease, in *Future Directions in Ophthalmic Research* (M. L. Sears, ed.), Yale University Press, New Haven, Connecticut.

Young, R. W., 1982, Biological renewal. Applications to the eye, *Trans. Ophthalmol. Soc. UK* **102**:42–75.

Young, R. W., 1986, Solar radiation and age-related macular degeneration, *Surv. Ophthalmol.* **32**:252–269.

Young, R. W., 1987, Pathophysiology of age-related macular degeneration, *Surv. Ophthalmol.* **31**:291–306.

Zigman, S., 1978, Spectral transmittance of intraocular lenses, *Am. J. Ophthalmol.* **85**:879.

CHAPTER 6

Ocular Thermal Injury from Intense Light

Ralph G. Allen and Garret D. Polhamus

USAF School of Aerospace Medicine
Brooks Air Force Base, Texas 78235

1. HISTORY

Ocular injury produced by intense light sources has troubled man for a long time, solar eclipse burns, snow blindness, glassblower's cataracts, and corneal injury from arc welders being examples of these early concerns. Flash blindness from large naval guns has also been of concern to the Navy. The potential for corneal injury from solar radiation, specifically the UV, during extravehicular activities required NASA to provide helmets with adequate attenuation properties for the lunar landings. The potential for retinal damage and flash blindness from the thermal emission of nuclear detonations also required eye protection for observers at the first detonation in 1945—with studies of retinal injury and flash blindness undertaken early in the weapons testing program that followed. Concurrently, Meyer-Schwickerath developed the Zeiss Xenon photocoagulator for treatment of eye disorders—particularly detached retina and diabetic retinopathy. Currently, the argon laser and the Nd YAG laser are very effective light sources for the treatment of retinal problems.

Systematic laboratory study of chorioretinal burns began around the mid-1950s at the USAF School of Aerospace Medicine and at the Medical

College of Virginia under Air Force contract. The purpose of these studies was to explore the eye hazard associated with nuclear detonations. The subjects were rabbits, and a Zeiss photocoagulator was the source. The primary end point in these early studies was a "minimum" visible retinal lesion with a threshold generally identified as the average of the lowest retinal irradiance that did not produce a visible lesion and the highest retinal exposure that did not produce a visible lesion—within 5 min. Also considered were such things as image size, blood flow, ocular transmission, enzyme inactivation, and light microscopy, etc. The Air Force, NASA, and their contractors extended many of these studies to primates (rhesus) in the mid-1960s. Unfortunately, limitations on the radiation sources available prevented systematic examination of spectral effects, exposure durations less than approximately 1 ms, and retinal image sizes less than 75–100 μm. During this period, the primary mechanism for retinal injury was almost universally considered to be thermal, or, for higher exposure levels, the production of a steam bubble or explosion and resultant mechanical trauma.

Shortly after the invention of the ruby laser in 1961, studies of the biomedical effects of laser radiation began at the Medical College of Virginia and the University of Cincinnati under contracts with the US Army Surgeon. Soon the Air Force also began laser effects studies through its Oculo-Thermal Groups at the USAF School of Aerospace Medicine —later to become the Laser Effects Branch of USAFSAM. In large part, the research performed by, or sponsored by the DOD through these Army and Air Force programs was the source of the ANSI Z-136.1 Laser Safety Standard, which is the basis for the individual services laser safety standards and the BRH Laser Product Performance Standard. Although these standards applied primarily to injury to the retina, lens, and cornea, potential for damage to skin from intense thermal radiation prompted studies of skin injury and consideration of thermal mechanisms for skin burns.

The complexities of these standards stem directly from the nature of laser radiation, the kinds of interactions possible between biological tissues and laser radiations, and the range of exposure conditions that can be encountered in the variety of applications for which lasers have been found useful. In spite of the fact that the current standards are broad in scope, they are based largely on empirical data, and there are still areas in which data are needed.

2. RETINAL THERMAL INJURY

Probably the most thoroughly studied mechanism of retinal injury is that described as thermal. This injury results from the absorption of energy by the retina and choroid after it has been focused by the cornea and lens.

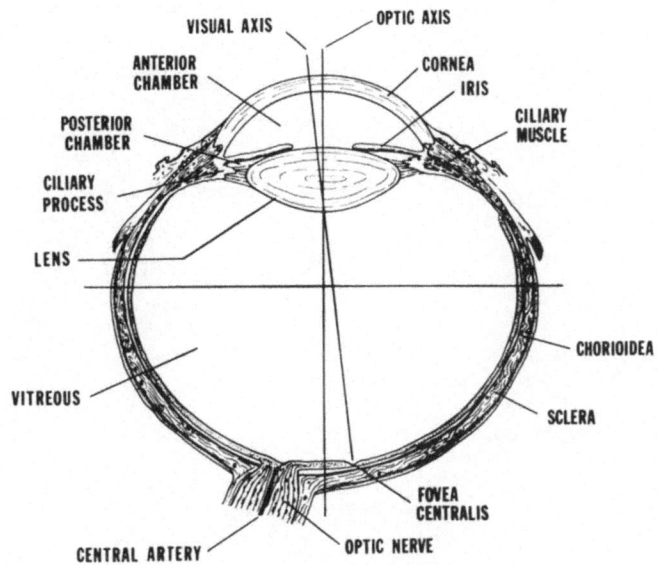

FIGURE 1. Schematic drawing of eye.

Heat generated in the retina and adjacent structures in time diffuses away from the area in which the image is focused. Heat will also be conducted away by the flow of blood in the vascular bed. If the rate of heat dissipation from an area is less than the rate of heat generation in that area, the temperature will increase. If the temperature exceeds the biologic tolerances for the area involved, injury to the photoreceptors (rods and cones), optical nerve tissue, and other structures in the retina and choroid will result—with a possible permanent loss of vision (Fig. 1).

The degree of visual impairment caused by a retinal burn will be

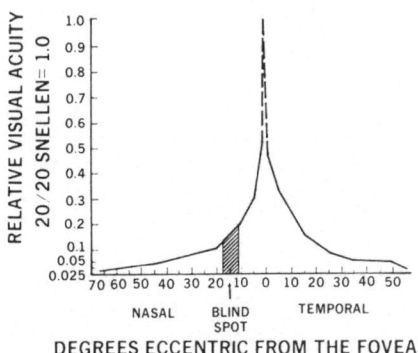

FIGURE 2. Visual acuity distribution at retina.

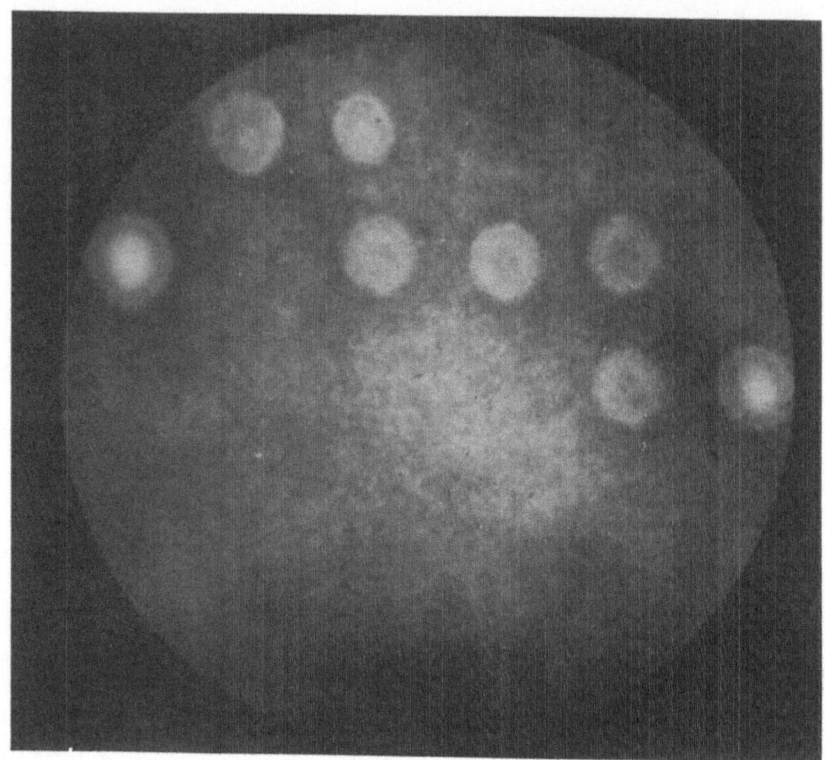

FIGURE 3. Typical thermal lesions.

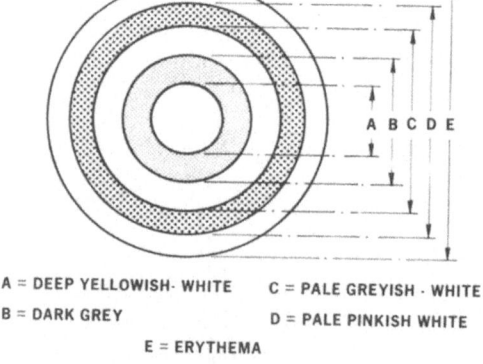

A = DEEP YELLOWISH- WHITE   C = PALE GREYISH · WHITE

B = DARK GREY   D = PALE PINKISH WHITE

E = ERYTHEMA

FIGURE 4. Typical characteristics of thermal lesions. A, Deep yellowish-white; B, dark gray; C, pale-greyish-white; D, pale pinkish-white; E, erythema.

dependent on the size, severity, and location of the burn. The size of the image, which influences the size of a burn, in general, depends on the size and distance of the object from the observer. The severity of a burn depends, in general, upon the amount by which the exposure exceeds the threshold exposure. The location of the burn will determine the function affected. For example, a large burn centered on the fovea would seriously affect visual acuity and color vision. A small burn in the periphery would have significantly less effect on visual acuity and, barring complications, could result in a scotoma or blind spot that would not normally be noticed.

From Fig. 2, a burn exactly centered on the fovea and large enough to include the central 2.5-degree visual field would reduce visual acuity to about 57% of normal (20/35 on the Snellen scale). In theory, if the central 10° visual field were destroyed, the acuity would be 29% (20/70). If the central 20° of vision were destroyed (an unlikely occurrence), visual acuity would be reduced to approximately 20% (about 20/100). Figure 3 shows typical thermal lesions made by an image of uniform intensity and a half millimeter in diameter, while Fig. 4 depicts schematically the typical characteristics of a thermal lesion.

2.1. Threshold Retinal Irradiance

The prediction model for thermal retinal injury used by Allen and Richey in 1958 at the USAF School of Aerospace Medicine and revised and refined in the mid-sixties by Allen (1968) had as its base the very early experimental data developed using rabbits as subjects (Ham et al., 1958). This data base was later updated with primate data (Allen, 1967; Richey and Lof, 1975). The experimental data consisted of threshold retinal exposures versus exposure durations for the production of "minimum" visible lesions viewed with an ordinary ophthalmoscope (Fig. 5). The light sources used in this early work, which preceded the development of lasers, were Zeiss photocoagulators with infrared radiation above 900 nm filtered out. The experimental results indicated an image size dependence and, in general, agreed with the widely held view that the mechanism involved in retinal injury was mainly thermal.

For this model, calculation of the retinal irradiance used the following equation:

$$H_r(t) = \frac{\pi}{[2f(t)]^2} \int_{\lambda_{\min}}^{\lambda_{\max}} N_S(\lambda, t) \, T_A(\lambda) \, T_E(\lambda) \, d\lambda \tag{1}$$

where $f(t)$ is the f number of the eye, $\lambda_{\min, \max}$ are the minimum and maximum wavelengths radiated by the source which reach the retina with energy sufficient to contribute to heating, t is time, $N_S(\lambda, t)$ is the radiance of

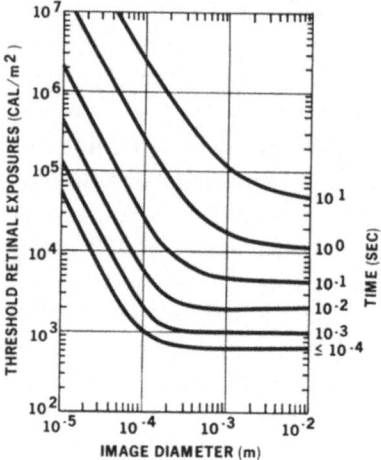

FIGURE 5. Threshold retinal exposure for visible.

the source (cal/m^2 s sr), $T_A(\lambda)$ is the atmospheric transmission, and $T_E(\lambda)$ is the transmission of the clear media of the eye. Integrating the retinal irradiance over the exposure duration yields the retinal exposure:

$$Q_r = \int_0^t H_r(t)\, dt \tag{2}$$

This model simply compared calculated retinal exposures as a function of exposure duration and image size with corresponding measured retinal threshold exposures. It still serves well for many situations; however, the empirical basis of this model limits its application to certain classes of exposure conditions and provides little insight into more complicated exposure conditions.

2.2. Threshold Temperature

In the early 1960s, Wray of DASA (1962) and Vos of the Institute for Perception, Netherlands (1962) independently developed "temperature" models based on solutions of the heat conduction equation. In cylindrical coordinates, and assuming cylindrical symmetry, this equation takes the form

$$\frac{\partial T}{\partial t} = \frac{1}{\rho C} S(r, z, t) + \frac{K}{\rho C}\left(\frac{1}{r}\frac{\partial T}{\partial r} + \frac{\partial^2 T}{\partial r^2}\right) + \frac{K}{\rho C}\frac{\partial^2 T}{\partial z^2} \tag{3}$$

where $T(r, z, t)$ is the temperature rise (°C), t is the time (s), ρC is the volumetric specific heat (cal/cm^3 °C), K is the thermal conductivity (cal/cm sec °C), and

$$S(r, z, t) = \int_{\lambda_{min}}^{\lambda_{max}} h(r) \, H_0(t, \lambda) \left[-\frac{dX}{dz}(\lambda, z) \right] d\lambda \tag{4}$$

with $\lambda(min, max)$, the minimum and maximum wavelengths radiated by the source; $h(r)$, the normalized intensity profile of the image at the retina; $H_0(t, \lambda)$ the center retinal spectral irradiance (cal/cm^2 s); and $(dx/dz)(\lambda, z)$, the derivative of the transmission through the tissue (cm^{-1}). Assuming Lambert–Beers absorption, the derivative of the transmission introduces different rates of absorption in the pigment epithelium and choroid $[\beta_2(\lambda)]$ as follows:

$$\left[-\frac{dX}{dz}(\lambda, z) \right] = \beta_1(\lambda) \, e^{-\beta_1(\lambda)z} \tag{5}$$

for the pigment epithelium (PE) and

$$\left[-\frac{dX}{dz}(\lambda, z) \right] = \beta_2(\lambda) \exp\{d_1[\beta_2(\lambda) - \beta_1(\lambda)] - \beta_2(\lambda)z\} \tag{6}$$

for the choroid, where d_1 is the thickness of the PE, and $z > d_1$. This model provided a more satisfying picture of the thermal process produced by

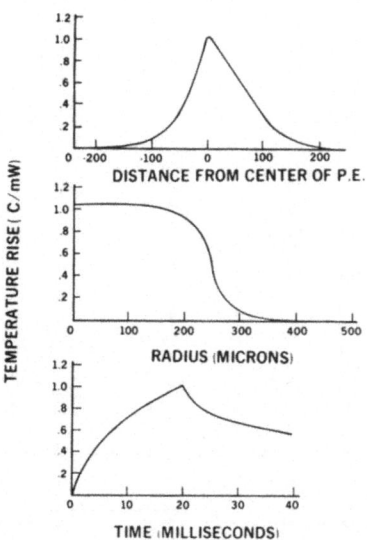

FIGURE 6. Typical results using temperature model.

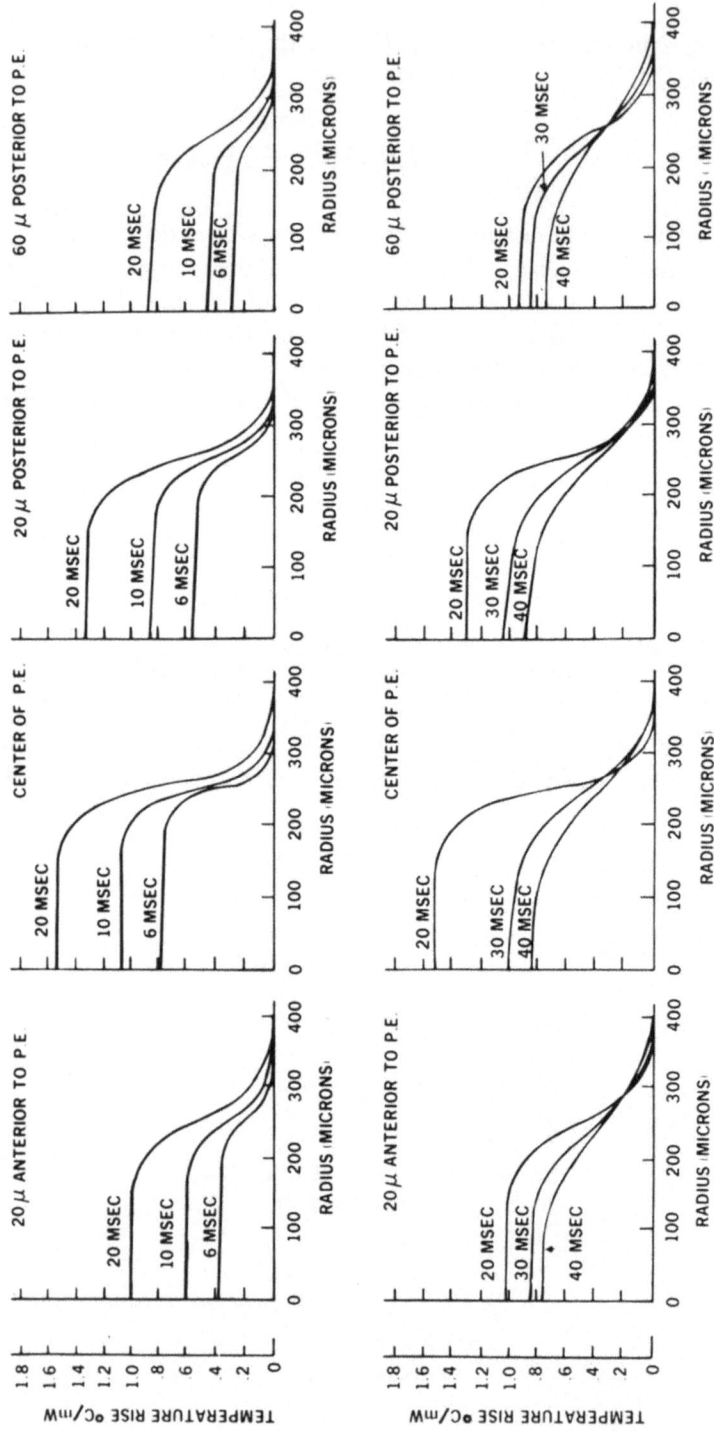

FIGURE 7. Temperature calculations at different axial positions.

the absorption of radiant energy at the retina and choroid and provided the opportunity to examine systematically more complicated exposure situations. Basically, the model provided a time-history of the volume temperature distribution in the retina and choroid. The solutions represent the temperature rise above an arbitrary ambient temperature. Figures 6 and 7 show the temperature response from a 20 ms exposure with a disk image 500 μm in diameter.

In 1967–1968, Clarke *et al.* (1969) provided a uniform absorption model with an analytical solution to the steady state diffusion equation. Later, in the early 1970s, the time-dependent temperature model was revised, improved, and adapted to symmetrical noncoherent sources by Mainster, White, and Allen of Technology Incorporated (1970). Shortly afterwards, Wissler (1976) obtained an analytical expression for the transient temperature distribution produced in the retina during laser radiation. His analysis included tissue layers with different physical properties, energy absorption in the sclera as well as PE and choroid, removal of heat from the tissue by blood circulation, and variation in the light source with time. However, the results were still limited in that they permitted only the prediction of a time-history of temperature distribution with retinal injury examined in terms of a maximum-temperature damage criterion.

2.3. Temperature and the Damage Integral

In the mid-1970s, Takata *et al.* (1974) at the Illinois Institute of Technology Research revised and generalized the model adapting it to coherent radiation for the Air Force. At this point in the evolution of the model, the geometry employed was as shown in Fig. 8.

The model used an exponentially stretched grid in a cylindrical geometry, and variable internal sources, and thermal properties. Boundary conditions for Eq. (3) were as follows:

$$T(R, z, t) = 0$$

$$T(r, \pm Z, t) = 0$$

$$T(r, z, 0) = 0$$

The values of R and Z were equal to 1 cm. The irradiance at the front surface of the PE was equal to the radiant energy entering the eye multiplied by a wavelength-dependent transmission coefficient for the cornea, lens, ocular fluids, and retina.

This version of the model had a heat sink to simulate convection of heat

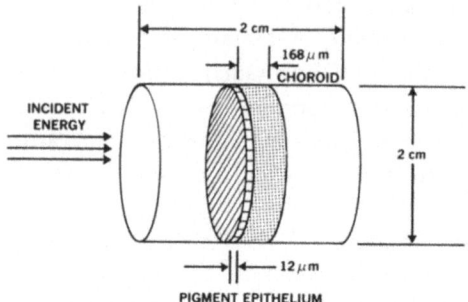

FIGURE 8. Geometry of heat conduction model of the ocular fundus.

by blood flow, nonuniform absorption in a homogenous PE through absorption by a distribution of discrete pigment granules in the PE, and an option to specify the distribution of energy at the retina or use a distribution of energy at the retina that approximates that produced by a highly collimated coherent source focused by the eye. In addition, Henriques's damage integral (1947) was coupled to the temperature calculation to allow systematic assessment of injury in terms of the temperature rise history. This version of the model is represented schematically in Fig. 9. The figure emphasizes the interrelation between the source term, heat conduction, and rate process portions of the model.

This model of the thermal mechanism of retinal injury appears adequate for our basic knowledge of the anatomy of the eye, its thermal properties, blood flow, pigment distribution, and absorption coefficients. This is not to say that all of these properties are adequately established—or that the limits of biological variation are known. In particular, questions still exist concerning the value of the spectral absorption coefficients of the PE and the choroid, the distribution of energy on the retina produced by a highly collimated laser beam, and the values of the rate process damage coefficients.

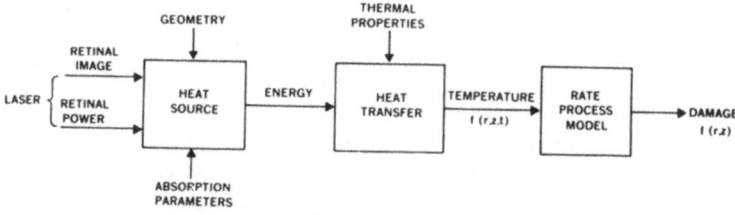

FIGURE 9. Schematic model of thermal damage.

2.3.1. Absorption Coefficient

Coogan *et al.* (1974) and Gabel *et al.* (1976) provided the only two sets of measured absorption coefficients of the PE and separate choroid in rhesus and human retinal tissue (Fig. 10). The thicknesses of the rhesus PE samples were, respectively, 12 and 4 μm from the studies of Coogan and Gabel. Choroid thicknesses were approximately equal, with a value of 170 μm.

2.3.2. Retinal Energy Distribution

Takata *et al.*'s version of the model used the following equation developed by Weigandt to calculate the normalized irradiance at the retina:

$$\frac{H(r)}{H(0)} = \left| \frac{1}{\lambda f'} \int_0^a [P_p(\rho)]^{1/2} \, J_0 \left(\frac{2\pi r\rho}{\lambda f'} \right) F_1(\rho) \, F_2(\rho)\rho \, d\rho \right|^2 \qquad (7)$$

where a is the pupil radius (cm), $f' = f_0 - p$ (cm), f_0 is the second principal focal length at a reference wavelength of λ_0 (cm), $F_1(\rho)$ is the function accounting for defocusing (chromatic and geometric), $F_2(\rho)$ is the function accounting for spherical aberration, J_0 is the zero-order Bessel function of the first kind, p is the distance of the pupil from second principal plane (cm), r is the radial distance in the retinal plane (cm), λ is the wavelength (cm), ρ is the radial distance in the pupil plane (cm), $P_c(R)$ is the beam profile at the cornea, $P_p(p) = (1/\phi_2) \, P_c(R/\phi)$ is the profile at the pupil plane, where R is the radial distance at the cornea, $\phi = 1 - P_c/zf$, P_c is the distance of the pupil from the cornea, $z_f = nz_0 f_c/(nz_0 - f_c)$, when $nz_0 > f_c$, z_c is the distance from the anterior apex of the cornea to the image plane in the eye in the

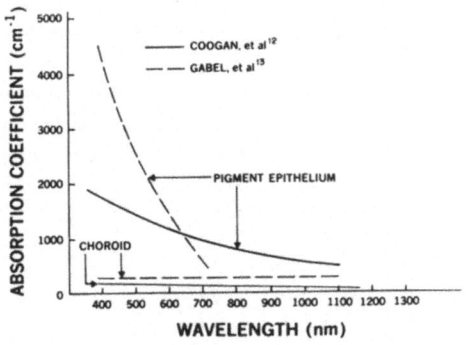

FIGURE 10. Absorption coefficients for rhesus monkey fundus.

absence of the lens, and f_c is the local length of the cornea. The functions $F_1(P)$ and $F_2(p)$ were given by

$$F_1(\rho) = \exp(iC_0\rho^2) \tag{8}$$

$$F_2(\rho) = \exp(iC_2\rho^4) \tag{9}$$

while the constants C_0 and C_2 were given by

$$C_0 = \frac{2\pi n}{\lambda a^2} W = \frac{2\pi n}{\lambda a^2} \left[-f' - \Delta z(1 - \cos\alpha) + (f'^2 - \Delta z^2 \sin^2\alpha)^{1/2} \right] \tag{10}$$

$$C_2 = -3 \times 10^6/\lambda$$

where α is the angle between the refracted beam at the cornea and the axis of the eye, and n is the index of refraction at laser wavelength. α, Δz, and f could be obtained from the following equations:

$$f = f_0 n(n_0 - 1)/n_0(n - 1) \tag{11}$$

$$\tan\alpha = a/(f' + \Delta z) \tag{12}$$

$$\Delta z = \frac{nz_0(f/f_0)}{n(z_0/f_0) - (f/f_0)} - f_0 \tag{13}$$

where f is the focal length at the laser's wavelength (cm), n_0 is the index of refraction at a reference wavelength λ_0 (cm), and z_0 is the distance of the pupil from the waist of the laser beam (cm).

Constants required for the optical analysis are presented in Table I. The equation accounts for scalar diffraction (using the Fresnel approximation), spherical aberration, a defocusing function, and chromatic aberration.

Flamant (1955) was the first of several (Westheimer and Campbell,

TABLE I. Physiological Parameters Required for Optical Analysis[a]

Parameter	Values of parameters (cm)	
	Human eye	Monkey eye
a_{max}	035	035
f_0 (at $\lambda_0 = 500$ nm)	2.24	1.68
p	0.135	0.12
f_c	3.12	2.43
p_c	0.31	0.29

[a] Takata et al. (1974).

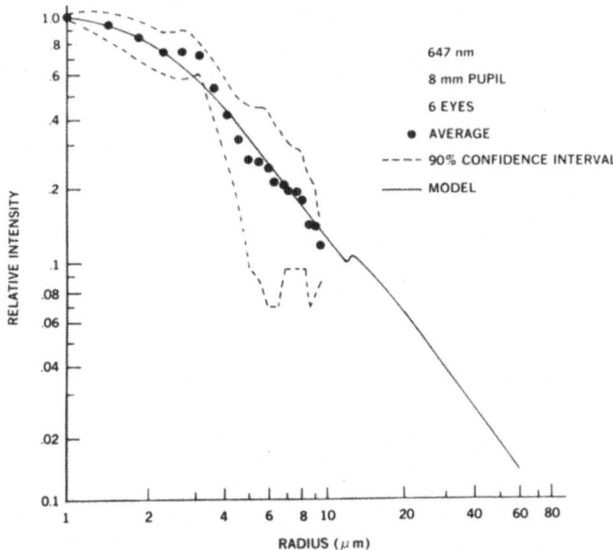

FIGURE 11. Measured images and predicted minimal profiles of intensity for 647-nm laser light.

1962; Campbell and Gubisch, 1966) to measure the image reflected from the human retina. In the late 1970s, Robson and Enroth-Cugell (1978) used a fiber optic probe to measure the retinal image directly at the retinal plane in anesthetized cats. Soon afterwards, Polhamus and Allen (1981) measured the *in vivo* irradiance profile in the rhesus monkey eye (Fig. 11). Measurements were made for three laser wavelengths (530, 647, and 1060 nm) and two pupil sizes (2 and 8 mm). For the visible wavelengths the 1/2-power radii were about 6 μm, though the images contained significant structure within the profiles.

2.3.3. Rate Process Model

For long exposure durations (1 ms to 10 s), the most widely used model of thermal tissue damage is the rate process model applied first by Henriques (1947) to porcine skin burns. He hypothesized that tissue burns resulted in thermal denaturation of protein and based his model on Arrhenius's reaction rate equation. Johnson *et al.* (1974) examined this equation and the assumptions required for its use in detail. Tissue damage was predicted from the following equation:

$$\Omega(r, z) = A \int_0^t \exp[-E/RT(r, z, t)]\, dt \qquad (14)$$

where A is a preexponential constant, E is the energy of activation (reflects species sensitivity to heat) (cal/mol), $T(r, z, t)$ is the tissue temperature, R is the universal gas constant, and Ω is an arbitrary criterion for damage (reflects percent of species that is damaged).

With the assumption that radiation produces a step increase in tissue temperature and a step decrease when the exposure stops, Eq. (14) simplifies to the following:

$$\Omega(r, z) = At_0 e^{-E/RT}(r, z, t_0) \tag{15}$$

where t_0 is exposure duration(s).

2.4. Temperature Measurements

To validate the temperature calculations, Cain and Welch (1974) at the University of Texas at Austin developed microthermocouples specially designed for measuring transient temperatures in tissue. Priebe, Cain, and Welch (1975) compared predicted temperatures to measurements in rhesus monkeys for laser exposures of 0.1–10 s. The average ratio of the model temperatures to their measured values was 0.77.

In the late 1970s, Polhamus (1980) at the School of Aerospace Medicine measured retinal temperatures from long-term laser exposures of 9, 100, and 1000 s and compared measured to predicted temperatures —measured at its center of the image and at various radial distances (Figs. 12 and 13). The measured temperatures associated with threshold lesions were 15% lower than the predicted temperatures (Fig. 14)

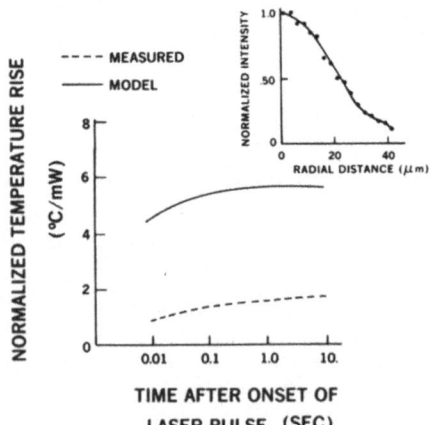

FIGURE 12. Temperature-rise history at the image center normalized with respect to corneal power. Retinal irradiance profile is shown in the insert.

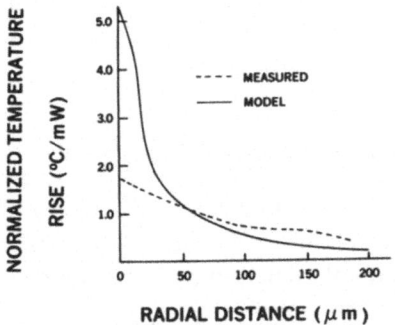

FIGURE 13. Radial temperature rise profiles at 9.0 s normalized with respect to corneal power.

—temperatures associated with the lesion radii. Thus, the model predicted the extent of damage reasonably well despite significant differences between predicted and measured temperatures at the center of an image.

2.5. Blood Flow

Eichler *et al.* (1978) tested the significance of blood flow on tissue temperature profiles in the late 1970s. Using several different visible wavelengths of laser radiation, they radiated living and dead tissue while measuring steady-state temperature rise as a function of depth. Only a 10% difference in the temperature rise between live tissue being supplied with blood and dead tissue was found. Shortly thereafter, Welch, Wissler, and Priebe (1980) used a dimensionless solution of the retinal temperature model to determine

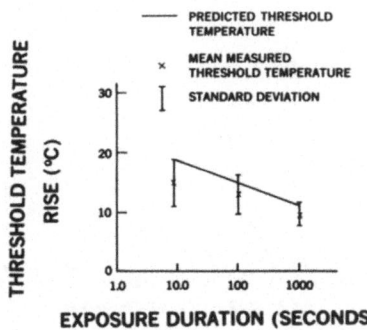

FIGURE 14. Maximum temperature rise at the lesion radius for combined macular and paramacular thresholds for various exposure duration. The predicted threshold temperature using Takata *et al.*'s recommended parameters are represented by the solid line.

that for exposures less than 8 s blood perfusion reduced temperature rise by only 10%. Therefore, the prediction of retinal temperature rise and injury could safely neglect the effects of blood flow.

2.6. Asymptotic Retinal Thermal Injury

The thermal model is normally solved numerically; however, the asymptotic nature of the solution lends itself to a graphical approach. To predict damage, one need only determine the energy required to produce a "threshold" temperature rise profile at a given lesion radius. That "threshold" temperature rise is one that will, when subjected to Henriques's damage integral, predict irreversible injury. The threshold exposure is then

$$Q_t = T_t/V_r A \qquad (16)$$

where Q_t is the threshold retinal exposure (J/cm^2), T_t is the threshold temperature rise at the lesion radius (°C), V_r is the normalized temperature rise at the lesion radius (°C/J), and A is the aperture area of a 7-mm pupil, $A = 0.38 \, cm^2$.

In the absence of a plot of normalized radial temperature profiles from Gaussian images from which to obtain V_r, minor modification to Eq. (16) yields

$$Q_t = T_t R_r/V_c A \qquad (16.1)$$

for the same definitions of Q_t, T_t, and A as for Eq. (16) and where R_r is the ratio of temperature rise at the image center to that at the lesion radius, and V_c is the normalized maximum center temperature rise (°C/J).

Thus, to obtain threshold temperature rise, use can be made of the asymptotic plot by Priebe and Welch (1978) of lesion radius threshold temperatures (T_t) as shown in Fig. 15. Figure 15 was derived with rate coef-

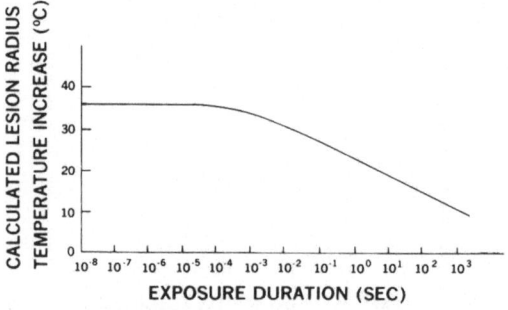

FIGURE 15. Asymptotically predicted threshold temperature at the lesion radius using the thermal model coupled with the rate process model—the coefficients in the rate process model are $P = 3.1 \times 10^{98} \, s^{-1}$ and $E_0 = 150,000 \, cal \, mole^{-1}$ (Priebe and Welch, 1978).

ficients obtained from porcine skin burns. A better set of damage coefficients should be those mentioned above generated from the data of Fig. 14; however, the difference between the two sets is only 5%. The plot shows the maximum temperature at the radius for irreversible injury ($\omega = 1$ for this plot). For instance, the maximum temperature at the lesion radius for a 1-ms threshold exposure is 33°C. Priebe and Welch generated this curve with experimental threshold data from the literature using an assumed image size of 25 μm $1/e^2$ radius.

Model solutions show that 40% changes in image size do not affect maximum temperature rise at the lesion radius by more than 3% (Welch *et al.*, 1978). We would expect this, since the exponential nature of the integrand in Henriques' rate process model makes the results highly sensitive to peak temperature instead of the general shape of the temperature–time profile.

The normalized temperature rise at the center of the image was tabulated by Mainster *et al.* (1970), and Priebe and Welch (1978) later emphasized its asymptotic nature (Fig. 16). The curves of Fig. 16 were calculated using absorption coefficients of $\beta_1 = 1425$ cm^{-1}, and $\beta_2 = 163$ cm^{-1}, which are the values of Coogan *et al.* (1974) for $\lambda = 530$ nm. However, Fig. 17 was obtained using $\beta_1 = 310$ cm^{-1} and $\beta_2 = 53$ cm^{-1}. Though the temperatures will be different owing to different absorption coefficients (the center temperature from Fig. 16 for $25 = 10$ is $100 \times$ larger than that of Fig. 17), the ratios of center to lesion radius temperature should be relatively unchanged (see Fig. 18 for definition of β). Model calculations with $\beta_1 = 108$ cm^{-1} and $\beta_2 = 145$ cm^{-1} compared to those with $\beta_1 = 360$ cm^{-1} and $\beta_2 = 260$ cm^{-1} show that their ratios of center tem-

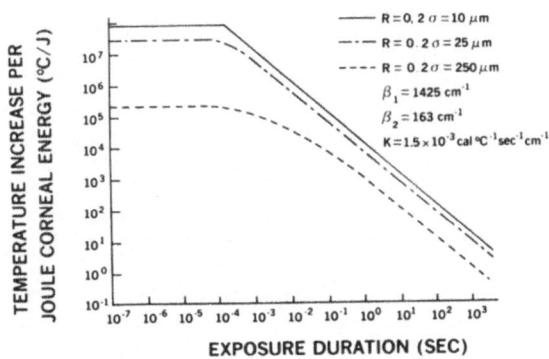

FIGURE 16. Temperature increase per J total intraocular energy versus exposure duration (Priebe and Welch, 1978).

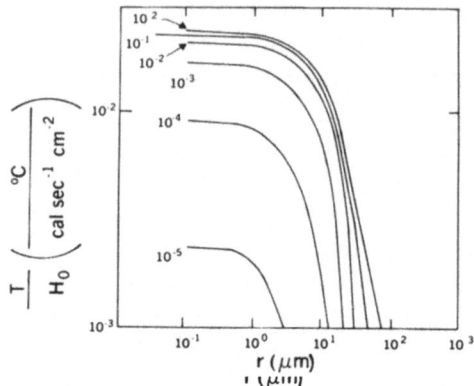

FIGURE 17. Normalized temperature for a constant exposure and the point-spread distribution associated with a 6.6-mm pupil diameter. Radial ($z = 1$ μm) temperature distributions are given (Mainster *et al.*, 1970).

perature to lesion radius temperature (averaged over exposures from 10^3 to 10^{-5} s) differ only by 19% (Polhamus, 1982).

This analysis and particularly the use of Fig. 17 is valid only for small images (~ 10 μm $1/e^2$ radius). To estimate threshold exposures from images larger than 10 μm $1/e^2$ radius, the dependence of radial temperature profiles on image size must be known. Priebe and Welch (1979) have normalized information for uniform profiles as shown in Fig. 18.

Using information from Figs. 15, 16, and 17, threshold exposures for

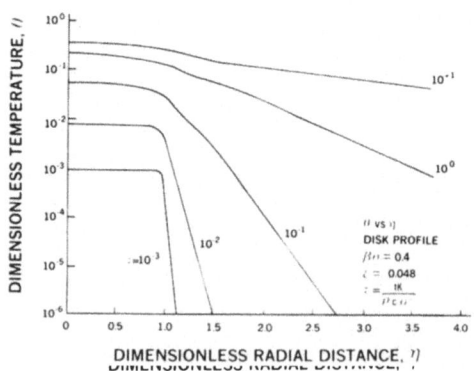

FIGURE 18. Dimensionless temperature versus dimensionless radial distance for uniform irradiance profile; $0 = Tk/H_0^2$, $n = r/\sigma$, and $\xi - \beta z$; T, temperature rise (°C); K, thermal conductivity (cal/cm s °C); β, absorption coefficient at wavelength λ for a single layer of the retinal-choroid complex (cm^{-1}); H_0, center retinal spectral irradiance (cal/cm^2 s); σ, radius of a Gaussian image at the $1/e$ point or the radius of a uniform image; r, radial distance in the image (cm); z, axial distance (cm); t, time (s); and pC, volumetric specific heat (cal/cm^3 °C) (Priebe and Welch, 1979).

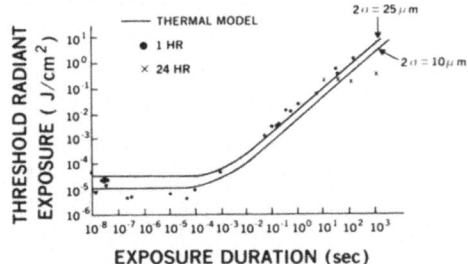

FIGURE 19. Model prediction for 25- and 10-µm image diameters compared to experimental observation of a minimum visible lesion of 25 µm $1/e^2$ diam (with Henriques's rate constants). Data points are from Landers *et al.* (1969); Bresnick *et al.* (1970); Lappin (1970); Vassiliadis *et al.* (1970); Vassiliadis *et al.* (1971); Beatrice and Steinke (1972); Skeen (1972); Gallagher (1975); Goldman *et al.* (1977); Gibbons and Allen (1977); Connolly *et al.* (1978); Zuclich *et al.* (1979); and Reed *et al.* (1980).

minimal size were calculated and compared to literature data in Fig. 19. The image was assumed to be either 25 µm or 10 µm $1/e^2$ in radius, and the minimum ophthalmoscopically visible lesion to be 12.5 µm in radius. For the particular example of 1 ms, the threshold calculated by Eq. (16) was $2.9 \times 10^{-5}/\text{cm}^2$. The calculated values are about 50 below experimental data for exposures shorter than 0.1 ms.

This approach shows agreement between a model of thermal injury and the general trend of the experimental data between 10^{-8} and 10 s exposure duration. Unfortunately, the model appears to be limited to single exposures since Hemstreet *et al.* (1978) showed that the model did not predict injury from multiple pulse exposures. Despite this limitation, the model has been used fairly successfully for the prediction of damage to tissues other than the retina.

3. CORNEAL THERMAL INJURY

In the far-IR region, i.e., for wavelengths greater than 1400 nm, the mechanism of interaction is generally considered to be thermal—at least at the threshold level—and effects are produced primarily in the cornea and lens, since very little energy is transmitted to the retina at these wavelengths. Although a minimal epithelial corneal lesion is not particularly significant in terms of a permanent effect on visual function, the production of stromal injury with associated edema and subsequent collagen shrinkage is viewed as

serious, since extended or permanent loss of visual capacity can be involved
—as shown by Gallagher (1975).

Safety standards for corneal exposure to laser infrared (IR) radiation
have been developed on the basis of the production of a minimal epithelial
corneal lesion visible within about one hour using a slit lamp for the
examinations. Typically, such a lesion would appear as a greyish cloudy
area, a small opacity, or a "stippling" located within the irradiated area.
Maximum permissible exposure levels (MPE) were established using a
suitable safety factor, generally around 10, below the value of the exposure
(in terms of energy per unit area) that produced the minimal lesion. A
relatively small body of experimental data was available upon which to
formulate the standards, but a well-developed understanding of thermal
heating phenomena was available to guide and assist in estimating safe
exposure levels. As a result of the wide range of wavelengths potentially
producible by lasers in the infrared region and the relatively broad and
irregular infrared absorption spectra of the cornea, aqueous, and lens, an
ANSI standard was developed that was independent of wavelength in the
spectral region 1.4–10^6 μm. This standard, which still stands, has only one
exception and that is for a wavelength of 1.54 μm—the wavelength
produced by an erbium laser (Fig. 20).

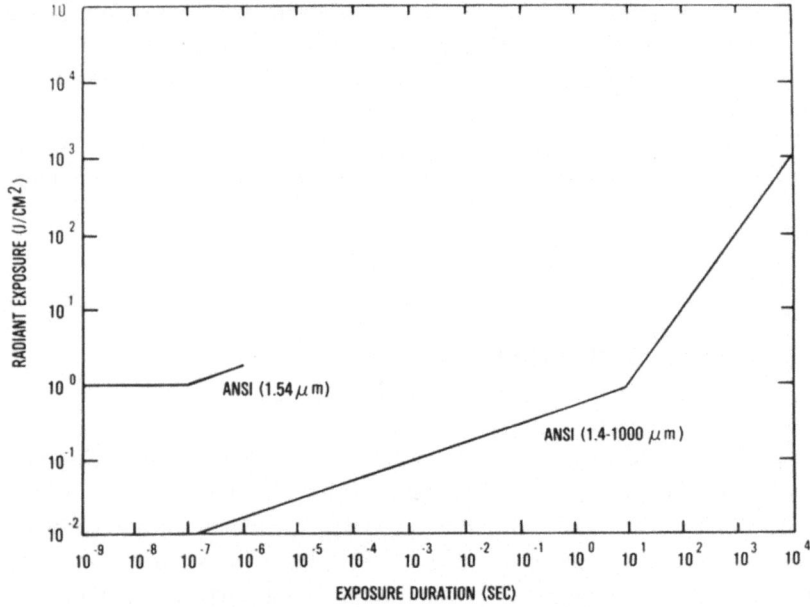

FIGURE 20. Safety standards for the cornea.

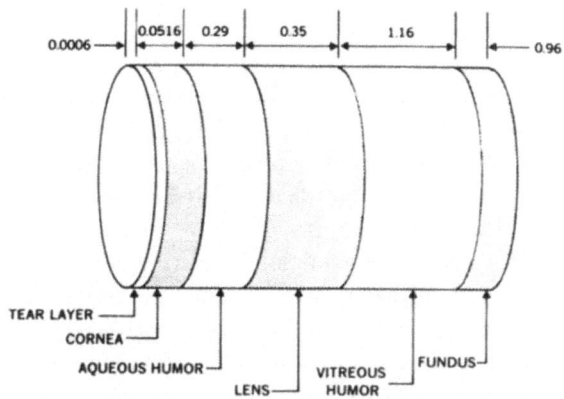

FIGURE 21. Cornea and lens model geometry (distances in cm).

The work of Gallagher stimulated an interest in predicting different degrees of corneal injury. Since the retinal thermal model was structured to accommodate various layers of absorbing tissues, it was rather easily adapted to calculating temperatures in the cornea and the other clear media of the eye. Takata *et al.* (1974) accomplished this by introducing anatomical and physical properties appropriate to the various layers of the clear media (Fig. 21) in place of those for the retina and choroid (Table II). This model assumed an air interface with the tear layer but considered transfer of heat at this boundary only by conduction. Absorption coefficients for the clear media were available from the work of Boetnner (1967) and Maher (1978)

TABLE II. Physiological Parameters Used for Clear Media Analysis[a]

	Thicknesses (cm)	
	Monkey	Man
Tear layer	6×10^{-4}	6×10^{-4}
Cornea	5.16×10^{-2}	5.86×10^{-2}
Aqueous humor	2.9×10^{-1}	3.1×10^{-1}
Lens	3.5×10^{-1}	3.6×10^{-1}
Vitreous humor	1.157	1.697
Pigment epithelium	1.2×10^{-3}	1.4×10^{-3}
Chorio-capillaris	1.0×10^{-3}	1.2×10^{-3}
Choroid	1.68×10^{-2}	1.42×10^{-2}
Sclera	1.0×10^{-1}	1.0×10^{-1}
Corneal surface to second principal plane	1.70×10^{-1}	1.75×10^{-1}
Corneal surface to pupil	2.9×10^{-1}	3.1×10^{-1}

[a] Takata *et al.* (1974).

(Fig. 22). This corneal model also included provision for injury assessment using the Henriques damage assessment technique.

In 1975–1977, Egbert and Maher (1979) conducted an extensive analysis of the relationship between experimentally determined corneal injury "thresholds" and corneal temperature predictions using the corneal thermal model. Egbert considered threshold production for three types of thermal corneal injury: (1) minimal epithelial lesion, defined as the appearance of a relatively faint greyish-white stippled area within 1–2 h; (2) stromal opacity observable as a clouding or an opacity within one hour postexposure; and (3) epithelial vaporization or perforation with severe stromal injury—crater formation, bubbles in the stroma, or perforation into the aqueous observable immediately following the exposure. He found that experimentally determined "thresholds," more specifically the exposure doses for which the probability of lesion formation was 50% (ED_{50}) for the various types of corneal injury, could be fairly well characterized by a "critical peak temperature." From his calculations using the model, the various thresholds were found to vary inversely with the absorption coefficient at lower values of the absorption coefficient but appeared to be independent of them for large values of the absorption coefficient, i.e., 1000 cm^{-1}. He also observed that thresholds were independent of the radius of the incident beam for radii greater than about 10^{-2} cm. Further, for exposures less than about 10^{-3}–10^{-4} s, thresholds were independent of exposure durations. For exposure durations greater than about 1 s, the thresholds appeared to be directly proportional to exposure duration.

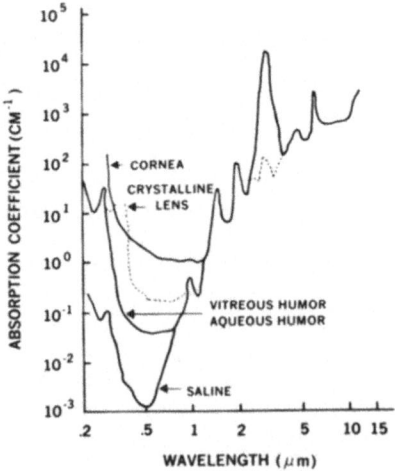

FIGURE 22. Absorption coefficients for the rhesus ocular media (Maher, 1978).

In summary, Egbert found the various threshold estimates made with the thermal model were within a factor of 0.3–10 of the mean experimental values. Of specific interest, he found that calculations using the damage integral technique for epithelial lesion prediction were, on the average, within 3% of the average experimental value, and that 90% of these predictions were within a factor of 2 of the experimental values.

During 1976–1978, Mikesell (1979) examined the production and "repair" of corneal injury produced by CO_2 laser radiation (10.6 μm). In this study, he used Dutch-belted rabbits and a CW laser having a beam diameter at the $1/e$ point of 2.6 mm. All exposures were of 0.5-s duration. He considered four types of injury: (1) minimum visible lesions observable with a bioslit lamp within 30 min of exposure; (2) visible lesions at specified times, up to one year; (3) changes in corneal curvature during the one year observation period; and (4) corneal heterogeneity, i.e., opacities or heterogenous indices of refraction within the cornea. The results are summarized in Table III for the presence or absence of a visible lesion. He found that, with time, corneal curvature changes (measured in diopters) occurred in all groups including the control group as expected. Further, he found that the changes that occurred consisted of both decreased and increased corneal curvatures—including both the exposed and control groups. However, the average change occurring in the exposed group showed a slow decrease or flattening of the curvature with time up to one year—with the more pronounced average flattening associated with corneas that had received the higher exposures. The number of corneas showing changes greater than expected (relative to the control animals) was significant, and the size of many of the measured changes indicated that significant changes in visual acuity could be expected.

Finally, Mikesell determined the threshold for corneal heterogeneity at

TABLE III. ED_{50} Values of Average Power for Corneal Lesions after Exposure to 10.6 μm Radiation[a]

Postexposure time	Number of eyes	ED_{50} (W)	95% Confidence limits
30 min	144	2.1	1.8– 2.3
1 day	138	2.3	1.9– 2.6
7 days	138	2.8	2.3– 3.2
30 days	134	4.9	4.2– 5.6
90 days	132	8.6	6.8–15.2
180 days	131	11.1	8.2–32.0
270 days	127	11.1	8.2–34.0
365 days	120	9.9	7.7–21.3

[a] Laser beam radius = 2.65 mm at $1/e$ point; duration of all exposures = 0.5 s.

one year to be 7.25 J/cm² by examining the fundus reflection for a lack of homogeneity using a band-held ophthalmoscope. The measured thresholds for each type of injury were more than a factor of 10 above the corresponding American National Standards Institute (ANSI) maximum permissible exposure (MPE). Interestingly, the ratio of ED_{50} for corneal heterogeneity to the Ed_{50} for a minimum visible lesion within 30 min was 1.5. If, as Mikesell recommends, the indication of permanent injury is taken as corneal heterogeneity at one year, the ED_{50} for just visible corneal stippling which is repairable within a day or two is not far removed from that for permanent injury. However, the effect on visual function of corneal heterogeneity as it was observed in these experiments has not been shown.

Based on thermal models, Reeds (1978) and Vos (1978) independently developed equations describing standards for eye safe exposures in the IR region between 1.4 and 10^6 μm which take into account the wavelength dependence of energy absorption in the corneal tissues. Such a modification of the laser safety standards appears quite attractive, since the current standards do not consider the rather significant wavelength dependent effects.

Figure 23 shows curves developed by Reed predicting minimum epithelial corneal lesions along with experimental data cited by Egbert (1979) and Maher (1978). The two data points indicated by large diamonds were obtained by Muller (1976), one at 1.4 ns for CO_2 radiation and one at 100 ns for HF radiation, and correspond to the observation of light stippling at 48 h postexposure—visible only with fluorescein staining of the cornea.

Figure 24 presents standards as proposed by Reed that can be com-

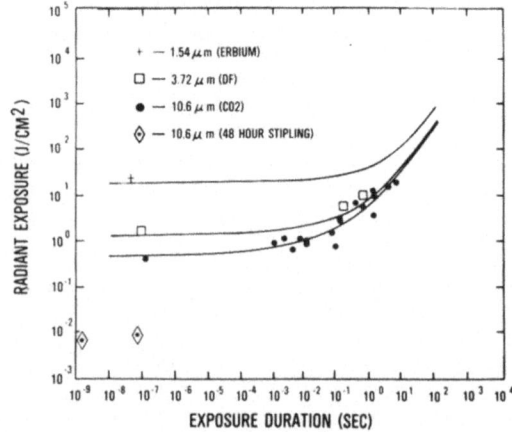

FIGURE 23. Comparison of experimental and theoretical [Eq. (3)] lesion thresholds for three absorption coefficients. Data points are as cited by Egbert and Maher (1979) for 10^{-8}–10^{-2} s.

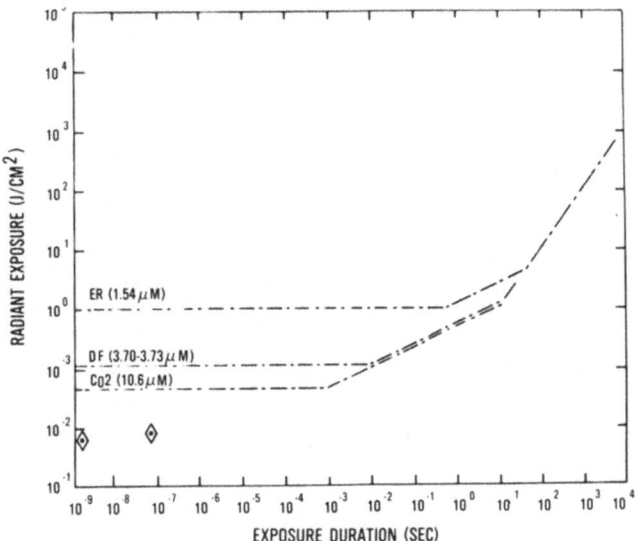

FIGURE 24. Proposed corneal safety standards (Ree, 1978; Vos, 1978).

pared with ANSI standard (Fig. 20). The two observations by Muller are also shown on this figure. Sliney (1978) has taken the position that these two values may be the result of acoustic phenomena, or some other mechanism operating at the very short exposure times involved. Also fluorescein uptake is known to be a more sensitive detector of injured tissue, especially 48 h postexposure. Taboada, Mikesell, and Reed (1981) found that detectable changes in the cornea using fluorescein could be observed at about 1/20 of the dose that produced changes detectable without fluorescein at 2 h postexposure for 248 nm, 50 ns radiation. In the author's view, had Muller used the same criterion for a minimum epithelial lesion that was commonly used in most of the earlier work, i.e., the appearance of a light stippling or cloudiness within one hour, visible without fluorescein staining, the threshold values could have been more nearly in line with other CO_2 data.

Subsequently, Zuclich *et al.* (1984), using facilities at the Air Force School of Aerospace Medicine and the Los Alamos National Laboratory, measured corneal thresholds (Ed_{50s}) for rabbits using single short pulses of CO_2 laser radiation. The values obtained were 0.182 J/cm^2 for a 250-ns pulse, 0.54 J/cm^2 for 25 ns, 0.33 J/cm^2 for 1.7 ns, and correspond to an area on the cornea with a whitish granular appearance having the shape of the beam detectable with a slit lamp within one hour postexposure. These experimental results are in accord with expectations based on a thermal damage mechanism, and are shown in Fig. 25 along with threshold values

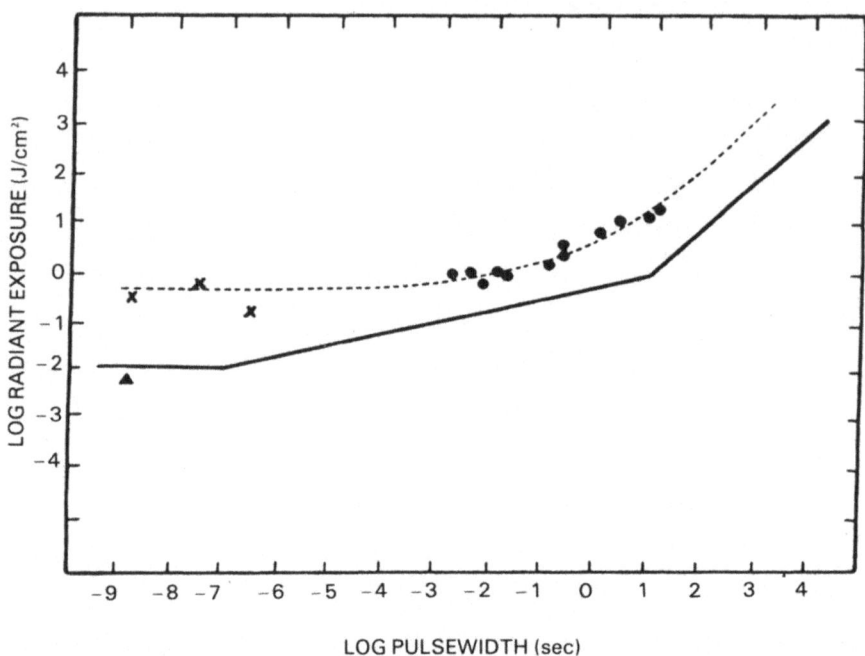

FIGURE 25.   Thresholds versus width plot of laser induced corneal damage. ●, CW CO_2 laser exposure; ×, Zuclich *et al.* (1982); ▲, Mueller and Ham (1976). The solid line is the MPE fn wavelengths from 1400 to 10^6 nm. The dashed curve is predicted by a model based on a purely thermal damage mechanism.

for CO_2 exposures at other pulse widths. Examinations of these lesions at 24 and 48 h postexposure, with and without fluorescein dye, showed no signs of the production of trauma at levels lower than the ED_{50}—in contrast to the results obtained by Mueller and Ham.

These results reinforce the case for modifying laser safety standards for IR radiation in the region between 1.4 and 10^6 μm to take into account the wavelength dependence of energy absorption in corneal tissue (Reed, 1978). Further, they tend to support other evidence that *threshold effects* may be accounted for by a simple thermal mechanism for pulses as short as 10^{-8} s.

In any event, additional attention should be given to this portion of the spectrum, and revision of the IR standards should be considered to either provide additional safety factors, as would be appropriate if the Muller data are considered in the same light as previous "thresholds," or to spread the standards with respect to wavelength, as has been done in the visible spectrum and as indicated appropriate in the IR by Reed and Vos.

4. BEYOND THERMAL INJURY

In the last few years, research has shown that a pure thermal model has a limited application for predicting injury. If exposures longer than a few seconds are involved, experimental evidence suggests a different mechanism of interaction—generally referred to simply as "photochemical" for want of a more definitive understanding of the detailed processes taking place. Undoubtedly, there is a region in which the photochemical process is thermally moderated to some extent. Recognition of these findings has caused a lowering of the safety standards for prolonged exposures to visible light and an introduction of a spectral dependence which reflects higher sensitivities of the retina as the wavelength approaches the blue end of the spectrum (Fig. 26). The evidence strongly suggests that the photopigments of the detector cells are involved, and a photochemical process appears to be a likely candidate for the effects observed following low-level chronic exposures (Gibbons and Allen, 1977; Ham *et al.*, 1979; Harwerth and Sperling, 1971).

At short exposure, less than approximately 10^{-6}–10^{-8} s, another mechanism appears to become important. This mechanism is the production of acoustic transients, which, if sufficiently large, can produce mechanical damage to the delicate retinal structures as well as the underlying choroidal tissues. Cleary of the Medical College of Virginia (1977) treated this subject in detail. Again, there presumably is a region in which near-threshold injury is produced by a combination of thermal and acoustic effects or perhaps by a simple thermal effect.

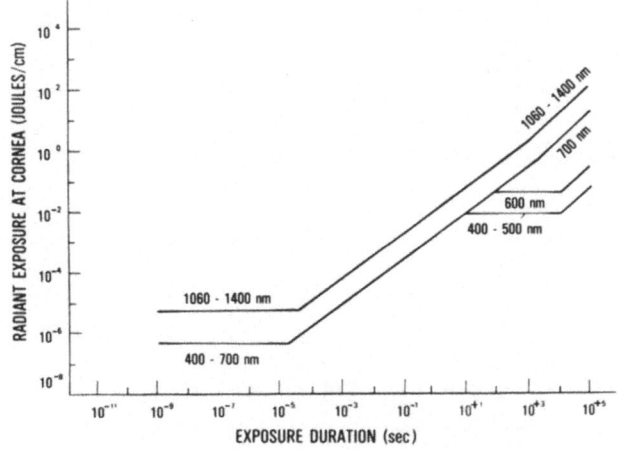

FIGURE 26. Safety standards for the retina.

TABLE IV. Bright Light Sources and Some Possible Consequences of Overexposure

Source	Site of effect	Possible effects	Consequences
Sun	Cornea (UV) Retina (visible)	Kerato conjunctivitis Solar retinal burn, flashblindness[a]	Pain, inflammation, recovery in 24–48 h No pain, small scotoma, possibly some noticeable loss of visual acuity at high acuities for bilateral burns
Arc weider Intense arc lights with glass envelopes (xenon)	Cornea (UV) Retina (visible and near IR)	Kerato conjunctivitis Retinal burn	Pain, inflammation, recovery in 24–48 h No pain, small scotoma, possibly some permanent loss of visual acuity
Intense arc lights with no UV filtering envelope	Retina (visible and near IR)	Retinal burn	No pain, small scotoma, possibly some permanent loss of visual acuity
Ophthalmic surgical lamps	Corneal (UV) Retina	Kerato conjunctivitis Retinal burns	Pain, inflammation, recovery in 24–48 h No pain, possible small scotoma, possible color response shift
Nuclear detonation	Retina	Retinal burn, retinal hemorrhage, flashblindness[a]	Effects may range from minor to severe, temporary or permanent, involving loss of visual acuity and scotomas
Lasers	Cornea (UV, IR)	Kerato conjunctivitis, epithelial cell injury, stromal cell injury	Effects may range from minor kerato conjunctivitis to severe stromal scarring and change of corneal curvature with associated loss of visual acuity
	Lens (UV, IR)	Cataracts	Small opacities in the lens producing increased light scattering and/or obstructions to vision in the eye. Prompt or delayed cataractogenesis
	Retina (near UV, visible, near IR)	Detector cell damage, retinal burn, retinal hemorrhage, neural damage	Effects on vision may range from no significant effect to permanent effects involving loss of visual acuity, scotomas, or shifts in color response
Photoflash bulbs	Retina	Flashblindness[a]	No permanent effects, recovery from flashblindness in a few seconds

[a] Flashblindness, which may occur following exposure to any visible "bright" light, is a partial and temporary loss of visual function that can last from several seconds to minutes or hours, depending upon the extent of the exposure.

Finally, when the retina is exposed to extremely high powers such as can occur in extremely short pulses (less than about a nanosecond), new or different mechanisms such as dielectric breakdown or multiple photon events may occur. These mechanisms are generally reflected in a power-dependent relationship and are suspected to involve cellular membranes as well as other structures (1975).

Retinal effects are produced primarily by wavelengths in the region between 400 and 1400 nm. As indicated earlier, primary effects produced by ultraviolet radiation occur in the cornea and lens. However, at 325–335 nm, Zuclich and Taboada (1978) observed retinal lesions (discolorations) at corneal exposures approximately a factor of 100 below the threshold for corneal effects observable with a slit lamp. Histological examinations implicate the detector cells carrying the photopigments—again suggesting a photochemical process. Although specific photochemical reactions may be difficult to identify, in general, they should display a cumulative effect, obey reciprocity, and be characterized by a wavelength dependence with a long-wavelength cutoff.

Possible ocular injury resulting from the interaction of radiant energy from "bright" lights of various wavelengths with varying parts of the eye have been briefly discussed. Table IV lists some of the bright light sources that can cause eye injury with possible consequences to vision.

5. CONCLUSION

Man's susceptibility to injury from intense light has required development of safety criteria—primarily for the retina but also for the lens, cornea, and skin. Historically, investigators have developed retinal safety data for worst-case situations based on minimum visible type lesions (visible using an ophthalmoscope or slit lamp) and the estimated exposure for 50% probability of lesion occurrence within one hour—suitably adjusted with safety factors developed using the best information available at the time. This approach apparently was and is reasonable and appropriate for thermal and thermal/acoustic interactions. However, studies dealing with long exposures or low-level repeated exposures require other end points and procedures.

Systematic laboratory study of ocular damage began in the early 1950s and has progressed more or less continuously ever since. Probably the most understood mechanism of injury is that described as thermal. Relatively thorough models of this mechanism exist and have been validated reasonably well within the limits of their applicability. However, other mechanisms of injury such as acoustical shock waves and photochemical interactions have been identified and have received considerable attention in the past decade. The results of the research efforts of many investigators over a considerable span of time have been incorporated into numerous Laser Safety Standards typified by the American National Standards Institute Z-136.1 Standard for the Safe Use of Lasers. These standards, although carefully conceived and based upon a large body of empirical information, are neither complete nor final and do not explicitly consider potential effects on visual function, the possibility of accelerated retinal aging, or latent effects. Further, they currently do not extend beyond 10^{-9} s. They must obviously be updated as additional information is obtained, and as pulse-shortening technology expands.

REFERENCES

Allen, R. G., 1967, Research on ocular effects produced by thermal radiation, Report for the USAF School of Aerospace Medicine by Technology Incorporated, Life Sciences Division, San Antonio, Texas.

Allen, R. G., et al., 1968, The calculation of retinal burns and flashblindness safe separation distances, Life Sciences Division, Technology Incorporated, USAF-TR-106.

Beatrice, E. S., and Steinke, C., 1972, Q-switched ruby retinal damage in rhesus monkey, Frankfort Arsenal Report R-2051.

Boettner, E. A., 1967, Spectral transmission of the eye, final report of USAF, School of Aerospace Medicine Contract AF 41 (609)–2966.

Bresnick, G. H., Frisch, G. D., Powell, J. O., Landers, M. B., and Holst, G. C., 1970, Ocular effects of argon laser radiation, Invest. Ophthalmol. 9:901–910.

Cain, C. P., and Welch, A. J., 1974, Thin film temperature sensors for biological measurements, IEEE Trans. Biomed. Eng. 21(4):421–423.

Campbell, F. W., and Gubisch, R. W., 1966, Optical quality of the human eye, J. Physiol. 186:558–578.

Clarke, A. M., Geeraets, W. J., and Ham, W. T., Jr., 1969, An equilibrium thermal model for retinal injury from optical sources, Appl. Opt. 8:1051–1054.

Cleary, S. G., 1977, Laser pulse and the generation of acoustic transients in biological material, in Laser Applications in Medicine and Biology, Plenum Press, New York, Vol. 3, Chap. 2, pp. 175–219.

Connolly, J. S., Hemstreet, H. W., Jr., and Egbert, D. E., 1978, Ocular hazards of picosecond and repetitive-pulsed lasers, Vol. II: Argon-ion laser (514.5 nm), Report No. SAM-TR-78-21.

Coogan, P. S., Hughs, W. F., and Mollsen, J., 1974, Histologic and spectrophotometric comparisons of the human and rhesus monkey retina and pigmented ocular fundus, final report of USAF School of Aerospace Medicine, contract No. AF41609-71-C-0006.

Egbert, D. E., and Maher, E. F., 1979, Corneal damage threshold for infrared laser exposure: Empirical data, model predictions and safety standards, USAFSAM-TR-77-29.

Eichler, J., Knof, J., Lenz, H., Salk, J., and Schafer, G., 1978, Temperature distribution in tissue during laser irradiation, *Rad. Environ. Biophys.* **15**:277–287.

Flamant, F., 1955, Repartition dans l'image retinienne d'une fente, *Rev. Opt.* **34**:433–459.

Gabel, V. P., Birngruber, R., and Hillenkamp, F., 1976, Die Lichtabsorption om Augenhintergrund, Gesellschaft fur Strahlen-und-Univehforschung Munchen, GSF Bericht A55.

Gallagher, J. T., 1975, Corneal curvature chages due to exposure to a carbon dioxide laser: A preliminary report, USAFSAM-TR-75-44.

Gibbons, W. D., and Allen, R. G., 1977, Retinal damage from long-term exposure to laser radiation, *Invs. Ophthalmol. Vis. Sci.* **16**:521–529.

Goldman, A. I., Ham, W. T., Jr., and Mueller, H. A., 1977, Ocular damage thresholds and mechanisms for ultrashort pulses of both visible and infrared laser radiation in the rhesus monkey, *Exp. Eye Res.* **24**:45–56.

Ham, W. T., Jr., Wiesinger, H., Schmidt, F. H., Williams, R. C., Ruffin, R. S., Schaffer, M. C., and Guerry, D., III, 1958, Flash burns in the rabbit retina, *Am. J. Ophthalmol.* **46**:700–723.

Ham, W. T., Jr., Mueller, H. A., Ruffolo, J. J., Jr., and Clarke, A. M., 1979, Sensitivity of the retina to radiation damage as a function of wavelength, *Photochem. Photobiol.*, **29**:735–743.

Harwerth, R. S., and Sperling, H. G., 1971, Prolonged color blindness induced by intense spectral lights in Rhesus monkeys, *Science* **174**:520–523.

Hemstreet, H. W., Jr., Conolly, J. S., and Egberts, D. E., 1978, Ocular hazards of picosecond and repetitive-pulsed laser, 1:ND: YAG laser (1064 nm), USAF School of Aerospace Medicine SAM-TR-78-20.

Henriques, F. C., 1947, Studies of thermal injury, *Arch. Pathol.* **23**:489–502.

Johnson, F. H., Eyring, H., and Stover, B. J., 1974, *Theory of Rate Process in Biology and Medicine*, John Wiley and Sons, New York.

Landers, M. B., Beatrice, E. S., Byer, H. H., Powell, J. O., Chester, J. E., and Frisch, G. D., 1969, Determination of visible threshold of damage in retina of Rhesus monkey by Q-switched ruby laser, U.S. Army, Frankfort Arsonal, Memo Report M69-26-1.

Lappin, P. W., 1970, Ocular damage thresholds for the helium–neon laser, *Arch. Environ. Health* **20**:177–183.

Maher, E. F., 1978, Transmission and absorption coefficients of ocular media of the Rhesus monkey, USAFSAM-TR-78-32, USAF School of Aerospace Medicine, Brooks, AFB TX.

Mainster, M. A., White, T. J., and Allen, R. G., 1970a, Spectral dependence of retinal damage produced by intense light sources, *J. Opt. Soc. Am.* **60**:848–855.

Mainster, M. A., White, T. J., Tips, J. H., and Wilson, P. W., 1970b, Retinal-temperature increases produced by intense light sources, *J. Opt. Soc. Am.* **60**:264–270.

Mikesell, G. W., Jr., Richey, E. O., Taboada, J., McNee, R. C., Anderson, B. R., and Bower, J. L., 1979, Lesion duration and curvature change in the cornea following exposure to a carbon dioxide laser, USAFSAM-TR-79-26.

Mueller, H. A., and Ham, W. T., Jr., 1976, The ocular effects of single pulses of 10.6-μm and 2.5–3.0-μm Q-switched laser radiation, report to the Los Alamos Scientific Laboratory, L-Division.

Polhamus, G. D., and Allen, R. G., 1981, *In vivo* measurement of the point spread function in the Rhesus monkeys, *Invest. Ophthalmol. Vis. Sci.* **20**(3):144.

Polhamus, G. D., 1980, *In vivo* measurement of long-term laser induced retinal temperature rise, *IEEE Trans. Biomed. Eng.* **27**:617–622.

Priebe, L. A., Cain, C. P., and Welch, A. J., 1975, Temperature rise required for production of minimal lesions in the Macaca Mulatta retina, *Am. J. Ophthalmol.* **79**:405–413.

Priebe, L. A., and Welch, A. J., 1978, Asymptotic rate process calculations of thermal injury to the retina following laser irradiation, *J. Biomech. Eng.* **100**:49–54.

Priebe, L. A., and Welch, A. J., 1979, A dimensionless model for the calculation of temperature increase in biologic tissues exposed to nonionizing radiation, *IEEE Trans. Biomed. Eng.* **26**:244–250.

Reed, R. D., 1978, Format of revised safety standards for infrared laser exposure, USAF-SAM-TR-78-29.

Reed, R. D., Taboada, J., and Griess, G. A., 1980, Thresholds and mechanisms of retinal damage from a white-light laser, *Health Phys.* **39**:33–39.

Richey, E. O., and Lof, N. E., 1975, Predicting safe distances to prevent retinal burns from nuclear detonations, USAFAM-TR-75-30.

Robson, J. G., and Enroth-Cugell, C., 1978, Light distribution in the cat's retinal image, *Vision Res.* **18**:159–173.

Skeen, C. H., Bruce, W. R., Tips, J. H., Jr., Smith, M. G., and Garza, G. G., 1972, Ocular effects of repetitive laser pulses, Technology, Inc., San Antonio, Texas, USAF Contract F41609-71-C-0018 (30 June, 1972) (AD746795).

Sliney, D. H., 1978, personal communication.

Taboada, J., Mikesell, G. W., Jr., and Reed, R. D., 1981, Response of the corneal epithelium to KrF eximer laser pulses, *Health Phys.* **40**:677–683.

Takata, A. N., Goldfinch, L., Hinds, J. K., Kuan, L. B., Thomopoulis, N., and Wiegandt, A., 1974, Thermal model of laser induced eye damage, final technical report IITRI, I-TR-74-6324, IIT Research Institute, Chicago (DDC AD/AO17-201).

Vassiliadis, A., Zweng, H. C., and Dedrick, K. G., 1971, Ocular laser threshold investigations, Stanford Research Institute, SRI Report #8209, contract No. F41609-70-C-0002, USAF School of Aerospace Medicine, Final Report.

Vassiliadis, A., Zweng, H. C., Peppers, N. A., Peabody, R. R., and Honey, R. C., 1970, Thresholds of laser eye hazards, *Arch. Environ. Health* **20**.

Vos, J. J., 1962, A theory of retinal burns, *Bull. Math. Biol.* **24**:115–128.

Vos, J. J., 1978, personal communication.

Welch, A. J., Wissler, E. H., and Priebe, L. A., 1980, Significance of blood flow in calculations of temperature in laser irradiated tissue, *IEEE Trans. Biomed. Eng.* **BME-27**:164–166.

Welch, A. J., Priebe, L. A., Forster, L. D., Gilbert, R. G., Lee, C., and Drake, P., 1979, Experimental validation of thermal retinal models of damage from laser radiation, Final Report of USAF School of Aerospace Medicine, Contract No. F33615-76-C-0605.

Westheimer, G., and Campbell, F. W., 1962, Light distribution in the image formed by the living human eye, *J. Opt. Soc. Am.* **52**:1040–1045.

Wissler, E. H., 1976, An analysis of chorioretinal thermal response to intense light exposure, *IEEE Trans. Biomed. Eng.* **BME-23**:207–215.

Wray, J. L., 1962, Model for prediction of retinal burns, Technical Staff Study, DASA 1281, Headquarters DASA, Washington, D.C.

Zuclich, J. A., and Taboada, J., 1978, Ocular hazard from UV laser exhibiting self-modelocking, *Appl. Opt.* **17**:1482–1484.

Zuclich, J. A., Griess, G. A., Harrison, J. M., and Brakefield, J. C., 1979, Research on the ocular effects of laser radiation, report No. SAM-TR-79-4.

Zuclich, J. A., Blankenstein, M. F., Thomas, S. J., and Haneson, R., 1982, Corneal damage induced by pulsed CO_2 laser radiation, Technology Incorporated Life Sciences Division Quarterly Progress Report No. 9 (Contract F33615-80-C-0610), USAF School of Aerospace Medicine, Brooks AFB, Texas, 1 October to 31 December 1982, pp. I-1–14.

Zuclich, J. A., Blankenstein, M. F., Thomas, S. J., and Harrison, R. F., 1984, Corneal damage induced by pulsed CO_2 laser radiation, *Health Phys.* **6**:829–835.

AUTHOR INDEX

SUBJECT INDEX